SOFT ERRORS

Devices, Circuits, and Systems

Series Editor
Krzysztof Iniewski
CMOS Emerging Technologies Research Inc.,
Vancouver, British Columbia, Canada

SOFT ERRORS

FROM PARTICLES TO CIRCUITS

Jean-Luc Autran

AIX–MARSEILLE UNIVERSITY
MARSEILLE, FRANCE

Daniela Munteanu

CENTRE NATIONAL DE LA RECHERCHE SCIENTIFIQUE (CNRS)
MARSEILLE, FRANCE

Krzysztof Iniewski MANAGING EDITOR

CMOS EMERGING TECHNOLOGIES RESEARCH INC.
VANCOUVER, BRITISH COLUMBIA, CANADA

CRC Press
Taylor & Francis Group
Boca Raton London New York

CRC Press is an imprint of the
Taylor & Francis Group, an **informa** business

To our beloved sons,

Alexandre and Florian.

To our families.

"Nous comprenons la Nature en lui résistant."

(We understand nature by resisting it.)

Gaston Bachelard (1884–1962)

Contents

Section I Environments: Definition and Metrology

Section III Soft Errors: Modeling and Simulation Issues

Foreword

It is a daunting task to describe a field that is constantly in motion and that, up to now, has followed Moore's Law in the rate of its evolution, i.e., with rapid and continuous changes.

To peer into the future and outline the probable developments expected requires intimate knowledge and a deep understanding of the subject, a clear mind and a sound background, and, most of all, a critical maturity in judgment. These are the qualities of the authors of this book.

Professor Jean-Luc Autran and Dr. Daniela Munteanu have mastered the past, absorbed the present, and captured the trends of the future in one of the most important technologies of our time. Their book is the culmination of a lifelong drive to understand and contribute to the field of radiation physics and electronics.

Soft Errors: From Particles to Circuits covers all aspects of the design, use, application, performance, and testing of parts, devices, and systems and addresses every perspective from an engineering, scientific, or physical point of view.

As the authors have mentioned in their Preface, many good texts have been written on similar subjects, but none as thorough, as clear, and as complete as this volume by Professor Autran and Dr. Munteanu. Their book provides vital information to casual readers, as well as to dedicated professionals in the field, on the design and development of safe and reliable electronic systems used widely in applications in industry, engineering, and science, e.g., in automobiles, airplanes, satellites, and medical facilities and instruments.

Drawing on over 50 years of research, experiments, and testing, from the radiation physics community, from failures and successes, and from discoveries and inventions, the authors have condensed an enormous wealth of information into this outstanding book that is well organized, comprehensive, and easy to read.

This great accomplishment has resulted in an extremely useful text that has succeeded in presenting with clarity a complex reality in accessible, understandable, and helpful terms. We are grateful to our colleagues, Jean-Luc and Daniela, for their excellent work, which is highlighted in this edition, and we look forward to their ongoing research and discoveries in the exciting years to come.

Dr. Epaminondas G. Stassinopoulos
Astrophysicist, Emeritus
NASA Goddard Space Flight Center

Preface

This book addresses the problem of soft errors in digital integrated circuits subjected to the terrestrial natural radiation environment, one of the most important primary limits for modern digital electronic reliability. Circumscribed for several decades to the domains of space and avionics applications, and known for their severe radiation constraints, soft errors have been recognized in the last decade as a major threat for electronics at terrestrial ground level, since the miniaturization and complexity of circuits have rendered them more and more sensitive to the tenuous ground-level radiation environment.

Soft errors are a multifaceted issue at the crossroads of applied physics and engineering sciences. They are by nature a multiscale and multiphysics problem that combines not only nuclear and semiconductor physics, material sciences, circuit design, and chip architecture and operation but also cosmic-ray physics and their interaction with the earth's atmosphere, natural radioactivity issues, particle detection, and related instrumentation.

This book originated from different invited presentations, lectures, and short courses we have authored or coauthored in the last decade on different facets of soft-error issues. It includes a large part of our personal contributions on these subjects. Its primary ambition is to give an overview on this complex subject with comprehensive and, as far as possible, self-consistent knowledge on the complete chain of the physics of soft errors, from particles to circuits. Another objective is to address not only the fundamental aspects but also some engineering considerations and technological aspects of soft errors, with different levels of reading according to the centers of interest of the reader:

- Fundamentals and basic knowledge on the natural radiation environment, particle interactions with matter, and soft-error mechanisms
- Practical knowledge and overview of instrumentation developments in the fields of environment characterization, particle detection, and real-time and accelerated tests
- Detailed review and engineering solution analysis concerning the most recent computational developments, modeling, and simulation strategies for the soft-error-rate estimation in state-of-the-art digital circuits
- Trends for future technological nodes and emerging devices

This book should thus be read and used in different ways and by a varied public concerned with the physics of soft errors: as an introductory course for graduate students, a specialized overview for academic researchers, and a practical guide for semiconductor industry engineers or application engineers. We would like to mention that excellent books on the same subject already exist. Without trying to be exhaustive, we would like to cite *Single Event Phenomena* by Messenger and Ash; *Handbook of Radiation Effects* by Holmes-Siedle and Adams; *SER—History, Trends and Challenges* by Ziegler and Puchner; *Radiation Effects and Soft Errors in Integrated Circuits and Electronic Devices*, edited by Schrimpf and Fleetwood; *Terrestrial Neutron-Induced Soft Errors in Advanced Memory Devices* by Nakamura, Baba, Ibe, Yahagi, and Kameyama; *Architecture Design for Soft Errors* by Mukherjee; and *Soft Errors in Modern Electronic Systems*, edited by Nicolaidis. The ensemble of these books covers all aspects of soft errors, from the environment to circuit and system levels. With respect to these important contributions in the domain, our objective was to propose a book that is clearly complementary in its content and that provides a state-of-the art over-view in several domains, including

- The models and computer codes used to estimate the composition and particle flux for terrestrial cosmic rays
- The detection and characterization of high-energy atmospheric neutrons using a neutron monitor and the related physics and operation of this instrument
- The metrology, modeling, and simulation of alpha-particle emissivity in micro-electronics materials
- The interaction of charged particles with matter, with a special emphasis on low-energy protons and muons
- A detailed review about real-time testing, with the most recent experiments and results in the domain completed by a comparison and a simulation-based analysis with accelerated tests
- A complete survey of the different modeling and simulation approaches at both device and circuit levels, with a methodical description of transport models, emerging physical effects in scaled devices, device numerical simulations, and compact models
- An accurate presentation of a Monte Carlo radiation transport code, detailing all the computational aspects of the simulation chain, including the analytical mod-els implemented to evaluate the response of various circuit architectures (static random-access memory and flash memories) to radiation
- A complete review of the evolving risks of single-event effects for current and future complementary-metal-oxide-semiconductor technologies, covering from bulk, silicon-on-insulator, and fin-shaped field-effect transistor circuits to multi-ple-gate, nanowire, and junctionless devices

Structure of the Book

This book consists of twelve chapters in four sections and is organized as follows:

Section I: Environments—Definition and Metrology

Section I focuses on the natural radiation environment and its metrology in the particular context of microelectronics.

Chapter 1 introduces the terrestrial radiation environment in the atmosphere and at ground level. It provides a short description of galactic cosmic rays and focuses on particle radiation within the earth's atmosphere and at ground level. The sea-level and mountain-altitude environments are described in terms of particle composition, absolute flux values, and flux variations as a function of time and various environmental param-eters. Shielding effects of terrestrial cosmic rays are also briefly discussed. The chapter is concluded by a short review of several models and computer codes used to estimate the composition and particle flux of the atmospheric environment, from high altitude to ground level.

Chapter 2 deals with the detection and characterization of high-energy atmospheric neutrons. The chapter focuses on the neutron monitor, a ground-level instrument used to measure the atmospheric neutron flux and to characterize its real-time variations. We describe the operation of this instrument and the evaluation of its response function for the different components of the atmospheric flux. The second part of the chapter is dedicated to the Plateau de Bure Neutron Monitor, an instrument specially designed and operated to characterize the magnitude of the neutron background. We successively describe its construction, installation, and operation over more than 7 years as well as its complete modeling and simulation.

Chapter 3 details the implications for microelectronics of the natural radioactivity present in all terrestrial materials. The chapter begins with the definition of several general aspects and quantities concerning radioactive elements. Primordial, cosmogenic, and human-produced radionuclides present in nature are next described, with a special emphasis on uranium and thorium due to the presence of numerous alpha-particle emitters in their radioactive decay chains. The implications of radon gas for microelectronics are also discussed, together with several important issues concerning the radioactive contamination of advanced complementary-metal-oxide-semiconductor technologies. Finally, an analytic model of the alpha-particle emissivity from single-layer materials and multilayer stacks is presented.

Chapter 4 introduces the alpha-particle detection techniques used as standard tools in microelectronics to detect, quantify, and identify the ultratraces of alpha emitters present in circuit materials. After introducing the basic definitions and important notions concerning alpha-radiation detection, the chapter focuses on the description of an ultralow-background alpha-particle counter recently introduced as a new standard of measurement at wafer level. The design, operation, modeling, and simulation of this metrology tool are illustrated from different contributions and recent works. Other detection techniques are briefly described in the last part of the chapter.

Section II: Soft Errors—Mechanisms and Characterization

The second section of the book deals with the soft-error mechanisms in semiconductor devices and the different experimental methods used to estimate the soft-error rate at circuit level.

Chapter 5 describes the interactions of both alpha and atmospheric particles with matter and their impact on the production of soft errors in semiconductor circuits. The interactions of neutrons and charged particles with matter are addressed, and different phenomena or quantities of interest are introduced. Interactions of protons, pions, and muons with matter are also detailed. The second part of the chapter introduces the basic mechanisms of creation of single-event effects and their impact on the operation of microelectronic devices. Finally, single-event upsets in advanced static random-access memory and single-event-effect mechanisms in logic circuits are considered. The production and propagation of digital single-event transients in sequential and combinational logic are also addressed.

Chapter 6 constitutes a short introduction to accelerated-soft-error-rate tests. We detail several types of accelerated-soft-error-rate experiments using intense beams of different types and flavors of particles: high-energy neutrons, thermal neutrons, monoenergetic protons, and muons. Accelerated tests using alpha-particle solid sources are also presented. The last part of the chapter details a recent simulation study that analyzes in-depth the differences between several artificial broad-spectrum sources of atmospheric-like neutrons in terms of recoils produced in the interactions of neutrons with silicon.

Chapter 7 gives an overview of recent real-time soft-error rate experiments, conducted at altitude, underground, or both, and investigating modern complementary-metal-oxide-semiconductor logic technologies down to the 40 nm technological node. The overview also includes several contributions by the authors, conducted during the last decade to characterize soft-error mechanisms in advanced static random-access memory. Finally, the chapter discusses the specific advantages and limitations of this approach as well as its comparison with accelerated tests using intense particle beams or sources.

Section III: Soft Errors—Modeling and Simulation Issues

The third section of this book overviews the crucial issues of soft-error modeling and simulation at both device and circuit levels.

Chapter 8 presents the different approaches of modeling and simulation of single-event effects in microelectronic devices and integrated circuits. The interest of simulating radiation effects is discussed in the first section, and a brief overview of the simulation at device and circuit levels is presented and illustrated. Device-level simulation is addressed next, including a detailed description of transport models, emerging physical effects in scaled devices, device numerical simulations, and compact models. Finally, circuit-level simulation is introduced, with a particular emphasis on the advantages and drawbacks of various approaches that can be used to simulate single-event effects at circuit level.

Chapter 9 gives an overview of different Monte Carlo computational methods applied to the analysis of single-event effects in semiconductor devices. In the first part, this chapter provides a brief inventory of Monte Carlo-based radiation transport tools to simulate a variety of effects that result from particle interactions with matter. We also provide a short description of a few recent simulation codes specially designed to support the analysis of single-event effects in semiconductor devices. In the second part of the chapter, we describe in more detail a complete general-purpose simulation platform we have developed in recent years for the numerical evaluation of the sensitivity of digital circuits subjected to natural radiation at ground level. Applications of this simulation platform are illustrated through different case studies.

Section IV: Soft Errors in Emerging Devices and Circuits

Finally, the fourth section of the book explores the important consequences of the evolution of microelectronics for the single-event susceptibility of current and future devices and circuits.

Chapter 10 discusses the major consequences of the scaling down of complementary metal-oxide semiconductors for the single-event susceptibility of devices and circuits. Several factors impacting their soft-error rate are surveyed, notably the reduction of device geometry, the increase of operation frequency, and the reduction of the critical charge/energy deposition necessary to cause a single event. The combination of these factors and their consequences for transistor/circuit operation are carefully examined. The impact of all these factors on the soft error rate is also explained and illustrated for sub-45 nm bulk and silicon-on-insulator complementary-metal-oxide-semiconductor technologies.

Chapter 11 addresses an important reliability issue concerning the sensitivity of non-volatile flash memories to the terrestrial radiation environment. We report a recent study based on a new type of experiment conducted at wafer level that combined characterization at mountain altitude and at sea level of several tens of gigabits of flash memories subjected to natural radiation. This experiment evidenced a limited impact of terrestrial radiation at ground level on the memory soft-error rate. Experimental values are compared with estimations obtained from computational simulations using a Monte Carlo radiation-transport code combined with a physical model for the charge-loss mechanism in such floating-gate devices.

Chapter 12 provides a detailed overview of single-event effects in ultrathin fully depleted silicon-on-insulator transistors, fin-shaped field-effect transistor devices, and multiple-gate technologies. We first present silicon-on-insulator technology and discuss specific phenomena that could impact its radiation sensitivity, such as parasitic bipolar amplification. Multiple-gate devices are next introduced, and experimental and simulation results concerning their response to radiation are presented. Finally, we review simulation results concerning the radiation sensitivity of several types of nonconventional devices, such as multichannel nanowire metal-oxide-semiconductor field-effect transistors, multiple-gate devices with independent gates, metal-oxide-semiconductor field-effect transistors without junctions, and tunnel field-effect-transistor devices.

Acknowledgments

Epaminondas G. Stassinopoulos, "Stass" to his friends and colleagues, has agreed to write the foreword for this book. As well as our thanks, we would like to show him our deepest gratitude, profound admiration, and professional and personal friendship. We would also like to honor the memory of his wife Effie.

We would like to express our fond memory of our late colleague, Jean-Claude Boudenot, who passed away too early, in 2008. Jean-Claude encouraged us for many years to start writing this book. Always present in our memories, we would also like to dedicate this book to him.

Writing our first book was a challenge, and without the support of many colleagues, including our PhD and postdoctoral students, the completion of this book would not have been possible.

Our first acknowledgments are for our close colleagues, with whom we have worked during the last decade on various soft-error-related issues. We would like to express our gratitude to Philippe Roche and Gilles Gasiot (STMicroelectronics) for the special relationship we constructed between our two groups, based on a permanent and solid link around scientific questions, successful collaborations, and exciting technical challenges. All of our colleagues, previously or currently involved in this common effort of research between STMicroelectronics and our laboratory, are also gratefully acknowledged: Jean-Pierre Schoellkopf, Sylvain Clerc, Dimitri Soussan, Thierry Parrassin, and Jean-Marc Daveau. We are also indebted to our past and current PhD and postdoctoral students involved in soft-error-related work: Damien Giot, Slawosz Uznanski (now at the European Organization for Nuclear Research [CERN]), Maximilien Glorieux (now at iROC Technologies), Mehdi Saligane, Martin Cochet, Victor Malherbe, Sébastien Martinie (now at French Atomic Energy Commission–Electronics and Information Technology Laboratory [CEA-LETI]), Sébastien Serre, Guillaume Just, Tarek Saad Saoud, and Soilihi Moindjie. Special thanks are due to Tarek and Soilihi for their contribution to several illustrations and graphics included in this book. We are also particularly indebted to all authors who provided original materials (figures, photos, or data) included in this book: Philippe Roche, Gilles Gasiot, and Damien Giot (STMicroelectronics), Brendan D. McNally (XIA LLC), Helmut Puchner (Cypress), Brian Sierawski (Vanderbilt University), Mélanie Raine (French Atomic Energy Commission), John Clem (University of Delaware), Nicolas Fuller (Paris Observatory), Ewart Blackmore (Tri-University Meson Facility accelerators, Vancouver, Canada), Jean-Paul Goglio (EASII-IC), and Jim Hinton (University of Leicester).

Because this book is essentially the result of a "long immersion" in the radiation effects community, the authors would like to show their gratitude to Jean-Luc Leray (French Atomic Energy Commission) and Jean Gasiot (University of Montpellier 2) for their professional friendship and always valuable advice during these last 20 years. We are also grateful to many colleagues in the community: Véronique Ferlet-Cavrois (European Space Agency), Philippe Paillet (French Atomic Energy Commission), Frédéric Saigné, Frédéric Wrobel, Antoine Touboul (University of Montpellier 2), Pete Truscott (Kallisto Consultancy Ltd), Rémi Gaillard (Gaillardremi), Philippe Calvel (Thales Alenia Space France), Christian Chatry (Test and Radiation, TRAD), Richard Sharp (Aeroflex RAD Europe), Robert Baumann (Texas Instruments), Simon Platt (University of Central Lancashire), Helen

Mavromichalaki (University of Athens), and François Mauger (Laboratory of Corpuscular Physics-Caen).

In addition, we would like to acknowledge all the persons who have participated, since 2004, in the effort to develop both the Altitude Single event effects Test European Platform (ASTEP) and Modane Underground Laboratory (LSM) platforms and related real-time experiments: Joseph Borel (JB R&D), Sébastien Sauze (formerly Institute for Materials, Microelectronics and Nanosciences of Provence, now at Sauze Ingénierie), David Gauthier (formerly iRoC Technologies and EASII-IC), Marc Derbey (formerly iRoC Technologies), Jean-Paul Goglio, Mehdi Naceur, Nadine Onesti (EASII-IC), Pierre Cox, Bertrand Gautier (Millimetric Radioastronomy Institute [IRAM] Observatory), Fabrice Piquemal, Michel Zampaolo, Guillaume Warot, Pia Loaiza (Modane Underground Laboratory [LSM, Modane]), Evgeny Yakusev (Joint Institute of Nuclear Research, Dubna), Sergey Rosov (Joint Institute of Nuclear Research, Dubna), and Sergey Semikh (formerly Joint Institute of Nuclear Research, Dubna).

The writing of this book indirectly benefited from a few long-term projects in which we were involved. We would like to mention the MEDEA+ Project #2A704 ROBIN (Robust Chip Design Initiative), the CATRENE Project #CA303 OPTIMISE (OPTImisation of MItigations for Soft, firm and hard Errors), and the DGA project EVEREST (evaluation of the soft-error rate of FDSOI technologies for strategic defense applications). The support of the French Ministry of Economy, Finances and Industry through research conventions #052930215, #062930210, #092930487, and #132906128 is acknowledged.

We cannot forget to mention the support of our laboratory, the Institute for Materials, Microelectronics and Nanosciences of Provence (IM2NP, UMR CNRS 7334), during the writing of this book.

Finally, we would like to particularly acknowledge Kris Iniewski, editor of the series "Devices, Circuits, and Systems"; Nora Konopka, publisher for engineering and environmental sciences; Jessica Vakili, project coordinator, Michele Smith, senior editorial assistant, and John Gandour, designer, from CRC Press/Taylor & Francis; and Michelle van Kampen, project manager from Deanta Global Publishing Services for their cooperation and support during the whole process of editing this book.

Last but not least, we would like to thank our children, our parents, and our families for their permanent understanding and continuous encouragement.

Jean-Luc Autran
Daniela Munteanu
Marseille

Authors

Jean-Luc Autran is distinguished professor of physics and electrical engineering at Aix-Marseille University and honorary member of the University Institute of France (IUF). He is deputy director of the Institute for Materials, Microelectronics and Nanosciences of Provence (IM2NP, UMR CNRS 7334) and the principal investigator of the Altitude Single event effects Test European Platform (ASTEP), a permanent mountain laboratory created in 2004 in the French Alps for the study of soft errors in electronics. Having worked for over 20 years in the field of semiconductor-interface defects, physics of advanced complementary-metal-oxide-semiconductor devices, and radiation effects in microelectronics, his current research interests focus on the physics of soft errors, from the characterization of natural radiation to Monte Carlo radiation-transport simulation. Jean-Luc Autran is the author or a coauthor of more than 300 papers published in international journals and conferences. He has supervised 28 PhD theses. He is a senior member of the Institute of Electrical and Electronics Engineers and a fellow of the Société de l'Electricité, de l'Electronique et des Technologies de l'Information et de la Communication (SEE).

Daniela Munteanu is director of research at the National Center for Scientific Research (CNRS). She is a fellow researcher at the Institute for Materials Microelectronics and Nanoscience of Provence (IM2NP, UMR CNRS 7334) and has 15 years of experience in characterization, modeling, and simulation of semiconductor devices. Her current research interests include emerging complementary-metal-oxide-semiconductor devices, compact modeling, numerical simulation in the domains of nanoelectronics, and radiation effects on components and circuits. Daniela Munteanu is the author or a coauthor of more than 200 papers published in international journals and conferences. She has supervised 12 PhD theses.

Introduction

Radiation-induced soft errors, and more generally single-event effects, are a recent reliability issue in the history of electronics; their discovery dates to the 1970s, and they have been seriously studied for less than 40 years. Several authors (in certain cases, the discoverers themselves) have published in the past detailed chronologies of the history of single-event effects, soft errors, and related subjects (Ziegler et al. 1996; Messenger and Ash 1997; Ziegler and Puchner 2004; Slayman 2010; Heijmen 2011; Pease 2013; Petersen et al. 2013). From these different texts, and, in particular, from the remarkable chronology published in the preface of Messenger and Ash (1997), we have extracted a few milestones and key works that marked the early history of soft errors.

To begin, we can cite two dates in the "prehistory" of soft errors (Cressler and Mantooth 2012): the period 1954–1957, during which the first random and unexplained failures (i.e., anomalies) in digital electronics following surface nuclear bomb tests were reported (Wang and Agrawal 2008); and, in 1962, the first paper pointing to the role of cosmic rays in electronics, published by Wallmark and Marcus (1962). These authors predicted that cosmic rays would start upsetting microcircuits due to heavy ionized-particle strikes and cosmic-ray reactions when the feature size became small enough (Messenger and Ash 1997).

During the 1970s, the effects of radiation on electronics received more and more attention, and an increasing number of research groups examined the physics of these phenomena (Wang and Agrawal 2008). There is a general consensus in the literature that the history of single-event effects really began at the end of the 1970s with the publication of two major papers:

- In 1975, Binder, Smith, and Holman reported for the first time soft errors in digital flip-flop circuits detected in certain satellites in space (Binder et al. 1975).
- In 1978, May and Woods published the first paper on soft errors observed in dynamic random-access memories at ground level (May and Woods 1978). These fails were induced by alpha particles emitted from radioactive impurities in the package materials. This paper marks the beginning of the soft-error issue at terrestrial level.

Other important papers of this pioneering period are summarized in Table I.1. They often correspond to the first contributions in several key domains of soft errors: prediction models (Pickel and Blandford 1978; Ziegler and Lanford 1979), accelerated tests (Guenzer et al. 1979; Wyatt et al. 1979), single-event latchup (Kolasinski et al. 1979), real-time tests (O'Gorman et al. 1996), and the boron-10 issue (Baumann et al. 1995).

This "introductory tableau" of soft errors should be completed by some historical examples of severe reliability problems caused by soft errors in electronic systems. We can cite the case of the Intel 2107-series 16 kbit dynamic random-access memories at the origin of the paper by May and Woods (1978), the "Hera" problem at IBM (Ziegler et al. 1996), and more recently the occasional crash of "Entreprise" servers of Sun in 2000 or the major *soft-error rate* issue in CISCO routers in 2003 (Heijmen 2011). These notorious cases correspond to the manifestation of soft errors at circuit or system level, and illustrate the difficulty

TABLE I.1

A Few Milestones and Pioneering Studies in the Recent History of Soft Errors

Year	Authors/Company (Affiliation)	Contribution/Soft-Error Issue	References
1962	Wallmark and Marcus (RCA Labs)	First prediction that cosmic rays would start upsetting microcircuits when the feature size became small enough, due to both heavily ionizing tracks and cosmic-ray spallation reactions	Wallmark and Marcus (1962)
1975	Binder, Smith, and Holman (Hughes Aircraft Company)	First report of soft errors in digital flip-flop circuits observed in certain satellites. First soft-error-rate estimations based on a diffusion-collection model	Binder et al. (1975)
1978	May and Woods (Intel Corporation)	First paper on soft errors in dynamic random-access memories at ground level induced by alpha particles emitted from radioactive impurities in package materials. Introduction of the concept of critical charge Q_{crit}	May and Woods (1978)
	Pickel and Blandford (Rockwell Int.)	Model developed to predict the cosmic-ray bit-error rate in dynamic-metal-oxide-semiconductor random-access memories	Pickel and Blandford (1978)
1979	Guenzer et al. (Naval Research Laboratory)	First accelerated test identifying neutron- and proton-induced nuclear reactions as the cause of upsets in dynamic random-access memories	Guenzer et al. (1979)
	Ziegler (IBM) and Lanford (State University of NY)	First quantitative predictions that cosmic rays would cause a major electronic-reliability problem at terrestrial sites and at aircraft altitudes	Ziegler and Lanford (1979)
	Kolasinski et al.	Discovery of heavy-ion-induced latchup in static random-access memories and first observation of cosmic-ray-induced latchup	Kolasinski et al. (1979)
	Wyatt et al. (Clarkson University)	Proton-induced soft errors investigated using accelerated tests. Proton-induced nuclear reactions identified as the cause of the errors	Wyatt et al. (1979)
1983–1988	O'Gorman et al. (IBM)	First real-time tests to measure the effects of cosmic rays on semiconductor memory chips	O'Gorman et al. (1996)
1995	Baumann et al. (Texas Instruments)	First study showing that the interactions of thermal neutrons with the boron-10 isotope are a nonnegligible source of soft errors	Baumann et al. (1995)

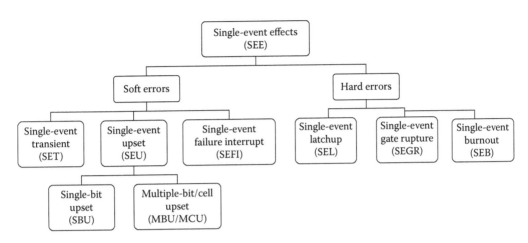

FIGURE I.1
Classification of single-event phenomena.

of identifying soft-error mechanisms at the origin of such sudden malfunctioning of the circuit equipment, in particular due to their stochastic nature.

During this pioneering period, in the early 1980s, most of the terminology used in the field of single-event effects was introduced and then more rigorously defined or specified in different standards (JESD89A 2006; EIA-JESD57 1996; ESCC25100 2002). We summarize in the following the most important terms used in this book.

Single-event effects indicate any measurable or observable change in state or performance of a microelectronic device, component, subsystem, or system (digital or analog) resulting from a single, energetic particle strike. As summarized and hierarchized in Figure I.1, single-event effects include single-event upset, multiple-bit upset, multiple-cell upset, single-event functional interrupt, single-event latchup, single-event hard error, single-event transient, single-event burnout, and single-event gate rupture. The soft-error rate indicates the rate at which soft errors occur. The term *soft-error rate* has been adopted by commercial industry, while the more specific terms *single-event upset, single-event functional interrupt*, and so on are typically used by the avionics, space, and military electronics communities.

Soft errors are a particular subset of single-event effects (see Figure I.1) that can affect digital circuits. Soft errors are related to the most benign form of radiation effects on the circuitry, where radiation directly or indirectly induces a localized ionization capable of upsetting internal data states (Degalahal et al. 2004). At circuit level, a *soft error* corresponds to an erroneous output signal from a latch or memory cell that can be corrected by performing one or more normal functions of the device containing the latch or memory cell. As commonly used, the term refers to an error caused by radiation or electromagnetic pulses and not to an error associated with a physical defect introduced during the manufacturing process. A soft error in logic occurs when the result of a transient fault in logic propagates to a storage element and is latched. A soft error in a memory element occurs when sufficient electrical charge is generated to invert the value stored in the memory element. Historically, the term *soft error* was first introduced (for dynamic random-access memories and integrated circuits) by May and Woods in their 1978 International Reliability Physics Symposium paper (May and Wood 1978), whereas the term "single-event upset" was introduced by Guenzer, Wolicki, and Allas in their 1979 Nuclear and Space Radiation Effects Conference paper (Guenzer et al. 1979).

Other important terms and related definitions are defined as follows.

- Single-event upset: A soft error caused by the transient signal induced by a single energetic particle strike. Another definition, given by the National Aeronautic and Space Administration Thesaurus, is: "Radiation-induced errors in microelectronic circuits caused when charged particles (usually from the radiation belts or from cosmic rays) lose energy by ionizing the medium through which they pass, leaving behind a wake of electron-hole pairs."

- Single-event upset cross section: The number of events per unit fluence. For device SEU cross section, the dimensions are sensitive area per device. For bit SEU cross section, the dimensions are area per bit.

- Single-event upset rate: The rate at which single-event upsets occur.

- Single-event transient: A momentary voltage excursion (voltage spike) at a node in an integrated circuit caused by a single energetic particle strike.

- Single-event latchup: An abnormal high-current state in a device caused by the passage of a single energetic particle through sensitive regions of the device structure and resulting in the loss of device functionality. Single-event latchup may cause permanent damage to the device. If the device is not permanently damaged, power cycling of the device (off and back on) is necessary to restore normal operation. An example of SEL in a complementary-metal-oxide-semiconductor device is when the passage of a single particle induces the creation of parasitic bipolar (p-n p-n) shorting of power to ground.

- Single-event gate rupture: Total or partial damage of the dielectric gate material due to an avalanche breakdown.

- Multiple-cell upset: A single event that induces several cell upsets in an integrated circuit. The struck cells are adjacent or topologically connected (in contrast to the corresponding error bits, which are not always adjacent).

- Multiple-bit upset: A multiple-cell upset in which two or more error bits occur in the same word data (a multiple-bit upset cannot be corrected by a simple single-bit error-code correction).

- Single-event functional interrupt: A soft error that causes the component to reset, lock up, or otherwise malfunction in a detectable way, but does not require power cycling of the device (off and back on) to restore operability, unlike single-event latch up, or result in permanent damage, as in single-event burnout. Note that a single-event functional interrupt is often associated with an upset in a control bit or register.

- Hard error: An irreversible change in operation that is typically associated with permanent damage to one or more elements of a device or circuit (e.g., gate-oxide rupture, destructive latch up events). The error is "hard" because the data are lost and the component or device no longer functions properly even after power reset and reinitialization. The generic term "single-event hard error" is also used in literature.

References

Baumann, R., Hossain, T., Murata, S., and Kitagawa, H. 1995. Boron compounds as a dominant source of alpha particles in semiconductor devices. In *Proceedings of the IEEE International Reliability Physics Symposium*, pp. 297–302. 4–6 April, Las Vegas, NV: IEEE.

Binder, D., Smith, E.C., and Holman, A.B. 1975. Satellite anomalies from galactic cosmic rays. *IEEE Transactions on Nuclear Science* NS-30:2675–2680.

Cressler, J.D. and Mantooth, H.A. 2012. *Extreme Environment Electronics*. Boca Raton, FL: CRC.

Degalahal, V., Vijaykrishnan, N., Irwin, M.J., Cetiner, S., Alim, F., and Unlu, K. 2004. SESEE: A soft error simulation and estimation engine. In *Proceedings of the 7th Annual MAPLD International Conference*. Available at: http://klabs.org/mapld04/abstracts/degalahal_a.pdf.

EIA-JESD57. 1996. *Test Procedures for the Measurement of Single-Event Effects in Semiconductor Devices from Heavy Ion Irradiation*. Arlington, VA: Electronic Industries Alliance (EIA).

ESCC25100. 2002. Single event effects test method and guidelines, ESCC basic specification No. 25100. European Space Agency.

Guenzer, C.S., Wolicki, E.A., and Allas, R.F. 1979. Single event upset of dynamic RAMs by neutrons and protons. *IEEE Transactions on Nuclear Science* NS-26:5048–5052.

Heijmen, T. 2011. Soft errors from space to ground: Historical overview, empirical evidence, and future trends. In M. Nicolaidis (ed.), *Soft Errors in Modern Electronic Systems*. New York: Springer.

JESD89A. 2006. Measurement and reporting of alpha particle and terrestrial cosmic ray-induced soft errors in semiconductor devices, JEDEC Standard JESD89A, October 2006. Available at: http://www.jedec.org/download/search/jesd89a.pdf.

Kolasinski, W.A., Blake, J.B., Anthony, J.K., Price, W.E., and Smith, E.C. 1979. Simulation of cosmic ray induced soft errors and latchup in integrated circuit computer memories. *IEEE Transactions on Nuclear Science* NS-26:5087–5091.

May, T.C. and Woods, M.H. 1978. A new physical mechanism for soft errors in dynamic memories. In *Proceedings of the 16th Annual IEEE International Reliability Physics Symposium*, pp. 33–40. April, San Diego, CA: IEEE.

Messenger, G.C. and Ash, M.S. 1997. *Single Event Phenomena*. Dordrecht: Springer.

O'Gorman, T.J., Ross, J.M., Taber, A.H., et al. 1996. Field testing for cosmic ray soft errors in semiconductor memories. *IBM Journal of Research and Development* 40:41–50.

Pease, R.L. 2013. A brief history of the NSREC. *IEEE Transactions on Nuclear Science* 60:1668–1673.

Petersen, E., Koga, R., Shoga, M.A., Pickel, J.C., and Price, W.E. 2013. The single event revolution. *IEEE Transactions on Nuclear Science* 60:1824–1835.

Pickel, J.C. and Blandford, I.T. 1978. Cosmic ray induced errors in MOS memory cell. *IEEE Transactions on Nuclear Science* NS-30:1166–1171.

Slayman, C. 2010. Soft errors—Past history and recent discoveries. In *Proceedings of the IEEE International Integrated Reliability Workshop Final Report (IRW)*, pp. 25–30. 17–21 October, Stanford Sierra, CA: IEEE.

Wallmark, J.T. and Marcus, S.M. 1962. Minimum size and maximum packing density of non-redundant semiconductor devices. *Proceedings of the IRE* 50:286–298.

Wang, F. and Agrawal, V.D. 2008. Single event upset: An embedded tutorial. In *Proceedings of the 21st International Conference on VLSI Design*, pp. 429–434. 4–8 January, Hyderabad, IEEE.

Wyatt, R.C., McNulty, P.J., Toumbas, P., Rothwell, P.L., and Filz, R.C. 1979. Soft errors induced by energetic protons. *IEEE Transactions on Nuclear Science* NS-26:4905–4910.

Ziegler, J.F. and Lanford, W.A. 1979. Effect of cosmic rays on computer memories. *Science* 206:776–788.

Ziegler, J.F. and Puchner, H. 2004. *SER—History, Trends and Challenges*. San Jose, CA: Cypress Semiconductor.

Ziegler, J.F., Curtis, H.W., Muhlfeld, H.P., et al. 1996. IBM experiments in soft fails in computer electronics (1978–1994). *IBM Journal of Research and Development* 40:3–18.

Glossary

1D	One-dimensional
2D	Two-dimensional
3D	Three-dimensional
3D–IC	Three-dimensional integrated circuits
3T-DGFET	Three-terminal DGFET (connected gates)
3T-FinFET	Three-terminal FinFET (connected gates)
3T-GAA	Three-terminal GAA (connected gates)
3T-MC-NWFET	Three-terminals MC-NWFET (connected gates)
3T-Trigate	Three-terminal triple-gate (connected gates)
4T	Four-terminal
4T-DGFET	Four-terminal DGFET (independently biased gates)
4T-FinFET	Four-terminal FinFET (independently biased gates)
4T-MC-NWFET	Four-terminal MC-NWFET (independently biased gates)
6T	Six-transistor
AF	Acceleration factor
AIDA	Abstract Interfaces for Data Analysis
ANITA	Atmospheric-like Neutrons from thIck Target, a neutron facility of the Svedberg Laboratory at Uppsala University
ASER	Accelerated soft-error rate
ASTEP	Altitude Single event effects Test European Platform
ATE	Automatic test equipment
AWE	Atomic Weapons Establishment (United Kingdom)
Balmos3D	Home made quantum simulator for DGFET
BEOL	Back-end-of-line
BOX	Buried oxide (SOI technology)
BPSG	Borophosphosilicate glass
BREL	Boeing Radiation Effects Laboratory (United States)
BTE	Boltzmann transport equation
CAD	Computer-aided design
CCD	Charge-coupled device
CERF	CERN-EU high-energy Reference Field
CERN	European Organization for Nuclear Research
CG	Control gate
CGR	Galactic cosmic rays

CHIPS	Chiral Invariant Phase Space model
CME	Coronal mass ejection
CMOS	Complementary MOS
CNRS	French National Center for Scientific Research
CORSIKA	COsmic Ray SImulations for KAscade
CPU	Central processing unit
CR	Cosmic rays
CRY	Cosmic-Ray Shower Library
DG	Double-gate
DG-FET	Double-gate MOSFET
DNW	Deep N-well
DRAM	Dynamic random access memory
DSET	Digital SET
DUT	Device under test
EAS	Extensive air shower
ECC	Error correction code
EEPROM	Electrically erasable programmable read-only memory
EGN	Equivalent number of gates
EXPACS	EXcel-based Program for calculating Atmospheric Cosmic-ray Spectrum
FDSOI	Fully depleted SOI
FET	Field-effect transistor
FEOL	Front-end-of-line
FG	Floating gate
FinFET	Fin-shaped field-effect transistor
FIT	Failure in time
FLUKA	FLUktuierende KAscade, fully integrated particle physics Monte Carlo simulation package
FPGA	Field-programmable gate array
FWHM	Full width at half maximum
GAA	Gate-all-around
GANIL	Grand Accélérateur National d'Ions Lourds, Caen (France)
GCR	Galactic cosmic rays
GDS	Graphic Data System (database file format)
GEANT4	Toolkit for the simulation of the passage of particles through matter
GLE	Ground-level enhancement

GPS	General Particle Source (Geant4 class)
HTFET	Heterojunction tunnel FET
IC	Integrated circuits
ICP	Inductively coupled plasma
ICP-MS	Inductively coupled plasma mass spectrometry
IEC	International Electrotechnical Commission
IGY	International Geophysical Year
ILL	Institut Laue-Langevin
IM-DGFET	Inversion-mode DGFET
IQSY	International Quiet Sun Year
IRAM	Institut de Radioastronomie Millimétrique, Grenoble, France
IRPS	International Reliability Physics Symposium
IRT	Intel Radiation Tool
ISIS	ISIS neutron source at Rutherford Appleton Laboratory, Oxfordshire (United Kingdom)
ITRS	International Technology Roadmap for Semiconductors
JEDEC	Joint Electron Device Engineering Council
JEITA	Japan Electronics and Information Technology Industry Association
JL-DGFET	Junctionless (without junctions) DGFET
LA	Low alpha
LANSCE	Los Alamos Neutron Science Center, New Mexico (United States)
LET	Linear energy transfer
LLB	Laboratoire Léon Brillouin, Gif-sur-Yvette (France)
LSM	Underground Laboratory of Modane
MARS	Monte Carlo code that simulates the passage of particles through matter
MBU	Multiple-bit upset
MC	Monte Carlo
MCNP	Monte Carlo N-Particle Transport Code
MCNPX	Monte Carlo N-Particle eXtended
MC-NWFET	Multi-Channel Nanowire MOSFET
MC-ORACLE	Radiation transport code from University of Montpellier-2
MCU	Multiple-cell upset
MIM	Metal-insulator-metal
MOS	Metal-oxide semiconductor
MOSFET	Metal-oxide semiconductor field-effect transistor
MRED	Monte Carlo Radiative Energy Deposition code, Vanderbilt University

MS	Mass spectrometry
NAND	Negated AND
NBTI	Negative-bias temperature instability
NEL	Number of electron loss
NEST	NMDB Event Research Tool
nHTFET	N-channel HTFET
NIST	National Institute for Standards and technology (United States)
NM	Neutron monitor
NMDB	Neutron Monitor Database
NMOS	N-channel MOSFET
NOR	Negated OR
n-SER	Neutron-SER
NSREC	Nuclear and Space Radiation Effects Conference
NVM	Nonvolatile memories
N-well	Well of n-type
NWFET	Nanowire MOSFET
NYC	New York City, reference location for atmospheric neutron flux (sea level, outdoors, and quiet sun activity)
ONO	Oxide-nitride-oxide
PARMA	PHITS based Analytical Radiation Model in the Atmosphere
PdBNM	Plateau de Bure Neutron Monitor
PDSOI	Partially depleted SOI
PERALS	Photon-Electron Rejecting Alpha Liquid Scintillation
PHITS	Particle and Heavy Ion Transport Code System
PHITS-HYENEXSS	PHITS Hyper Environment for Exploration of Semiconductor Simulation
PLANETOCOSMICS	Simulation framework that computes the interactions of cosmic rays with Earth, Mars, and Mercury
PMOS	P-channel MOSFET
ppb	Parts per billion
PSI	Paul Scherrer Institute (Switzerland)
P-sub	P-type substrate
P-well	Well of p-type
QARM	Quotid (formerly QinetiQ) Atmospheric Radiation Model
QGSP	Quark-Gluon String Precompound model
RADECS	European Conference on Radiation and its Effects on Components and Systems
RAL	Rutherford Appleton Laboratory (United Kingdom)
RAM	Random-access memory

RCNP	Research Center for Nuclear Physics, Osaka (Japan)
ROI	Region of interest in the energy range
ROM	Read-only memory
ROOT	CERN data analysis framework (an object-oriented framework for large-scale data analysis)
RTSER	Real-time soft-error rate
SAA	South Atlantic Anomaly
SBGA	Super ball grid array
SBU	Single-bit upset
SEB	Single-event burnout
SEE	Single-event effects
SEGR	Single-event gate rupture
SEL	Single-event latchup
SEP	Solar energetic particle
SER	Soft-error rate
SET	Single-event transient
SEU	Single-event upset
SIMS	Secondary ion mass spectrometry
SoC	System-on-chip
SOI	Silicon-on-insulator
SPA	Single-photon absorption
SPICE	Simulation Program with Integrated Circuit Emphasis
SRAM	Static random-access memory
SRIM	Stopping and range of ions in matter
SRH	Shockley–Read–Hall
SSER	System SER
STI	Shallow trench isolation
SULA	Super ultralow alpha
TAT	Trap-assisted tunneling
TCAD	Technology computer-aided design
TCF	Transient carrier flux
TEM	Transmission electron microscopy
TFET	Tunneling FET
TIARA	Tool suite for rAdiation Reliability Assessment (radiation transport code from STMicroelectronics and Aix-Marseille University)
TIARA-G4	TIARA using Geant4 classes and libraries and compiled as a full Geant4 application
TIARA-G4 NVM	TIARA-G4 for nonvolatile memories
TID	Total ionizing dose

TNF	Neutron source facility at Tri-University Meson Facility accelerators, Vancouver, Canada
TO	Tunnel oxide
TRIUMF	Tri-University Meson Facility accelerators, Vancouver, Canada
TSL	The Svedberg Laboratory at Uppsala University, Sweden
TTF	Time-to-failure
TW	Triple well
ULA	Ultralow alpha
UTBB	Ultrathin body and box (SOI technology)
UT-FDSOI	Ultrathin FDSOI
VGM	Virtual geometry model
VLSI	Very-large-scale integration
VPD	Vapor-phase decomposition
WIPP	Waste Isolation Pilot Plant (WIPP) facility in Carlsbad (NM, United States)
WL	Wordline
α-SER	Alpha-soft-error rate
Ω-Gate	Omega-gate transistor
π-Gate	Pi-gate transistor

Section I

Environments
Definition and Metrology

1

Terrestrial Cosmic Rays and Atmospheric Radiation Background

1.1 Primary Cosmic Rays

The earth is continuously bombarded by cosmic radiation or cosmic rays (CRs) coming from all directions of deep space and from the sun. This extraterrestrial particle flux constitutes the so-called *primary cosmic rays*, a cocktail of charged energetic particles essentially composed of protons, helium nuclei, and heavy ions. The term *cosmic rays* may also include high-energy electrons, positrons, and other subatomic particles. The definition is wide since this generic term was historically introduced and then used overall to name the different components of this high-energy radiation that may hit the earth. We know now that these primary CRs have various origins and can be subdivided or classified into different categories (Dorman 2004), as reported in Table 1.1. Although the origin and properties of such deep-space radiation are one of the most intriguing problems in modern astrophysics (fundamental open questions concern their origin, acceleration, and propagation mechanisms) (Boezio and Mocchiutti 2012), its impact on the earth's atmosphere was discovered a century ago and has been the object of continuous research ever since. We provide, in the following, a short historical background before detailing the composition and main properties of these primary CRs.

1.1.1 Historical Background

CRs were discovered at the beginning of the twentieth century, although one of their physical manifestations was indirectly observed no later than the end of the eighteenth century by the French physicist and engineer Charles Augustin de Coulomb (1736–1806). Coulomb conducted a series of experiments describing the discharge of an electrically charged sphere perfectly isolated at the extremity of a silk wire (Capdevielle 1984). To explain the slow discharge observed over time, he assumed that dust particles in the surrounding air could be responsible for the discharge. But the problem remained unsolved for more than 100 years (Walter and Wolfendale 2012) because it could not be explained with the knowledge of the time: indeed, the role of CRs is fundamental in the continuous ionization of the air, directly contributing to this discharge phenomenon. One century later, the discovery of atmospheric ionization was demonstrated by Charles Thomson Rees Wilson (1869–1959), a Scottish physicist and meteorologist (his research work on atmospheric electricity, ionization, and condensation led him to design the first cloud chamber, an important discovery rewarded with the Nobel Prize in Physics in 1927). At first, Wilson speculated that the origin of this ionization might be natural radiation emitted by the earth. But scientists were

TABLE 1.1

Classification of Primary Cosmic Rays after Dorman

Origin of Cosmic Rays	Energies up to	Sources	Internal/External
Extragalactic	10^{21} eV	Radio-galaxies, quasars, powerful objects in the universe	*External* relative to our galaxy
Galactic	10^{15}–10^{16} eV	Supernovae explosions, in magnetospheres of pulsars and double stars, shock waves in interstellar space	*Internal* related to our galaxy and *external* to our heliosphere and the earth's magnetosphere
Sun	15–30 GeV	Powerful solar flares in the solar corona	*Internal* for the sun's corona and *external* for interplanetary space and the earth's magnetosphere
Interplanetary	10–100 MeV	Terminal shock wave at the boundary of the heliosphere	*Internal* to our heliosphere, *external* to the earth's magnetosphere
Magnetospheric	30 keV (Earth) 10 MeV (Jupiter and Saturn)	Generated inside the magnetospheres of rotating magnetic planets	*Internal* to the earth's magnetosphere

Source: Adapted from Dorman, L.I., *Cosmic Rays in the Earth's Atmosphere and Underground*, Kluwer Academic, Dordrecht, 2004.

soon intrigued by the excess of ions compared with the quantity normally induced by natural radiation at ground level. Several scientists then speculated about and studied the extraterrestrial nature of this underlying mechanism. In 1912, the Austrian scientist Victor Franz Hess (1883–1964) measured the ionization rate of air according to altitude using a gold-leaf electroscope in a balloon ascending up to 5 km of altitude. As illustrated in Figure 1.1, ionization was found to decrease up to 700 m and to increase above this height, with very little difference between day and night. Hess concluded that this so-called *penetrating radiation*

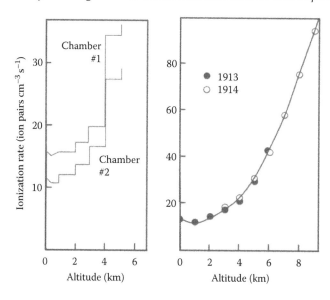

FIGURE 1.1

Increase of air ionization with height as measured by Hess in 1912 and by Kolhörster in 1913 and 1914.

has a cosmic origin (i.e., from beyond the solar system). In 1913–1914, Werner Kolhörster (1887–1946) completed Victor Hess's earlier results by measuring the increased ionization rate at an altitude of 9 km. These results were confirmed later by the American physicist Robert Millikan (1868–1953), who performed systematic measurements of ionization due to CRs from deep under water to high altitudes and around the globe. Millikan also introduced the name *cosmic radiation* in 1925. For "his discovery of cosmic radiation," Victor Francis Hess was awarded the Nobel Prize in physics in 1936.

CR identification was the basis of numerous works from the 1930s until after the Second World War. In 1927, the Dutch researcher Jacob Clay (1882–1955) characterized the variation of CR intensity with latitude, indicating that primary CRs are deflected by the geomagnetic field and must therefore be charged particles. This observation was confirmed in many other experiments. A wide variety of investigations also confirmed that primary CRs are mostly positive particles, identified as protons, and the secondary radiation produced in the atmosphere consists primarily of electrons, photons, neutrons, and muons (see Section 1.2).

1.1.2 Extragalactic and Galactic Cosmic Rays (GCRs)

A CR is a relativistic-speed particle (an atomic nucleus or an electron) that travels through deep space outside or inside our galaxy (the Milky Way), including the solar system. Some of these particles originate from the sun (see Section 1.1.3), but most of them come from sources outside the solar system and are known as galactic cosmic rays (GCRs). The particles that arrive at the earth's upper atmosphere are termed primary CRs; their collisions with atmospheric nuclei give rise to secondary or terrestrial CRs (see Section 1.2).

GCRs contribute to an energy density in our galaxy of about 1 eV cm^{-3}. They are mainly protons (hydrogen nuclei, 90%), with about 9% helium nuclei and smaller abundances (1%) of heavier elements (Valkovic 2000).

The flux of all nuclear components present in primary CRs is shown in Figure 1.2. This curve is known as the *all-particle spectrum*. At low energies (typically below ~30 GeV), the spectral shape bends down, as a result of the modulation imposed by the presence of a magnetized wind originating from the sun, which inhibits very-low-energy particles from reaching the inner solar system (Blasi 2013). The steepening of the spectrum at energy around 3×10^{15} eV is named the *knee*: at this point the spectral slope of the differential flux (flux of particles reaching the earth per unit time, surface, and solid angle, and per unit energy interval) changes from approximately −2.7 to −3.1. There is evidence that the chemical composition of CRs changes across the knee region, with a trend to becoming increasingly dominated by heavy nuclei at high energy, at least up to about 10^{17} eV. At even higher energies, the chemical composition remains a matter for debate (Blasi 2013).

1.1.3 Solar Wind and Solar Energetic Particles

The sun is a source of charged particles (plasma) ranging from a few tens of kiloelectronvolts to tens of gigaelectronvolts. Two main components can be distinguished: a continuous stream of particles, called the *solar wind*, and a second component related to burst emissions of solar energetic particles (SEPs).

The solar wind is a background stream of particles released from the upper atmosphere of the sun. It predominantly consists of electrons and protons with energies typically

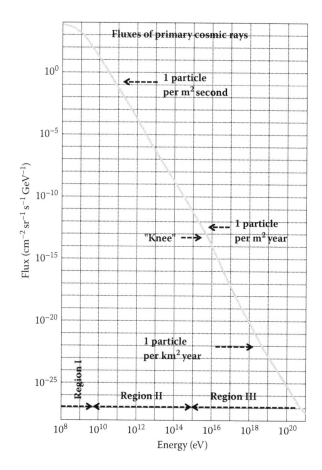

FIGURE 1.2
Flux of primary cosmic rays as a function of particle energy. The flux for the lowest energies (region I) is mainly attributed to solar cosmic rays, intermediate energies (region II) to galactic cosmic rays, and highest energies (region III) to extragalactic cosmic rays. (Data from Swordy, S., *Sci. Rev.* 99, 85–94, 2001.)

between 1.5 and 10 keV (Meyer-Vernet 2007). This stream of particles varies in density, temperature, and speed over time and over solar longitude. These particles can escape the sun's gravity because of their high kinetic energy and the high temperature of the corona. The solar wind flows outward supersonically to great distances, filling a region known as the heliosphere, an enormous bubble-like volume surrounded by the interstellar medium. The solar wind is divided into two components, respectively termed the *slow solar wind* and the *fast solar wind* (Meyer-Vernet 2007). The slow solar wind has a velocity of about 400 km s^{-1}, a temperature of 1.4–1.6 × 10^6 K, and a composition that is a close match to the corona. By contrast, the fast solar wind has a typical velocity of 750 km s^{-1} and a temperature of 8 × 10^5 K, and it nearly matches the composition of the sun's photosphere (Feldman et al. 2005). The slow solar wind is twice as dense as the fast solar wind, and more variable in intensity. The slow wind also has a more complex structure, with turbulent regions and large-scale structures.

With respect to the background solar wind, SEPs have higher energies, typically ranging from tens of kiloelectronvolts to tens of gigaelectronvolts. They are primarily ejected during solar flares or coronal mass ejections (CMEs). These latter are sometimes, but not always, associated with solar flares. Both types of events are the manifestation of sudden outbursts of magnetic energy from the sun's atmosphere; they eject enormous quantities of electrons, protons, and ions through the corona of the sun into space. When ejected in the direction of our planet, these particle clouds can reach the earth's vicinity in a relatively short time (hours, days) after the event, as a function of the speed of the emitted particles.

The intensity of the solar wind, as well as the frequency of solar flares and CME events, is directly linked to the solar activity, that is, the amount of radiation emitted by the sun. Solar activity variations have periodic components, the most remarkable being the 11-year solar cycle evidenced by the variations of the number of sunspots on the radiating surface of the sun (see Section 1.2.2.2).

Sunspots are the manifestation of intense magnetic activity; most solar flares and CMEs originate in magnetically active regions around visible sunspot groupings. Consequently, solar particle events are known to occur more frequently and with higher intensity during the solar maximum, and more precisely during the declining phase of the solar maximum. The impact of both solar wind and CME events on the production of secondary CRs in the atmosphere and at ground level is described in Section 1.2.2.

1.1.4 Magnetospheric Cosmic Rays

Energetic protons and electrons, associated with the magnetosphere, constitute a last category of energetic particles that can interact with the earth's upper atmosphere. These particles form two belts of energetic plasma around the earth, named after their discoverer, the American scientist James Van Allen (1914–2006). The Van Allen radiation belts are held in place around our planet under the action of the earth's magnetic field (see Figure 1.3). They are located in the inner region of the earth's magnetosphere and extend from an altitude of about 1000 to 60,000 km above the surface. Most of the particles that form the belts are thought to come from the solar wind, and other particles from CRs. The belts contain energetic electrons that form the outer belt and a combination of protons and electrons that form the inner belt. The radiation belts additionally contain lower amounts of other nuclei, such as alpha particles. The belts are known as a serious threat for satellites, which must protect their sensitive components with adequate solutions if their orbit spends significant time in such regions.

The Van Allen belts are symmetric about the earth's magnetic axis, which is tilted with respect to the earth's rotational axis by an angle of approximately 11°. The intersection between the magnetic and rotation axes of the earth is located not at the earth's center, but some 500 km further north. Because of this asymmetry, the inner Van Allen belt is closest to the earth's surface over the South Atlantic Ocean, where it dips down to 200 km altitude, and farthest from the earth's surface over the North Pacific Ocean. The area that precisely corresponds to the region where the inner radiation belt comes closest to the earth's surface is called the *South Atlantic Anomaly* (SAA, see Figure 1.3). This leads to an increased flux of energetic particles in this region and exposes orbiting satellites to higher-than-usual levels of radiation. If the SAA is primarily of great significance to satellites and other spacecraft that orbit the earth at several hundred kilometers of altitude, it is also directly responsible for an increase of particle flux (neutrons, muons) at ground level and for a modification of their energy distributions with respect to other locations on the earth (Augusto et al. 2008; Federico et al. 2010). The reason is that, in this area, the geomagnetic field is approximately 30% less than one would expect at similar latitudes and altitude in the world (see Section 1.2.2).

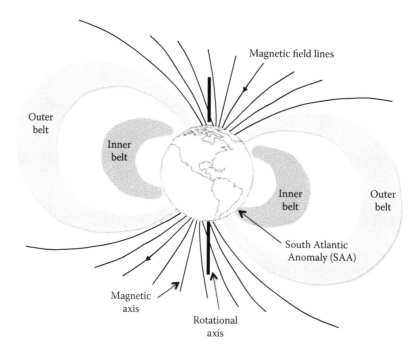

FIGURE 1.3
Schematic illustration of the earth's radiation belts with the South Atlantic Anomaly (SAA) indicated.

1.2 Secondary Cosmic Rays in the Atmosphere and at Ground Level

The interaction of primary CRs with the earth's upper atmosphere is the origin of atmospheric showers that produce secondary particles down to sea level. We detail in this section the development of these air showers and the different factors responsible for the modulation of particle production in the atmosphere and at ground level.

1.2.1 Development of Air Showers

An extended air shower (EAS) is a large cascade of ionized particles and electromagnetic radiation produced in the atmosphere when a primary CR enters the upper layers of the atmosphere (at about 40 km above sea level) and interacts with an air nucleus. EASs were discovered by the Italian American experimental physicist Bruno Rossi (1905–1993) in 1934 and independently, in 1937, by the French physicist Pierre Auger (1899–1993). By observing the CR with different detectors placed apart from each other, Rossi recognized that many particles arrive simultaneously at the detectors (Rao and Sreekantan 1998). Auger also found that the cosmic radiation events were coincident in time, meaning that they were associated with a single event.

The term *cascade* means that the incident high-energy (ultrarelativistic) particle strikes a molecule in the air so as to produce a cascade of new particles more or less in the same direction as the original primary particle. Each constituent of this secondary flux can in turn interact with air nuclei or decay, according to its mean lifetime. In both cases, new particles are created, which can be classified into three major components: the hadronic component (mostly mesons, such as pions and kaons), the muonic component (muons and neutrinos), and the electromagnetic component (electrons, positrons, and photons).

After the primary cosmic particle has collided with the air molecule, the first interactions primarily produce pions, and the secondary interactions produce kaons and baryons. Both pions and kaons are unstable; they decay into other particles, producing gamma photons, muons, neutrinos, and so on. As the air shower develops, the mean energy available to each particle in the shower front decreases. If the energy of the particle falls below the threshold for particle production, its energy is then gradually lost by ionization and other radiative processes. An EAS will thus reach a maximum number of particles as it develops as a function of atmospheric depth (see the definition in Section 1.2.2.3), after which it will deplete again by the loss of low-energy particles. A good approximation gives for an EAS a maximum of ~1–1.6 particles for every gigaelectronvolt (10^9 eV) of energy carried by the primary CR. In addition, the average depth in the atmosphere at which the shower maximum occurs varies logarithmically with the energy of the primary CR. The lateral extent of a shower also increases as a function of depth, reaching hundreds of meters or several kilometers at ground level, as a function of the incident primary. Figure 1.4 shows the typical development of an EAS obtained by numerical simulation for an incident primary CR of 10^{12} eV (1 PeV). For a better view, only a small fraction of secondary particles is shown. Figure 1.4 (*right*) also illustrates evolution of the total particle number with depth, as discussed above. For higher incident energies, the development of the EAS can be larger; for example, a vertical shower induced by an incident proton of 10^{19} eV will give, at ground level, about 10^{11} secondary particles with energy above 90 keV and a shower core of approximately 10 km. About 99% of the particles are photons, electrons, and positrons, with 90% of the primary particle energy being dissipated by the electromagnetic component.

1.2.2 Modulation Factors of Particle Production in the Atmosphere and at Ground Level

Different phenomena can modulate the primary flux of CRs impacting the earth and the production of EAS secondary particles in the atmosphere. We examine in the following a few important (natural) factors that directly determine the intensity of terrestrial CRs: location in the geomagnetic field, solar activity, and atmospheric depth (pressure and altitude).

1.2.2.1 Earth's Geomagnetic Field

To interact with the upper atmosphere and to initiate an EAS, a primary CR originating from interplanetary space must enter the magnetosphere surrounding the earth. This area of space around our planet is controlled by its dipolar magnetic field (resulting from electric currents in its core) and tends to deflect charged particles away via the Lorentz force. Figure 1.5 illustrates the structure of the earth's magnetosphere. Its distorted shape is the direct result of complex and dynamic interactions between the solar wind, the interplanetary magnetic field, and the geomagnetic field. The magnetosphere structure can be decomposed into different areas, as shown in Figure 1.5. The bow shock forms the outermost layer of the magnetosphere: it corresponds to the boundary between the magnetosphere and the interplanetary medium, where the solar wind is decelerating. In the latter, the pressure from the earth's magnetic field is balanced by the pressure from the solar wind. Between the bow shock and the magnetopause, the magnetosheath is mainly formed from shocked solar wind, though it contains a small amount of plasma from the magnetosphere. It is an area exhibiting high particle-energy flux, where the direction and

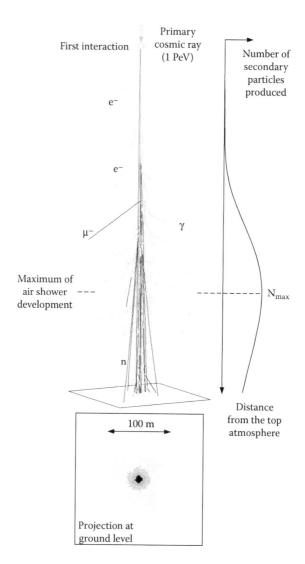

FIGURE 1.4
Development of an EAS induced by a primary cosmic ray of 1 PeV as obtained by numerical simulation. The lower figure shows the distribution of particles at ground level. (Adapted from Hinton, J.A., PhD. Thesis, University of Leeds, 1998. Courtesy of J.A. Hinton.)

magnitude of the magnetic field vary erratically. Finally, opposite the compressed magnetic field is the magnetotail, a region that extends far beyond the earth. It contains two lobes separated by a plasma sheet where the magnetic field is weaker and the density of charged particles is higher.

As stated, the magnetosphere plays the crucial role (in particular for life on Earth) of a natural magnetic shield or filter that deflects incoming charged particles more or less efficiently, depending primarily on their energy. To describe and quantify the efficiency of such a geomagnetic shield against the arrival of charged CRs from outside the magnetosphere, the concept of geomagnetic cutoff rigidity (R_c) has been introduced. This quantity is defined as the "minimum momentum per unit charge (magnetic rigidity) that an incident (often, vertically incident) particle can have and still reach a given location above

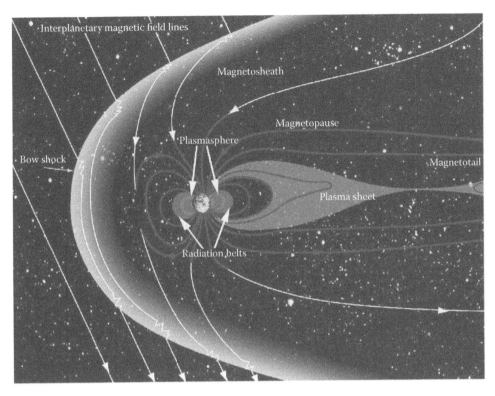

FIGURE 1.5
Schematic illustration of the earth's magnetosphere. This is a highly dynamic structure that responds dramatically to solar variations. (Credit: NASA/Goddard/Aaron Kaase.)

the Earth" (JEDS89A 2006). The effective vertical rigidity of geomagnetic cutoff is a good approximation of the value averaged over all directions. It depends primarily on the horizontal component of the earth's magnetic field. Its physical unit (SI) is the volt; R_c is near zero at the poles and a maximum of 15–17 GV at the equator (JEDS89A 2006). Consequently, the terrestrial CR flux is higher at the poles and lower at the equator, as we will show quantitatively in Section 1.3. Values of cutoff rigidity can be either calculated by different methods of particle trajectory tracing in the earth's magnetic field or obtained on the basis of experimental data on penetration of charged particles of solar CRs to spacecraft orbits (Nymmik et al. 2009). Figure 1.6 shows a projection on the earth's surface of the typical values of vertical cutoff rigidity computed for the International Geomagnetic Reference Field for 1995 (Smart and Shea 2007). These numerical values correspond to those reported in the JEDEC standard JESD89A (2006) and are frequently used to calculate the atmospheric neutron flux at a given location, as explained in Section 1.3.

1.2.2.2 Solar Activity

The sun is not only a source of energetic particles but also a modulator of terrestrial CRs through the variations of the solar wind and its magnetic field during the so-called *solar cycle* (ISES 2014). This dominant and most important timescale corresponds to an 11-year quasiperiodicity of solar activity (Potgieter 2013); it is clearly characterized by the number of sunspots and groups of sunspots present on the surface of the sun (the Wolf number).

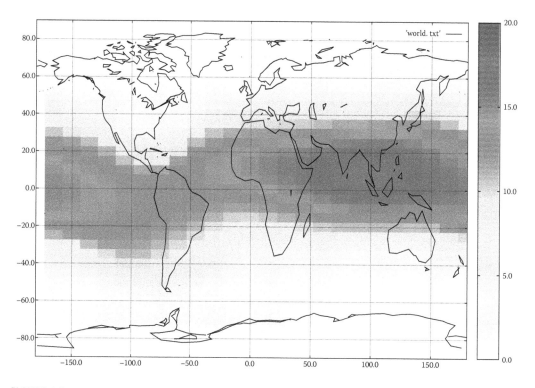

FIGURE 1.6
Mercator projection of the vertical cosmic-ray cutoff rigidity computed for the International Geomagnetic Reference Field for Epoch 1995.0. (Data from Smart, D.F. and Shea, M.A., *Proceedings of the International Cosmic Ray Conference*, 2007.)

Long-term monitoring with the worldwide network of neutron monitors (see Chapter 2) has revealed over decades that the flux of GCRs has a similar but opposite variation: when the number of sunspots is high, the GCR flux is low, and vice versa (Furet 2009). The main reason is that the magnetic field in the heliosphere is higher during periods of high activity, that is, when the number of sunspots is high. In other words, the active sun with its large solar wind creates an additional magnetic field around the earth, and this field increases the shielding against intragalactic CRs. The effect is to reduce sea-level CRs during the period of the active sun by about 30% (Ziegler 1996). Figure 1.7 shows the perfect anticorrelation experimentally observed between the CR flux monitored at earth level using a neutron monitor (deduced from pressure-corrected data; see Chapter 2) and the averaged sunspot number over the last six decades. In addition to these 11-year variations, other long-term solar cycles influence the GCR intensity and spectrum at the earth's level; for example, the 22-year solar-magnetic-field cycle is visible in CR data through maxima shape analysis of GCR intensity cycles (Usoskin et al. 2001).

Solar transient events occurring during geomagnetic storms can also influence the terrestrial flux of CRs. The most spectacular are solar flares and CMEs, defined in Section 1.1.3. When a CME impacts the heliosphere and the earth's magnetosphere, it temporarily changes their conjugated shielding efficiency against GCRs. This change may result in a sudden decrease followed by a gradual recovery of the observed GCR intensity, termed a Forbush decrease after the American astronomer Scott Forbush (1904–1984), who first evidenced and explained this effect. An example of a recent Forbush decrease, occurring on

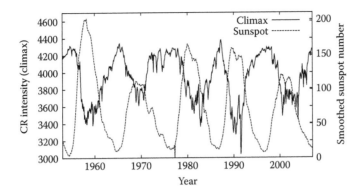

FIGURE 1.7
Anticorrelated variations between cosmic-ray intensity at earth level and the sunspot number at the surface of the sun. (Data from the University of Chicago [Climax neutron monitor] and from the National Geophysical Data Center [NGDC]).

February 15–17, 2011, is illustrated in Figure 1.8. This is the first Forbush decrease in solar cycle #24, after a very long solar minimum from December 2006. The sun was active, with an X-class flare and many M-class flares during this period. Figure 1.8 shows the responses of different neutron monitors located at Jungfraujoch (Switzerland), Rome (Italy), Athens (Greece), Kerguelen (French islands in the Southern Hemisphere), and Plateau de Bure (French South Alps; see Section 2.2). The coincidence between the different signals during this effect is spectacular, as well as the correlation between the signals in terms of amplitude variations before and after this decrease, demonstrating a global change of GCR flux at the earth's planetary level.

Finally, the sun emits particles (mainly protons) of sufficient energy and intensity to raise radiation levels on the earth's surface. These sudden emissions constitute a special class of events (solar proton events) in which ions are accelerated to relativistic energies, causing a significant sudden increase of CRs at ground level, mainly detected by neutron monitors (Papaioannou et al. 2014). To date, 71 such events have been identified since the initiation of reliable recordings, which began in the 1950s. The mean occurrence rate of ground-level enhancements (GLEs) is almost one per year, with a slight deviation due to intense solar activity around solar maximum. The most recent event was recorded on May 17, 2012, designated as GLE71 (Papaioannou et al. 2014).

1.2.2.3 Atmospheric Depth

EAS development is mainly determined by the amount of traversed air in the atmosphere.

For vertical EAS, a path integral of the air density, called atmospheric depth (X), is introduced:

$$X(h) = \int_{h}^{\infty} \rho(z)\,dz \qquad (1.1)$$

where ρ is the atmospheric density, which is a function of height (or altitude, h) above the earth's surface. X has the dimension of a mass divided by a surface. It is generally

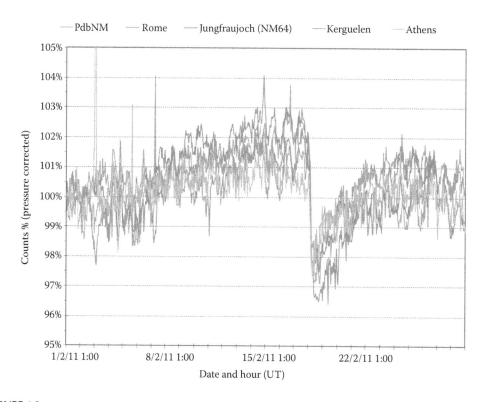

FIGURE 1.8

Comparison between pressure-corrected signals from five neutron monitors during the Forbush effect observed on February 15–17, 2011. (Data from Jungfraujoch, Rome, Athens, and Kerguelen neutron monitors courtesy of the Monitor DataBase [NMDB]).

expressed in grams per square centimeter. As given by Equation 1.1, X is the integral of density of the overlying air.

For vertical EAS, this quantity can be easily linked to the atmospheric pressure P(h) at altitude h, which is due to the weight of the air column above that point. Considering that the earth is so large, the atmosphere so thin, and the acceleration of gravity g is essentially constant (within 1%), we obtain

$$X(h) \approx \frac{P(h)}{g} \qquad (1.2)$$

Equation 1.2 indicates that, at a given altitude h, the atmospheric pressure (in Pa) divided by the acceleration of gravity g ($g = 9.8$ m s^{-2} at sea level) is a good estimation of the atmospheric path. At sea level and for the reference atmospheric pressure $P_0 = 101,325$ Pa, the atmospheric depth is $X_0 = 10,339$ kg m$^{-2} = 1034$ g cm^{-2} (1000 g cm^{-2} is a very good order of magnitude generally indicated in the literature). At an altitude of 15 km, which approximately corresponds to maximum muon production, this quantity drops to $X = 130$ g cm^{-2}.

For nonvertical EAS, the value of X is calculated by integrating the density of air from the entry point of the air shower at the top of the atmosphere, along the trajectory of the shower, to the point in question. In this case, the quantity X is termed *slant depth*. An inclined shower will thus traverse more than 1000 g cm^{-2} to reach sea level.

Following the above convention, the depth of shower maximum is denoted X_{max}. With a value of about 500 g cm^{-2} at 10^{15} eV, the average X_{max} for CR showers increases by about 60–70 g cm^{-2} for every decade of energy.

From Equations 1.1 and 1.2, it is now clear that the flux of terrestrial CRs at ground level, as the result of EAS development in the atmosphere, will be dependent on the altitude of the measurement location at the earth's surface and, for that location, will also be influenced by variations of the atmospheric pressure due to changes in weather conditions. An increase in altitude or a decrease of atmospheric pressure will result in an increase of the secondary particle flux at ground level (total flux for all produced particles), because the maximum of the EAS will be nearest the surface in this case. This point will be illustrated in Section 1.3 and details will be given for each type of secondary particle.

1.3 Radiation Environment at Ground Level (Particles, Flux, Variations, Shielding)

This section reports information and useful data concerning the different components of the terrestrial CRs that significantly interact with matter at ground level; the term *significantly* means, in the framework of this book, those that are capable of inducing single-event effects in terrestrial electronic circuits. The information detailed in this section is thus essentially restricted to atmospheric neutrons, protons, pions, and low-energy (<GeV) muons at sea level and at mountain altitudes. Other terrestrial cosmic ray components, such as electrons, positrons, photons, and neutrinos (Grieder 2001) are not able to induce significant effects in components; they will not be considered in the following.

We will frequently refer, in this section, to the EXPACS and QARM atmospheric radiation models; these models will be described and referenced in Section 1.4.

1.3.1 Particle Fluxes at Sea Level

The production of EAS secondary particles becomes significant at about 55 km of altitude in the atmosphere, with an intensity reaching a maximum (termed the *Pfotzer maximum*) at approximately 20 km. The intensity of secondaries then decreases from this maximum to the earth's surface as the particles lose their energy by additional collisions until the majority either decay or are absorbed. At ground level, the result of this production of secondaries in the atmosphere is a continuous flux of particles distributed in energy. Figure 1.9 shows typical energy distributions of particle flux (differential flux dϕ/dE) for high-energy (above 1 MeV) atmospheric neutrons, protons, muons, and pions. The location and conditions for these curves have been chosen as New York City (NYC) outdoors at sea level and at a time of average solar activity (*quiet sun*), denoted as *reference location* in the following. The same distributions are plotted in fluence rate per lethargy in Figure 1.10a; in such a lethargic representation (E×dϕ/dE on the vertical axis vs. E with a logarithmic scale on the horizontal axis), equal areas under the spectrum in different energy regions represent equal integral fluxes. The corresponding integral flux values above 1 MeV are shown in Figure 1.10b and reported in Table 1.2. These figures and table give a very good overview of the natural radiation environment of atmospheric origin at sea level. As we will show in Section 1.3.2, CR component intensities, characteristics of other locations around the world, and solar activity index can be roughly estimated from these reference values by

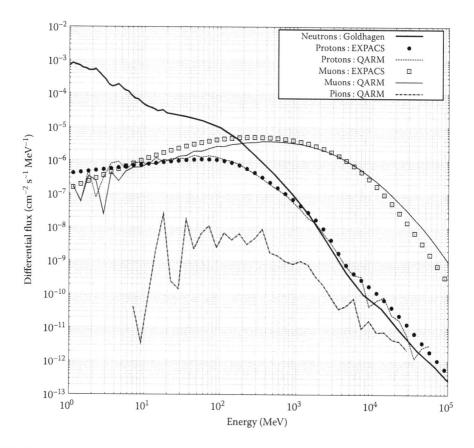

FIGURE 1.9
Differential flux for atmospheric neutrons, protons, muons, and pions as a function of particle energy. These spectra have been measured (neutrons) or computed (protons, muons, pions) for reference conditions (sea level, New York City, midlevel solar activity, outdoors). The data points correspond to the EXPACS model, the thin curves to the QARM model, and the bold curve to Goldhagen's data for neutrons.

applying simple analytical transformation factors. We discuss, in the following, the different spectra, flux values, and other particularities (angular dependence, etc.) per particle type.

1.3.1.1 Muons

Muons are the products of the decay of secondary charged pions ($p^{\pm}$) and kaons ($K^{\pm}$). Despite their lifetime of about 2.2 μs, most muons survive to sea level due to their ultra-relativistic character. They are the most abundant particles at sea level, with a total ($\mu^{+} + \mu^{-}$) integrated flux above 1 MeV of around 60 muons $cm^{-2}\,h^{-1}$, as estimated using the EXPACS or QARM models (Figure 1.10 and Table 1.2). High-energy physicists are familiar with an order of magnitude of one particle per square centimeter and per minute for horizontal detectors (Beringer et al. 2012). This corresponds to ~70 $m^{-2}\,s^{-1}\,sr^{-1}$ above 1 GeV.

An abundant literature exists for muons at ground level and underground, but the overwhelming majority of work concerns high-energy physics, or particle physics with energy ranges typically above the gigaelectronvolt range. With respect to the scope of this book and to the impact of atmospheric particles on terrestrial electronics, we will see later

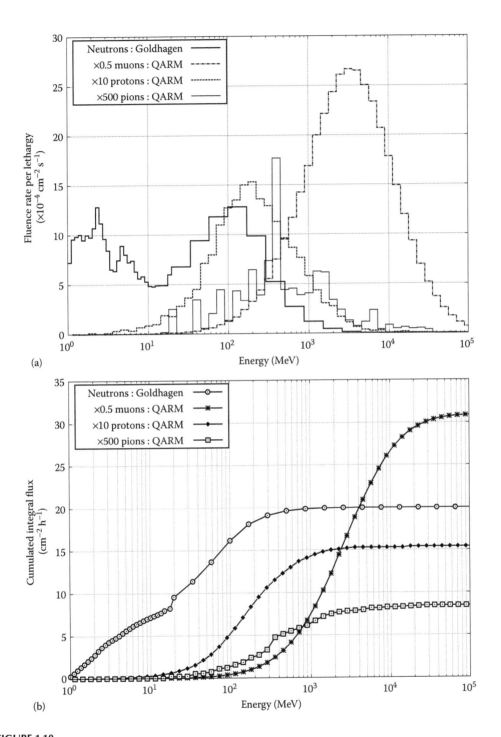

FIGURE 1.10
Fluence rate per lethargy (a) and integral flux (b) for high-energy (E > 1 MeV) atmospheric neutrons, protons, muons, and pions as a function of particle energy. Same data as used in Figure 1.9 for neutrons (Goldhagen) and for muons, protons, and pions (EXPACS model).

TABLE 1.2

Integrated Flux above 1 MeV for Different Terrestrial Cosmic-Ray Particles

Particles	Integrated Flux for E > 1 MeV (cm^{-2} h^{-1})		
	EXPACS Model	QARM Model	Measurements
Neutrons	23.0	39	20.0
Muons	61.8	63.3	–
Protons	1.54	1.53	–
Pions	–	1.69×10^{-2}	–

Note: Conditions: New York City outdoors, sea level, quiet sun. The experimental flux for neutrons above 1 MeV is deduced from Goldhagen's data.

(Chapter 5) that pertinent energy ranges for such effects are, rather, in the megaelectronvolt range and below, an energy domain not covered by the great majority of studies performed until now. This lack of knowledge in the characterization of low-energy muons (below 1 MeV) at ground level is a challenging issue for the prediction of radiation effects in microelectronics; this point will be discussed in Chapter 10.

Figure 1.9 shows that, between 1 MeV and a few tens of megaelectronvolts, the differential flux of muons is similar to that of protons; above 200 MeV it becomes higher than the flux of all other particles, including neutrons. With respect to the integral flux from 1 MeV (Figure 1.10b), muons dominate the high-energy particle spectra, typically above 1 GeV.

The overall angular (zenith angle θ) distribution of muons at the ground is $\propto \cos^2 \theta$, which is characteristic of muons with energy around 3 GeV. At lower energy, the angular distribution becomes increasingly steep, while at higher energy it flattens, approaching a sec θ distribution for $\theta < 70°$ (Beringer et al. 2012). In particular, muons with energy below 1 GeV fade fairly quickly with increasing zenith angle, with dependence $\propto \cos^n \theta$, where n ~ 2–3 (Cecchini and Spurio 2012).

Another important characteristic of muons at sea level is their charge ratio. It is well known from decades of observation that the number of positively charged muons exceeds the number of negatively charged muons over a wide range of energy. This muon charge ratio reflects the excess of π^+ over π^- and K^+ over K^- in the forward fragmentation region of proton-initiated interactions, together with the fact that there are more protons than neutrons in the primary spectrum. Typical muon charge ratio for muon momentum below 1 GeV c^{-1} is thus around 1.1–1.2 (Beringer et al. 2012).

1.3.1.2 Neutrons

After muons, the next most abundant particles at sea level are neutrons. Neutrons represent the most important part of the natural radiation constraint at ground level likely to impact electronics. Because neutrons are not charged, they are very invasive and can penetrate deeply into circuit materials. In contrast to muons, for which quantitative information is missing below the megaelectronvolt level, the energy distribution of neutrons at ground level is well characterized from thermal to high-energy values. In complement to Figures 1.9 and 1.10, Figure 1.11 shows this full energy distribution (lethargic representation) up to 1 GeV, as measured by Goldhagen (2003) and Gordon et al. (2004) using a Bonner multisphere spectrometer at the reference location.

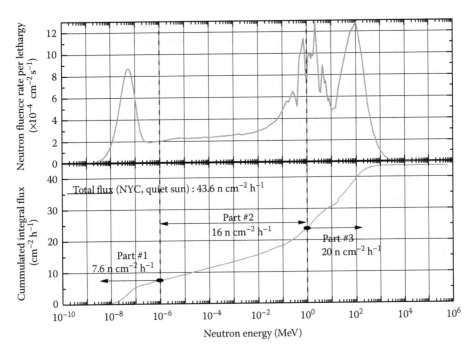

FIGURE 1.11

Top: Complete reference atmospheric neutron spectrum measured on the roof of the IBM Watson Research Center main building. (Data courtesy of Paul Goldhagen, U.S. Department of Homeland Security). *Bottom*: Cumulated integral flux corresponding to the above spectrum. The total neutron flux is 43.6 neutrons $cm^{-2} h^{-1}$. (Reprinted from Autran, J.L., et al. Soft-error rate induced by thermal and low energy neutrons in 40 nm SRAMs, *IEEE Trans. Nucl. Sci.* 59, 2658–2665. © 2012 IEEE. With permission.)

The spectrum has three broad peaks: a high-energy peak centered at about 100 MeV and extending up to about 10 GeV, a *nuclear evaporation* peak centered around 1 or 2 MeV, and a thermal peak. This latter corresponds to neutrons that have been slowed down by scattering until they are in thermal equilibrium with atoms in surrounding materials (Gordon et al. 2004). Between the evaporation peak and the thermal peak, there is a plateau region where $d\varphi/dE$ is approximately proportional to $1/E$. The evaporation peak has fine structure from nuclear resonances in the nitrogen and oxygen of the atmosphere and in materials in the concrete roof. As explained by Gordon et al. (2004), most of this structure is finer than the resolution of the spectrometer, and it appears in the measured spectrum because it is present in the calculated spectrum used as the default spectrum for the unfolding.

The integration of this spectrum, shown in Figure 1.11 (*bottom*), gives the total neutron flux expressed in neutrons per square centimeter per hour: this flux is equal to 7.6 neutrons $cm^{-2} h^{-1}$ for the lower part (thermal and epithermal neutrons below 1 eV), 16 neutrons $cm^{-2} h^{-1}$ for the intermediate part (between 1 eV and 1 MeV), and 20 neutrons $cm^{-2} h^{-1}$ for the upper part (high-energy neutrons above 1 MeV). Another value often reported and used in the soft-error-related literature is 13 neutrons $cm^{-2} h^{-1}$, which corresponds to the same flux but integrated above 10 MeV.

These two last values are frequently used in the radiation effects community for expressing the averaged intensity of the high-energy-neutron radiation background of a given location with respect to that of the reference location NYC. As we will see in Chapter 7, in the framework of real-time SER experiments, a dimensionless factor termed *acceleration factor* (AF) is defined as the ratio of the neutron integrated flux at a given test location (IF$_{test}$) divided by this reference value (see Equation 7.3). The term *acceleration factor* suggests that, for a given exposure to atmospheric neutrons, an experiment conducted at the location defined by the AF will be shorter by this factor compared with a similar experiment carried out at sea level. Evidently, NYC is characterized by a unit AF under mean sea-level atmospheric pressure (1013 hPa).

In their 2004 paper, Gordon et al. (2004) provided tabulated values for the reference NYC differential flux shown in Figure 1.11. They also proposed a simple analytical expression that has been fitted on this spectrum in the energy range from 0.1 MeV to 10 GeV:

$$\frac{d\Phi_0(E)}{dE} = \sum_{j=1}^{2} c_j \exp\left[-\beta_j\left(\ln(E)\right)^2 + \gamma_j \ln(E)\right] \quad (1.3)$$

where the numerical values of the fitting parameters are as follows: $\beta_1 = 0.3500$, $\beta_2 = 0.4106$, $\gamma_1 = 2.1451$, $\gamma_2 = -0.6670$, $c_1 = 1.006 \times 10^{-6}$, and $c_2 = 1.011 \times 10^{-3}$.

This analytical fit is reported in the standard JESD89A (2006). Figure 1.12 shows the comparison of this analytic fit with the reference neutron spectrum. The proposed fit is found to be very good above 10 MeV and reasonably good in the evaporation region down to about 0.4 MeV.

In complement to the previous model, we proposed another analytical description of the reference spectrum including the whole energy range, from thermal to high-energy

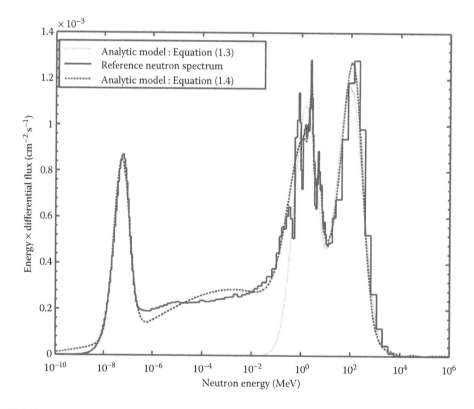

FIGURE 1.12

Reference neutron spectrum at New York City (same data as in Figure 1.11) and analytic fits from Equation 1.3 (From Gordon, M.S., Goldhagen, P., Rodbell, K.P. et al., *IEEE Trans. Nucl. Sci.* 51, 3427–3434. © (2004) IEEE.) for the upper-energy portion and from Equation 1.4 for the full energy range.

values. This model is based on power polynomial fitting functions well adapted to the description of the fluence rate per lethargy:

$$E \times \frac{d\Phi_0(E)}{dE} = \sum_{j=1}^{8} a_j \exp\left[-\left(\frac{\log(E) - b_j}{c_j}\right)^2\right] \tag{1.4}$$

The 24 numerical parameters of Equation 1.4 are given in Table 1.3. Figure 1.12 compares this additional fit with the reference spectrum; a very good agreement is found for the thermal peak and above 10^{-1} MeV. In the intermediate energy domain between 3×10^{-7} and 10^{-1} MeV, the fit is less accurate but reasonable in view of the overall agreement of this fit on the full spectrum.

The last point concerns the angular distribution of neutrons. Similarly to muons, the neutron flux at sea level has been found to be a function of the zenith angle θ, following a dependence $\propto \cos^n \theta$. Different values for the coefficient n can be found in past literature. Among the most recent studies, we can cite the results obtained by Moser et al. (2005) and Mascarenhas et al. (2007). Moser et al. used a neutron double-scatter telescope to characterize the neutron-flux spectrum and zenith angle distribution; their data were correctly fitted with $n = (2.9 \pm 0.3)$. Mascarenhas et al. used a neutron-scatter camera to measure the CR neutron background flux at Livermore, CA (approximately at sea level) as a function of angle. The measured neutron background (integrated over all angles) satisfactorily agreed with the reference spectrum from Gordon et al. (2004). In addition, they measured the angular dependence of the detected neutron flux, well described with a power exponent evaluated to $n = 2.7$.

1.3.1.3 Protons

Protons are several tens of times less numerous than muons and neutrons at ground level. The integral intensity of vertical protons above 1 GeV c^{-1} at sea level is ≈ 0.9 m^{-2} s^{-1} sr^{-1} (Beringer et al. 2012); both EXPACS and QARM models predict a total integrated flux of protons above 1 MeV equal to 1.5 protons cm^{-2} h^{-1} (Table 1.2 and Figure 1.10b).

Figure 1.9 shows that the differential flux for protons is similar to that of muons below a few tens of megaelectronvolts and close to that of neutrons above 1 GeV. But, as for

TABLE 1.3

Values of the Fitted Parameters for the Analytical Model of Equation 1.4 Reproducing the Reference Neutron Spectrum over Its Full Energy Range

j	a_j	b_j	c_j
1	0.000388	0.3811	0.07495
2	0.001175	2.07	0.5644
3	0.0007698	0.1013	0.8751
4	0.0007723	−7.277	0.4567
5	0.0001866	1.42	0.4537
6	0.0001992	0.7418	0.1034
7	0.000286	−2.766	4.07
8	-1.734×10^{-5}	4.395	0.6521

muons, an evident lack of data characterizes the low-energy domain, typically around and below a few megaelectronvolts. In this range, the EXPACS model predicts a differential flux more important for protons than for muons. The impact of both low-energy protons and muons on upcoming and future integrated circuit technologies will be discussed in Chapter 10.

1.3.1.4 Pions

Atmospheric charged pions are typically produced at an altitude of about 15 km and decay very fast (lifetime of 26 ns), producing mostly muons and neutrinos. Neutral pions have a much shorter lifetime (8.4×10^{-17} s) and produce gammas, electrons, and positrons. The very low intensity of pions observed at sea level is predominantly produced in local interactions. The QARM model (see Figures 1.9 and 1.10) predicts a total integrated flux above 1 MeV of 1.69×10^{-2} pions $cm^{-2} h^{-1}$ for charged pions. This is a hundred times lower than for other particles, which could explain why this component of natural radiation at ground level is not well characterized. The impact of pions on electronics will be discussed in Chapter 5.

1.3.2 Flux Variations

As introduced in Section 1.2.2, the intensity of all CR-induced particles in the atmosphere varies with location in the geomagnetic field, atmospheric depth, and solar magnetic activity. We quantitatively examine in this section the impact of such parameters on the terrestrial CR intensities at ground level.

1.3.2.1 Variations with Atmospheric Depth

Throughout most of the atmosphere, the altitude dependencies of secondary CR fluxes (I) are well described by Desilets and Zreda (2003) and Ziegler (1996):

$$-\frac{dI}{dX} = \frac{1}{\Lambda} = \beta I \tag{1.5}$$

where X is the atmospheric depth given in mass-shielding units ($g\ cm^{-2}$) and Λ is the atmospheric attenuation length ($g\ cm^{-2}$) and β its reciprocal, the attenuation coefficient ($cm^2\ g^{-1}$).

The solution to Equation 1.5 is a simple exponential relation:

$$I_2 = I_1 \exp\left(\frac{X_1 - X_2}{\Lambda}\right) = I_1 \exp\left(\beta(X_1 - X_2)\right) \tag{1.6}$$

where I_1 is the flux at some altitude (or pressure) X_1, and I_2 is the flux at altitude (or pressure) X_2, both altitudes being expressed in grams per square centimeter.

Considering a reference flux I_0 at sea level (altitude h = 0 m), Equation 1.6 can be rewritten as

$$F_A(h) = \frac{I(h)}{I_0} = \exp\left(\frac{X_0 - X(h)}{\Lambda}\right) \tag{1.7}$$

where $X_0 = 1034$ g cm^{-2} is the atmospheric depth at sea level and $F_A(h)$ is a modulation function describing the principal dependence of the terrestrial CR flux with altitude (JEDS89A 2006).

At low altitudes (the troposphere), the altitude dependence of X can be very easily introduced into Equation 1.7 considering one of the following analytic approximations (not exhaustive):

- For $h < 11$ km, $X(h) = \exp\{5.26 \times \ln[(44.34 - h)/11.86]\}$, where altitude h is in kilometers and atmospheric depth X is in grams per square centimeter (Stanev 2003)
- Using Equation 1.2 and considering an analytical expression for the altitude dependence of the atmospheric pressure, for example, $P(h) = ((44331.514 - h)/11880.516)^{5.255877}$, where P is in hectopascals and h is in meters (JEDS89A 2006)
- $X(h) = 1034 - (0.03648 \times h) + 4.26 \times 10^{-2} \times h^2$, where X is in grams per square centimeter and h is in feet (Ziegler 1996)

The atmospheric attenuation length Λ introduced in Equation 1.5 depends on the particle type and energy; a longer absorption length means slower attenuation, and hence less difference in flux when comparing locations with different altitudes (Ziegler 1996). Numerous experimental and simulated values have been reported in literature over decades in the framework of different studies (CR or EAS characterization and simulation, neutron-monitor operation, muon tomography, accelerator science, etc.). In the particular context of soft-error-related studies, Ziegler reported the following typical absorption lengths in the lower atmosphere (Ziegler 1996): $\Lambda_e = 100$ g cm^{-2} for electrons, $\Lambda_p = 110$ g cm^{-2} for protons and pions, $\Lambda_n = 148$ g cm^{-2} for protons, and $\Lambda_\mu = 520$ g cm^{-2} for muons.

Gordon et al. (2004) reported a value $\Lambda_n = 135$ g cm^{-2} deduced from measurements performed with a neutron spectrometer transported to different locations (see also Section 1.3.2.2). Moser et al. (2005) also performed itinerant measurements using a neutron-double-scatter telescope; they obtained an averaged value of $\Lambda_n = (134 \pm 7)$ g cm^{-2}. In the same way, a value of $\Lambda_n = 133.3$ g cm^{-2} can be deduced from data reported by Semikh et al. (2012) considering neutron-monitor measurements (see figure 4 of this reference) performed at sea level and at mountain altitude using the same instrument.

Numerical simulation can be an accurate alternate solution to Equation 1.6 to compute the intensity variation of CR components with altitude. Figure 1.13 shows the altitude dependence of the integral flux ($E > 1$ MeV) for atmospheric neutrons, protons, muons, and electrons. Data have been obtained using the EXPACS model for NYC location and quiet sun conditions.

This figure shows that, if muons dominate at sea level, neutrons rapidly become the most important component of atmospheric radiation when the altitude increases above 1.8 km. At avionics altitudes (10 km), the integral flux of neutrons is 300 times the value observed at sea level (about 6000 neutrons cm^{-2} h^{-1}). The proton flux intensity also dramatically increases with altitude, due to a relatively short attenuation length with respect to neutrons and muons.

A last important remark for this section concerns the impact of the daily and seasonal variations of atmospheric pressure at a fixed location. After Equations 1.2, 1.6, and 1.7, such pressure variations introduce atmospheric-particle-flux variations at ground level: a decrease of atmospheric pressure reduces the atmospheric depth and, consequently, increases the flux at the considered location. Inversely, an increase of pressure results in a corresponding decrease of the particle flux. In other words, the atmospheric particle flux and the atmospheric pressure, both related to a given location at ground level, show

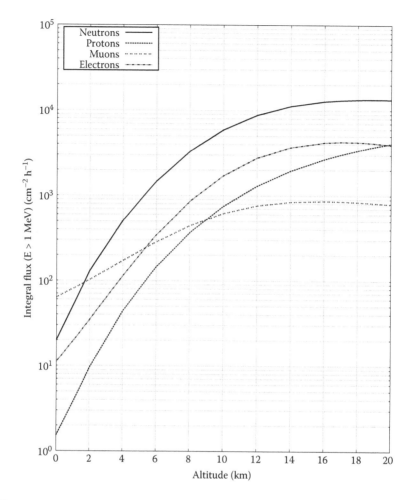

FIGURE 1.13
Integral flux (E > 1 MeV) for atmospheric neutrons, protons, muons, and electrons as a function of the altitude for the geographic coordinates (42° N 72° W) corresponding to New York City. Data obtained using the EXPACS model.

opposite variations, as is clearly shown by the signal of neutron monitors, for example, being perfectly anticorrelated with the local atmospheric pressure. This aspect will be more widely illustrated and discussed in Chapter 2.

1.3.2.2 Variations with Geomagnetic Location and Solar Activity

In addition to the atmospheric shielding described in the previous paragraph, the intensity of the different atmospheric particle fluxes also depends on the geomagnetic location and solar activity.

The impact of such factors has been qualitatively discussed in Sections 1.2.2.1 and 1.2.2.2 on the development of particle air showers. From a quantitative point of view, numerical (Monte Carlo) simulation is generally used to take into account the effects of these modulation factors on the particle fluxes. Numerical tools and codes, as described in Section 1.4, can be used to estimate the flux and the energy distribution of a given particle type at a given location and for a specific solar activity.

Because it is time consuming to systematically perform Monte Carlo simulation of terrestrial CRs for each flux calculation, alternate solutions consider analytic functions or models parameterized on experimental values. We can cite, for example, the considerable work performed by Sato and Niita (2006) and Sato et al. (2008). They proposed a series of analytic models for estimating the atmospheric CR spectra for neutrons, protons, He nuclei, muons, electrons, positrons, and photons applicable to any global conditions at altitudes below 20 km. The model was designated PHITS (Sato et al. 2013)-based Analytical Radiation Model in the Atmosphere (*PARMA*). PARMA is able to calculate CR doses and energy distributions instantaneously with precision equivalent to that of the Monte Carlo simulation, which requires much computational time. The EXPACS tool described in Section 1.4 is based on precisely this PARMA model.

For neutrons, considerable work has also accumulated during the last decades in the framework of the modeling and simulation of neutron-monitor operation and response functions. Based on this work, and similarly to Equation 1.7, Gordon et al. (2004) introduced a second parameterized function to take into account the location dependence of the neutron flux in terms of cutoff rigidity, solar modulation, and atmospheric depth. Considering that the shape of the neutron spectrum above 5 MeV does not change significantly with altitude, cutoff, or solar modulation, the neutron-fluence-rate spectrum outdoors at any location can be expressed as follows:

$$\frac{d\phi(E)}{dE} = \frac{d\phi_0(E)}{dE} \times F_A(h) \times F_D(R_c, h, I) \tag{1.8}$$

where:

$d\phi_0(E)/dE$ is the differential neutron spectrum at reference location given by Equation 1.3
$F_A(h)$ is the function describing the dependence on atmospheric depth given by Equation 1.7
$F_D(R_c, h, I)$ is a second modulation function describing the dependence on geomagnetic location and solar modulation

F_D corresponds to an atmospheric-depth Dorman function (Clem and Dorman 2000) parameterized by Belov, Struminsky, and Yanke to describe the location dependence of ground neutron monitors in terms of cutoff and solar activity (Gordon et al. 2004):

$$F_D(R_c, h, I) = N\left(1 - \exp\left(-\alpha/R_c^k\right)\right) \tag{1.9}$$

where:

R_c is the vertical cutoff rigidity
I is the relative count rate of a neutron monitor measuring solar modulation
N is a normalization factor
α and k are two parameters that depend on both atmospheric depth h and neutron-monitor signal I

Values of these parameters for solar minimum (S_{min}), maximum (S_{max}), and midlevel (S_{mid}) activities are reported in Gordon et al. (2004) with a normalization factor $N = 1.098$, which gives precisely $F_D = 1$ for midlevel solar activity at NYC ($R_c = 2.08$ GV, sea level, outdoors) and under standard pressure (101,325 Pa).

Using barometric pressure p (expressed in bar) instead of depth h (in g cm^{-2}), the two renormalized functions for S_{min} and S_{max} are given by the following expressions (Gordon et al. 2004):

$$F_{D,Smin}\left(R_c,p\right) = 1.098 \times \left(1 - \exp\left(-\alpha_1/R_c^{k1}\right)\right) \tag{1.10}$$

$$F_{D,Smax}\left(R_c,p\right) = 1.098 \times \left(1 - \exp\left(-\alpha_2/R_c^{k2}\right)\right)$$
$$\times\left(1 - \exp\left(-\alpha_1/50^{k1}\right)\right)\Big/\left(1 - \exp\left(-\alpha_2/50^{k2}\right)\right) \tag{1.11}$$

where parameters α and k are given by

$$\alpha_1\left(p\right) = \exp\left[1.84 + 0.094p - 0.09\exp\left(-1.1p\right)\right] \tag{1.12}$$

$$k_1\left(p\right) = 1.4 - 0.56p + 0.24\exp\left[-8.8p\right] \tag{1.13}$$

$$\alpha_2\left(p\right) = \exp\left[1.93 + 0.15p - 0.18\exp\left(-10p\right)\right] \tag{1.14}$$

$$k_2\left(p\right) = 1.32 - 0.49p + 0.18\exp\left[-9.5p\right] \tag{1.15}$$

Figure 1.14 shows the evolution of this F_D modulation factor with the vertical cutoff rigidity, as deduced from Equations 1.10 through 1.15 at sea level (p=1 bar) and for solar minimum, maximum, and midlevel activities. After Gordon et al. (2004), F_D is able to fit the observed cutoff dependence and monthly averaged solar modulation of the rates of many neutron monitors. This fit is reasonably good for all the observed solar cycles except for the extremely low rates between 1989 and 1991. Because both functions F_A and F_D are atmospheric depth dependent, the best fit of experimental data using Equation 1.8 was obtained with a neutron-attenuation length $\Lambda_n = (131.3 \pm 1.3)$ g cm^{-2}; this value is slightly different from the value reported in Section 1.3.2.1 by the same authors and extracted from the measured flux, not corrected in this case for F_D.

1.3.3 Shielding Issues

All building materials, snow, and water attenuate atmospheric particle flux as a function of their chemical composition, internal microstructure, density, and thickness.

Historically, this effect was first investigated and quantified for atmospheric neutrons, which represent the most important threat for electronics at ground level. Ziegler (1996) and Dirk et al. (2003) conducted a series of experiments and simulations for atmospheric neutrons, precisely in the framework of soft-error-related studies. For the portion of the spectrum above 10 MeV, a simple exponential absorption law with a mean attenuation length (Λ_n) for neutrons allows expression of an AF under the form

$$\phi = \phi_0 \exp\left(-L \times \rho / \Lambda_n\right) \tag{1.16}$$

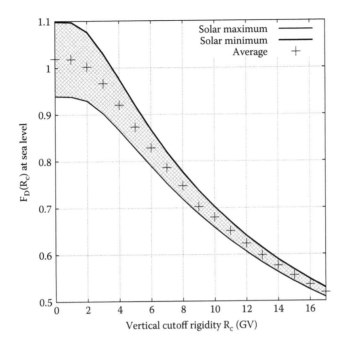

FIGURE 1.14
Scaling function $F_D(R_c)$ at sea level for solar minimum, maximum, and midlevel activities. (Numerical data after Gordon, M.S., Goldhagen, P., Rodbell, K.P. et al., *IEEE Trans. Nucl. Sci.* 51, 3427–3434. © (2004) IEEE; JEDS89A, Measurement and reporting of alpha particle and terrestrial cosmic ray-induced soft errors in semiconductor devices, JEDEC Standard JESD89A, October 2006.)

where:
ϕ and ϕ_0 = attenuated and initial fluxes, respectively
ρ = material density
L   = thickness of the shielding layer

This equation is strictly equivalent to Equation 1.6; the product $L \times \rho$ represents penetration depth (in g cm^{-2}) of the incoming particles into the considered shielding material.

The attenuation length for neutrons in concrete varies as a function of concrete composition (chemical elements, presence of steel reinforcement, aggregates, etc.) and neutron energy. A mean value of 216 g cm^{-2} was reported in Ziegler and Puchner (2004) and in Ziegler (1996), corresponding to an incident flux reduced to $\phi = 0.31 \times \phi_0$ for 1 m of concrete with a density of 2.54 g cm^{-3}. For steel, the mean attenuation coefficient for neutrons is about 266 g cm^{-2}, which signifies a lower absorption. In Dirk et al. (2003), other values of the attenuation lengths were reported for reinforced concrete: 155 and 102 g cm^{-2} for high-energy (above 100 MeV) and thermal neutrons, respectively. Similar experiments were performed in sea water (Dirk et al. 2003): values of 210 and 200 g cm^{-2} were measured for high-energy and thermal neutrons, respectively. Nakamura et al. (2008) monitored neutron dose rates both outside and inside a concrete building with a thick upper concrete ceiling of about 1 m. They performed continuous measurements from August to October, 2001 at Tohoku University, Sendai, Japan. The averaged neutron dose rate was 3.3 nSv h^{-1} outside and 1.1 nSv h^{-1} inside a concrete building, demonstrating a neutron attenuation of about two thirds after penetration through 1 m of concrete, in very good agreement with the above data. The neutron counter used has high sensitivity for low-energy neutrons and very low sensitivity for neutrons above 15 MeV.

It is also important to note that shielding not only attenuates but can also modify the natural radiation environment via the interactions of incoming neutrons with the shielding material. For example, the presence of snow on a building (a neutron-moderating material) will increase the production of thermal and low-energy neutrons, due to the scattering and slowdown processes of the atmospheric neutrons. But, apart from the shielding-related effects of snow and soils specifically studied in geosciences for cosmogenic nuclide-production mechanisms (Zweck et al. 2013) (atmospheric neutrons can also play the role of a gauge in various measurement techniques, in particular for measuring soil moisture [Hussein 2003]), these effects have been little studied so far for building materials, either experimentally or by simulation. Tosaka et al. (2008) performed very interesting work by simultaneously measuring the SER of 90 nm CMOS SRAM (see Chapter 7) and the neutron energy spectra at the summit of Mauna Kea (altitude 4200 m). Using a Bonner sphere spectrometer, they evidenced significant differences in both high-energy-neutron energy spectra (above 1 MeV) and integrated flux when performing measurements in the open air with respect to indoors.

For atmospheric muons, considerable literature exists to quantify the natural shielding of muons below the earth's surface (Dorman 2004); this generally corresponds to studies performed in the framework of high-energy-physics experiments conducted deep underground, under water, or under ice to protect the experiments from the terrestrial CR background. Regarding atmospheric proton shielding at ground level, the majority of data concerns cosmic radiation in commercial aviation, at avionics altitudes. Among all the references specifically focusing on flux metrology and shielding issues at sea level, we can cite a recent series of studies conducted at the Pacific Northwest National Laboratory (PNNL, Richland, WA). Aguayo et al. (2011) performed extensive Monte Carlo simulations to quantify and analyze the shielding properties of six commonly used shield materials (iron, lead, polyethylene, borated polyethylene, water, and concrete) against the proton, neutron, and muon components of CR showers. The simulations were performed using a simple slab of each material exposed to the simulated cosmic particle shower (from no material up to 1 m of thickness of material, in increments of 10 cm). These simulations showed that atmospheric neutrons and muons passing through concrete shielding can also generate a significant contribution of secondary neutrons to the total outbound flux that can represent a threat for electronics. It is clear that this point will require further future work to quantify exactly the importance of such effects on the modification of the natural radiation environment.

1.3.4 Synthesis

To conclude this section, Table 1.4 recapitulates the most important fixed and variable factors that can impact the atmospheric neutron flux at ground level.

Typical modulation factor amplitudes of this terrestrial neutron-flux intensity are reported. This table highlights the importance of radiation-environment (real-time) metrology in the framework of soft-error experiments, to correctly evaluate the AF and thus to express (normalize) the SER result with respect to a reference atmospheric flux.

TABLE 1.4

Main Natural Factors Impacting the Atmospheric Flux at Ground Level

Factors Impacting the Neutron Flux at Ground Level		Ratio (NYC=1)	Fixed/Variable
Geometric latitude		0.5–1.1	Fixed
Altitude (sea level to 4000 m)		1–20	Fixed
Solar activity	Solar cycle	0.7–1	Variable (11-year cycle)
	Forbush decrease	0.5–1	Variable (days) (rare events)
	Ground-level enhancement (GLE)	1–3	Variable (days) (rare events)
Shielding of building		0.3–1	Fixed
Atmospheric pressure (daily variations)		0.7–1.3	Variable (daily)
Atmospheric pressure (seasonal variations)		0.9–1.1	Variable (yearly)
Snow on building		0.7–1	Variable (winter)

1.4 Tools, Codes, and Models to Simulate Atmospheric and Terrestrial CRs

Different tools and numerical or analytical (fitting) models have been developed to estimate the terrestrial CRs in terms of composition, flux, or both. We present in the following a selection of useful approaches largely used in the radiation effects community to compute such radiation environments in the atmosphere and at ground level.

1.4.1 SEUTEST

Seutest.com (www.seutest.com) is a cooperatively managed noncommercial website providing links and support for soft-error testing compatible with the JEDEC standard JESD89A (JEDS89A 2006). The "flux calculation" menu proposes an applet to determine the relative neutron flux at a particular location. The output value is relative to the sea-level flux in NYC, according to the analytic model proposed by Gordon et al. (2004) and included in the JESD89A document. The user must enter the geographic coordinates (latitude, longitude) for the location of interest and either the elevation, barometric station pressure, or atmospheric depth of the location. The user can also enter the value of his or her choice for the solar modulation factor (from 0% to 100%). The station pressure corresponds to the mean, uncorrected barometer reading for the location. The values of geomagnetic vertical cutoff rigidity used to calculate the relative neutron flux correspond to data shown in Figure 1.6 (Smart and Shea 2007).

1.4.2 EXPACS (PARMA Model)

EXPACS is the acronym for EXcel-based Program for calculating Atmospheric Cosmic-ray Spectrum. The program was made for estimating the atmospheric CR spectrum based on the PARMA analytic model fitted on Monte Carlo calculations, as introduced in Section 1.3.2.2. EXPACS is capable of calculating not only neutron but also proton, He nucleus, muon, electron, positron, and photon spectra for anywhere in the atmosphere at altitudes below 20 km. It can also be used to estimate the ambient dose equivalent and the effective dose due to CR exposure. The calculated CR dose rates can be visualized on the map of Google Earth® using EXPACS-V. The software is open to the public via the website http://phits.jaea.go.jp/expacs/. The detailed description of the calculation procedure for neutron spectra was presented in Sato and Niita (2006), and for other particles in Sato et al. (2008).

Because EXPACS is an Excel-based program, the user must enter input data in different sheet cells before obtaining the calculated particle fluxes and dose estimations. The inputs include the geographic location (or the vertical cutoff rigidity), the altitude (or the atmospheric depth), the date (or the force-field potential, which is an index of the solar activity, based on the count rates of several ground-level neutron monitors on this date), and the surrounding environment and input local effect parameter (this parameter influences only the neutron spectra; for ground-level neutron spectra, the user can input fraction of water in the ground, or, for neutron spectra in aircraft, the mass of the aircraft). Finally, the user can specify the output units for both estimated doses and calculated fluxes.

1.4.3 QARM

QARM, initially the Qinetiq Atmospheric Radiation Model, is a comprehensive atmospheric radiation model constructed using Monte Carlo simulations of particle transport through the atmosphere. Its predictions are based on atmospheric response matrices containing the response of the atmosphere to incident particles on the upper atmosphere. The two main reference papers for this code are Lei et al. (2004, 2006).

QARM consists of two main parts: the web interface and the calculation engine, which is also available as a stand-alone tool upon request. QARM started its life around 2002 in a U.K.-government-funded research project at QinetiQ. Its first version was released to the public in 2005 on the site qarm.space.qinetiq.com. It was maintained and updated by the space radiation environment group in QinetiQ until mid-2012. Since then, its main developer, Dr. Fan Lei, has funded RadMod Research to carry on the support and development of the model. The current version, v3.0, is released by RadMod Research on the new site www.qarm.eu. Its name has been changed to Quotid Atmospheric Radiation Model, reflecting the change in the affiliation of its developer. The word "quotid" is derived from the Latin word "quotidie," meaning "daily," "everyday."

Currently, the QARM website provides access to four services: single-point rigidity calculations, single-point dose-rate calculations, flight-path dose calculations, and flight-path SEU calculations. The rigidity service calculates the vertical cutoff rigidity at any point near to the earth using CR-trajectory-tracing codes. The radiation and dose-rate service calculates the radiation spectra and dose rate at any point within the atmosphere in response to GCRs or solar proton events.

For single-point dose-rate calculation, the required user inputs for the execution of QARM are: (i) the date for which the atmospheric radiation environment is to be evaluated; (ii) the location (i.e., where in the atmosphere the radiation environment is to be predicted), defined by the latitude, longitude, and altitude (between 0 and 100 km); and (iii) the geomagnetic condition, specified by the K_p index of the geomagnetosphere (this index quantifies disturbances in the horizontal component of the earth's magnetic field with an integer in the range 0–9, with 1 being calm and 5 or more indicating a geomagnetic storm). In addition, QARM allows the user to specify his or her own incident spectrum at the upper atmosphere (100 km) above the specified location on the specified date. This will be used in calculating the radiation instead of the incident spectra determined internally by QARM. The results of QARM are output in three parts: (i) the summary (date, location, geomagnetic condition of the calculation, and dose rate predicted at the location for the given conditions), (ii) the energy-spectra plot (i.e., the calculated radiation energy spectra, downward, upward, and omnidirectional, of the particle type specified), and (iii) the data table (i.e., the numerical data of the energy spectra).

1.4.4 CORSIKA

COsmic Ray SImulations for KAscade (CORSIKA) is a detailed Monte Carlo program to study the evolution and properties of extensive air showers in the atmosphere. It was developed to perform simulations for the KASCADE experiment at Karlsruhe in Germany (Heck and Pierog 2013). This experiment measured the elemental composition of the primary cosmic radiation in the energy range 3×10^{14}–1×10^{17} eV, and after its upgrade to KASCADE-Grande it reached 10^{18} eV. The CORSIKA program allows interactions and decays of nuclei, hadrons, muons, electrons, and photons in the atmosphere up to energies of some 10^{20} eV to be simulated. It gives the type, energy, location, direction, and arrival times of all secondary particles that are created in an air shower and pass a selected observation level.

The CORSIKA program is a complete set of standard FORTRAN routines; it consists basically of four main parts. The first part is a general program frame handling the input and output, performing decay of unstable particles, and tracking the particles, taking into account ionization energy loss and deflection by multiple scattering and the earth's magnetic field. The second part treats the hadronic interactions of nuclei and hadrons with the air nuclei at higher energies. The third part simulates the hadronic interactions at lower energies, and the fourth part describes transport and interaction of electrons, positrons, and photons. CORSIKA contains several models for the latter three program parts that may be activated optionally with varying precision of the simulation and consumption of CPU time.

1.4.5 PLANETOCOSMICS

PLANETOCOSMICS is a simulation framework based on the toolkit Geant4 (Agostinelli et al. 2003) that allows the hadronic and electromagnetic interactions of CRs with Earth, Mars, and Mercury to be computed. PLANETOCOSMICS was developed by the University of Bern, under a European Space Agency/European Space Research and Technology Centre (ESA/ESTEC) contract. It is an extension of the MAGNETOCOSMICS and ATMOCOSMICS (Desorgher et al. 2005) codes that compute the propagation of CRs in the earth's atmosphere and magnetosphere. The code is available for download upon request at the following URL: http://cosray.unibe.ch/~laurent/planetocosmics/. Before installing the code, the user should install the Geant4 toolkit, different libraries, and visualization tools. Two different versions of the code have been developed, which differ in the analysis package used for histogramming. In the first version, the ROOT package is used to store the results of the code in histograms. In the second version, the analysis part of the code has been developed in compliance with the Abstract Interfaces for Data Analysis (AIDA) interface.

PLANETOCOSMICS takes into account, for each planet, the presence of the magnetic field, atmosphere, and soil. Different magnetic field models and atmospheric models are available. The addition of new models and new planets is possible. The code is well adapted to the computation of the flux of particles resulting from the interaction of CRs with the planet at user-defined altitudes and atmospheric depths and in the soil. In addition, the code can compute, for example, the energy deposited by CR showers in the planet's atmosphere and in the soil, the propagation of charged particles in the planet's magnetosphere, and the cutoff rigidity in function of position and direction of incidence. The study of quasi-trapped-particle populations is also possible, as well as the visualization of magnetic field lines and trajectories of primary and secondary particles in the planet's environment.

1.4.6 CRY

The Cosmic-Ray Shower Library (CRY) is free software produced by the Lawrence Livermore National Laboratory that is used to generate correlated CR particle showers as either a transport or a detector simulation code (Hagmann et al. 2012). It generates correlated CR particle-shower distributions at one of three elevations (sea level, 2100 m, and 11,300 m) for use as input to transport and detector simulation codes. Particles from CR showers over a wide range of energies (1 GeV–100 TeV primary particles and 1 MeV–100 TeV secondary particles) are generated from data tables. These tables are derived from full MCNPX simulation for muons, neutrons, protons, electrons, photons, and pions for several altitudes. The CRY software package generates shower multiplicity within a specified area (up to 300 m by 300 m) as well as the time of arrival and zenith angle of the secondary particles. Currently the code provides a latitude dependent geomagnetic cutoff of the primary cosmic-ray spectrum and modulation of the spectrum over time based on the average solar cycle. The software library and examples can be downloaded from http://nuclear.llnl.gov/simulation. CRY can run as a stand-alone program, or linked and run directly in Geant4, Monte Carlo N-Particle (MCNP), and Monte Carlo N-Particle eXtended (MCNPX).

References

Agostinelli, S., Allison, J., Amako, K., et al. 2003. Geant4—A simulation toolkit. *Nuclear Instruments and Methods Section A* 506:250–303. See also http://geant4.cern.ch.

Aguayo, E., Ankney, A.S., Berguson, T.J., Kouzes, R.T., Orrell, J.L., and Troy, M.D. 2011. Cosmic ray interactions in shielding materials. Pacific Northwest National Laboratory Report, PNNL-20693, 2011. Available at: http://www.pnnl.gov/main/publications/external/technical_reports/PNNL-20693.pdf.

Augusto, C.R.A., Dolival, J.B., Navia, C.E., and Tsui, K.H. 2008. Effects of the South Atlantic Anomaly on the muon flux at sea level. arXiv:0805.3166v1 [astro-ph].

Beringer, J., Arguin, J.-F., Barnett, R.M., et al. (Particle Data Group). 2012. The review of particle physics. *Physical Review* D86:010001.

Blasi, P. 2013. The origin of galactic cosmic rays. arXiv:1311.7346 [astro-ph.HE].

Boezio, M. and Mocchiutti, E. 2012. Chemical composition of galactic cosmic rays with space experiments. *Astroparticle Physics* 39–40:95–108.

Capdevielle, J.N. 1984. *Les rayons cosmiques [Cosmic Rays]*. Paris: Presses Universitaires de France.

Cecchini, S. and Spurio, M. 2012. Atmospheric muons: Experimental aspects. arXiv:1208.1171v1 [astro-ph.EP].

Clem, J.M. and Dorman, L.I. 2000. Neutron monitor response functions. *Space Science Reviews* 93:335–363.

Desilets, D. and Zreda, M. 2003. Spatial and temporal distribution of secondary cosmic-ray nucleon intensities and applications to *in situ* cosmogenic dating. *Earth and Planetary Science Letters* 206:21–42.

Desorgher, L., Fluckiger, E.O., Gurtner, M., Moser, M.R., and Butikofer, R. 2005. Atmocosmics: A GEANT4 code for computing the interaction of cosmic rays with the Earth's atmosphere. *International Journal of Modern Physics* A20:6802–6804.

Dirk, J.D., Nelson, M.E., Ziegler, J.F., Thompson, A., and Zabel, T.H. 2003. Terrestrial thermal neutrons. *IEEE Transactions in Nuclear Science* 50:2060–2064.

Dorman, L.I. 2004. *Cosmic Rays in the Earth's Atmosphere and Underground*. Dordrecht: Kluwer Academic.

Federico, C.A., Gonçalez, O.L., Fonseca, E.S., Martin, I.M., and Caldas, L.V.E. 2010. Neutron spectra measurements in the South Atlantic Anomaly region. *Radiation Measurements* 45:1526–1528.

Feldman, U., Landi, E., and Schwadron, N.A. 2005. On the sources of fast and slow solar wind. *Journal of Geophysical Research: Space Physics* 110(A7):2156–2202.

Furet, N. 2009. Solar wind, heliosphere, and cosmic ray propagation. Available at: http://www.nmdb.eu/?q = node/135, 2009.

Goldhagen, P. 2003. Cosmic-ray neutrons on the ground and in the atmosphere. *MRS Bulletin* 28:131–135.

Gordon, M.S., Goldhagen, P., Rodbell, K.P., et al. 2004. Measurement of the flux and energy spectrum of cosmic-ray induced neutrons on the ground. *IEEE Transactions on Nuclear Science* 51:3427–3434.

Grieder, P.K.F. 2001. *Cosmic Rays at Earth: Researcher's Reference Manual and Data Book*. Amsterdam: Elsevier.

Hagmann, C., Lange, D., and Wright, D. 2012. Monte Carlo simulation of proton-induced cosmic-ray cascades in the atmosphere. Lawrence Livermore National Laboratory, UCRL-TM-229452, 2012. Available at: http://nuclear.llnl.gov/simulation/doc_cry_v1.7/cry_physics.pdf.

Heck, D. and Pierog, T. 2013. Extensive air shower simulation with CORSIKA: A user's guide. Available at: https://web.ikp.kit.edu/corsika/usersguide/usersguide.pdf.

Hussein, E.M. 2003. *Handbook on Radiation Probing, Gauging, Imaging and Analysis. Volume II: Applications and Designs*. Dordrecht: Kluwer Academic.

International Space Environment Service (ISES). 2014. Solar cycle progression. NOAA Space Weather Prediction Center. Available at: (http://www.swpc.noaa.gov/SolarCycle/).

JEDS89A. 2006. Measurement and reporting of alpha particle and terrestrial cosmic ray-induced soft errors in semiconductor devices, JEDEC Standard JESD89A, October 2006. Available at: http://www.jedec.org/download/search/jesd89a.pdf.

Lei, F., Clucas, S., Dyer, C., and Truscott, P. 2004. An atmospheric radiation model based on response matrices generated by detailed Monte Carlo simulations of cosmic ray interactions. *IEEE Transactions on Nuclear Science* 51:3442–3451.

Lei, F., Hands, A., Clucas, S., Dyer, C., and Truscott, P. 2006. Improvement to and validations of the QinetiQ Atmospheric Radiation Model (QARM). *IEEE Transactions on Nuclear Science* 53:1851–1858.

Mascarenhas, N., Brennan, J., Krenz, K., Marleau, P., and Mrowka, S. 2007. A measurement of the flux, angular distribution and energy spectra of cosmic ray induced neutrons at fission energies. In *Proceedings of IEEE Nuclear Science Symposium Conference Record* N38-6, pp. 2050–2052. 26 October–3 November, Honolulu, HI: IEEE.

Meyer-Vernet, N. 2007. *Basics of the Solar Winds*. Cambridge: Cambridge University.

Moser, M.R., Ryan, J.M., Desorgher, L., and Flückiger, E.O. 2005. Atmospheric neutron measurements in the 10–170 MeV range. *Proceedings of the International Cosmic Ray Conference* 2:421–424. 3–10 August, Pune, India.

Nakamura, T., Baba, M., Ibe, E., Yahagi, Y., and Kameyama, H. 2008. *Terrestrial Neutron-Induced Soft Errors in Advanced Memory Devices*. Singapore: World Scientific.

Nymmik, R.A., Panasyuk, M.I., Petrukhin, V.V., and Yushkov, B.Y. 2009. A method of calculation of vertical cutoff rigidity in the geomagnetic field. *Cosmic Research* 47:191–197.

Papaioannou, A., Souvatzoglou, G., Paschalis, P., Gerontidou, M., and Mavromichalaki, H. 2014. The first ground-level enhancement of solar cycle 24 on 17 May 2012 and its real-time detection. *Solar Physics* 289:423–436.

Potgieter, M.S. 2013. Solar modulation of cosmic rays. *Living Reviews in Solar Physics* 10(3). Available at: http://www.livingreviews.org/lrsp-2013-3.

Rao, M.V.S. and Sreekantan, B.V. 1998. *Extensive Air Showers*. Singapore: World Scientific.

Sato, T. and Niita, K. 2006. Analytical functions to predict cosmic-ray neutron spectra in the atmosphere. *Radiation Research* 166:544–555.

Sato, T., Niita, K., Matsuda, N., et al. 2013. Particle and heavy ion transport code system PHITS, version 2.52. *Journal of Nuclear Science and Technology* 50:913–923.

Sato, T., Yasuda, H., Niita, K., Endo, A., and Sihver, L. 2008. Development of PARMA: PHITS based analytical radiation model in the atmosphere. *Radiation Research* 170:244–259.

Semikh, S., Serre, S., Autran, J.L., et al. 2012. The Plateau de Bure neutron monitor: Design, operation and Monte Carlo simulation. *IEEE Transactions on Nuclear Science* 59:303–313.

Smart, D.F. and Shea, M.A. 2007. World grid of calculated cosmic ray vertical cutoff rigidities for Epoch 1995.0. In *Proceedings of the 30th International Cosmic Ray Conference*, vol. 1, pp. 733–736. 3–11 July, Merida, Mexico.

Stanev, T. 2003. *High Energy Cosmic Rays*, 2nd edn. Berlin: Springer.

Swordy, S. 2001. The energy spectra and anisotropies of cosmic rays. *Science Reviews* 99:85–94.

Tosaka, Y., Takasu, R., Uemura, T., et al. 2008. Simultaneous measurement of soft error rate of 90 nm CMOS SRAM and cosmic ray neutron spectra at the summit of Mauna Kea. In *Proceedings of the IEEE International Reliability Physics Symposium*, pp. 727–728. 27 April–1 May, Phoenix, AZ: IEEE.

Usoskin, I.G., Mursula, K., Kananen, H., and Kovaltsov, G.A. 2001. Dependence of cosmic rays on solar activity for odd and even solar cycles. *Advances in Space Research* 27:571–576.

Valkovic, V. 2000. *Radioactivity in the Environment*. Amsterdam: Elsevier.

Walter, M. and Wolfendale, A.W. 2012. Early history of cosmic particle physics. *European Physical Journal H* 37:323–358.

Ziegler, J.F. 1996. Terrestrial cosmic rays. *IBM Journal of Research and Development* 40:19–39.

Ziegler, J.F. and Puchner, H. 2004. *SER—History, Trends and Challenges*. San Jose, CA: Cypress Semiconductor.

Zweck, C., Zreda, M., and Desilets, D. 2013. Snow shielding factors for cosmogenic nuclide dating inferred from Monte Carlo neutron transport simulations. *Earth and Planetary Science Letters* 379:64–71.

2

Detection and Characterization of Atmospheric Neutrons at Terrestrial Level: Neutron Monitors

2.1 Neutron Monitors (NM)

2.1.1 Historical Background

Terrestrial cosmic rays (i.e., atmospheric secondary radiation), which are related to primary cosmic rays (CRs) impacting the earth's upper atmosphere, have been measured by a wide variety of techniques since the 1930s. In particular, systematic and uninterrupted monitoring of atmospheric radiation was initiated in 1932 using ionization chambers and has continued since 1951 using neutron monitors (NMs) (Shea and Smart 2000). From this date, the NM rapidly became the detector of choice to monitor cosmic radiation variations, due to some limitations of the ionization chambers in detecting primary CRs below ~4 GeV and to potential long-term drift and calibration problems (Shea and Smart 2000).

The NM was invented by J.A. Simpson (University of Chicago) in 1948 (Simpson 1958, 2000). It is a ground-based instrument designed to detect predominantly the secondary neutrons produced in extensive atmospheric showers (EASs). They are sensitive to the hadronic component induced by CRs penetrating the earth's atmosphere with energies from about 0.5 to 20 GeV, that is, in an energy range that cannot be measured with detectors in space in the same simple, inexpensive, and statistically accurate way (Fuller 2009). The first NM was put in operation at Climax (Colorado, United States) in 1951. This is the longest continuously operated NM in the world (Shea and Smart 2000). The NM designed by Simpson became the standard cosmic radiation detector for the International Geophysical Year (IGY, July 1957–December 1958); it was called the IGY neutron monitor. Twelve IGY NMs were in operation by January 1957; 39 more were installed during the year, bringing the worldwide network to 51 monitors (Shea and Smart 2000).

About 10 years later, Carmichael designed a larger NM with an increased counting rate (Carmichael 1964). This instrument was the standard ground-based CR detector for the International Quiet Sun Year (IQSY) of 1964. This larger monitor, called the "super" monitor, NM64, or X-NM64 (where X denotes the number of detector tubes operating in the entire monitor), utilized larger-volume detectors than the IGY monitors and more modern data-collection techniques. Over the years, this monitor progressively replaced many of the initial IGY monitors. During the transition period, most groups operated both the IGY and the NM64 monitors for several months to establish a normalization factor for long-term studies. Although the majority of CR measurements since 1970 have been made using the NM64 NMs, there are still some well-maintained IGY monitors in operation. The data from these monitors, which have proven to be extremely stable detectors, are particularly valuable for long-term cosmic radiation studies (Shea and Smart 2000).

2.1.2 Neutron Monitor Design and Operation

An NM schematically consists of several gas-filled proportional counters surrounded by a moderator, a lead producer, and a reflector. Figure 2.1 shows the internal structure of both IGY and 3-NM64 NMs. For both types of NM, the principle of neutron detection is the same. The counter tubes of the instrument are largely sensitive to thermal neutrons resulting from: (i) the direct moderation of incident low/moderate-energy atmospheric neutrons and (ii) the production followed by the moderation of secondary neutrons produced in the lead by high-energy incident nucleons (primarily atmospheric neutrons, but also protons and muons) via spallation reactions. A spallation reaction is a process in which a light projectile (proton, neutron, or light nucleus) with kinetic energy from several hundreds of megaelectronvolts to several gigaelectronvolts interacts with a heavy nucleus (lead in this case) and causes the emission of a large number of hadrons (mostly neutrons) or fragments. The moderator slows these produced neutrons down to thermal energies; the proportional counter tubes finally detect a fraction of these moderated neutrons. In the NM64, for example, about 6% of these secondary neutrons are detected (Fuller 2009). We briefly detail in the following the different parts of a typical NM and the main physical mechanisms involved in the instrument response.

2.1.2.1 Counter Tubes

The counter tubes in an NM mainly detect thermal neutrons. The sensitivity of such a gas-filled detector is a function of the amount of sensitive gas: it increases with gas pressure

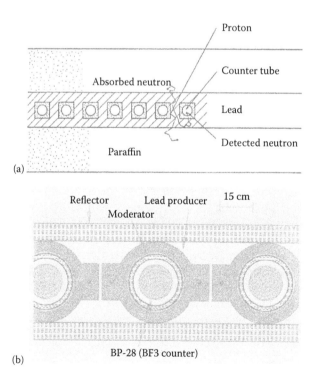

FIGURE 2.1
Schematic views of IGY (a) and NM64 (b) neutron monitors. ((a) Courtesy of N. Fuller, Neutron Monitor Database, http://www.nmdb.eu. (b) Courtesy of J.M. Clem.)

for a fixed volume. In the great majority of NMs in service around the world, the counter gas is boron trifluoride (BF_3), enriched to 96% of the ^{10}B isotope. Thermal neutrons are detected via the $^{10}B(n, \alpha)^7Li$ exothermic reaction (see Section 5.1.2 for the notation of nuclear reactions). In 93% of the reactions, the lithium nuclide remains in the first excited state ($Q = +2.31$ MeV); the other 7% of the reactions liberate $Q = +2.79$ MeV with the lithium nuclide in the ground state. The reaction products are detected by their ionization of the counter gas, creating a charge cloud in the stopping gas. The greater the energy deposited in the gas by these particles, the higher the number of primary ion pairs, the higher the number of avalanches, and the larger the pulse detected as the output signal by the electronic acquisition chain. The counter tube is operated as a proportional counter with an operating voltage of about 2–2.5 kV for a BF_3 tube.

Since the 1990s, counter tubes filled with ^3He gas instead of BF_3 have also been used in NMs. ^3He detectors respond to thermal neutrons by the exothermic $^3He(n, p)T$ nuclear reaction ($Q = +764$ keV). ^3He presents several advantages with respect to BF_3. It has a higher cross-section for thermal neutrons (5330 barns against 3840 barns), making ^3He detectors more sensitive. In addition, ^3He can be compressed to much higher pressures than BF_3, and ^3He tubes work at lower voltages (1–2 kV) than BF_3 counters. Nevertheless, the Q of the reaction $^3He(n, p)T$ is lower than for BF_3; it is thus easier to discriminate between neutron and gamma pulses with a BF_3 tube than with a ^3He detector.

In the last decade, unfortunately, the world has experienced a shortage of ^3He due to a sudden increase in demand for ^3He gas, primarily for homeland and national security, and also for medicine, industry, and science (Shea and Morgan 2010). After several years of demand exceeding supply, the price of ^3He has increased from typically $100 L^{-1} to as much as $2000 L^{-1} in recent years, encouraging the emergence of alternate solutions for thermal neutron detection, such as scintillation neutron detectors that include liquid organic scintillators, crystals, plastics, glass, and scintillation fibers.

2.1.2.2 Moderator

A neutron moderator is a medium that reduces the speed of fast neutrons, thus turning them into thermal neutrons. In an NM, the role of the moderator material is to decrease the energy of the incoming neutrons to as close as possible to the thermal energy in order to enhance their probability of interaction with ^{10}B or ^3He isotopes present in the gas-filled detectors. To slow down neutrons, the principle is to make them collide with other nuclei. Due to conservation of momentum, the neutron energy loss per elastic collision increases with decreasing atomic mass; therefore materials with a high concentration of hydrogen are most effective. Hydrogen-rich materials are thus used as moderators, such as paraffin wax in the IGY neutron monitor or polyethylene in the NM64 (Fuller 2009). Water and heavy water are also excellent neutron moderators and can be used in special dedicated setups.

2.1.2.3 Lead Producer

The function of the lead surrounding the moderator in the NM is to produce secondary neutrons via the interactions (spallation reactions) of the incident energetic nucleons with the lead. The evaporation neutrons produced have an energy distribution that shows a maximum at about 2 MeV and reaches energies up to about 15 MeV. The average number of evaporation neutrons per incident nucleon that undergoes a nuclear interaction in the lead is around 15 (Fuller 2009). The presence of lead in the NM structure thus increases the overall detection probability. Lead is chosen as producer because an element with a high atomic mass provides a large nucleus as a target for producing evaporation neutrons. In addition, lead has a relatively low absorption cross section for thermal neutrons (Fuller 2009).

2.1.2.4 Reflector

This last part of an NM constitutes a sort of outer box or shell in which the other parts of the instrument (counter tubes, moderator, and lead producer) are enclosed. Its role is to moderate the evaporation neutrons produced in the lead and reflect them into the counter tubes. The reflector also shields and absorbs low-energy neutrons that are produced in the surrounding materials outside the neutron monitor. This prevents changes of material in the environment of the detector (e.g., snow accumulation on the detector housing) causing a major change in the NM count rate (Fuller 2009). The reflector is made of proton-rich materials: paraffin wax for the early neutron monitors (IGY) and polyethylene for the more modern ones (NM64).

Table 2.1 recapitulates the main characteristics of both standard IGY and NM64 neutron monitors based on BF_3 gas-filled detectors. On the basis of such geometrical and detector characteristics, their respective response function to the different components of the atmospheric cosmic ray flux is described in Section 2.1.3.

2.1.3 Neutron Monitor Detection Response

How does an NM respond to terrestrial CRs? In order to answer this question, the relationship between the instrument counting rate at ground level and the flux of the different secondary particles such as neutrons, protons, muons, and pions in the atmosphere must be established. Because NMs are ordinarily used to study primary CRs via their manifestation in the atmosphere, different methods to estimate the relationship between the counting rate and the primary particle flux impacting the earth have been proposed in the literature (Clem and Dorman 2000). The response to the above question is thus included in the solution of this more general problem, which requires not only knowledge of detection efficiency but also correct modeling of primary composition and energy transport, atmospheric particle transport, and geomagnetic particle transport.

TABLE 2.1

Main Characteristics of IGY and NM64 Neutron Monitors

		IGY	6-NM64
Counters	Number of counters	12	6
	Type of counter	NW G-15-35A	BP-28
	Detection gas	BF_3 (96% ^{10}B)	
	Active length (cm)	86.4	191
	Diameter (cm)	3.8	14.8
	Pressure (bar)	0.60	0.27
Moderator	Material	Paraffin	Polyethylene
	Average thickness (cm)	3.2	2.0
Producer	Material	Lead	Lead
	Average depth (g cm^{-2})	153	156
Reflector	Material	Paraffin	Polyethylene
	Average thickness (cm)	28	7.5

Source: Adapted from Stoker, P.H. et al., *Space Science Reviews*, 93, 361–380, 2000; Fuller, N., Neutron monitors, Neutron Monitor DataBase website, http://www.nmdb.eu, 2009.

Clem (1999) determined the IGY and NM64 detection responses for secondary particles from extensive numerical simulation performed using FLUKA, a fully integrated particle physics Monte Carlo (MC) simulation package (Ferrari et al. 2005), combined with programs written by the author to simulate the proportional tube and electronics response to energy deposition in the gas. The standard dimensions and composition of materials of both IGY and NM64, given in Table 2.1, were used as input to the geometry. A parallel beam of monoenergetic particles at a fixed angle, 4 m in diameter, was considered to fully illuminate the modeled instrument. The irradiation was repeated for different incident beam angles, initial energy, and particle species, including neutrons, protons, and positive and negative pions and muons. Data were stored for every beam particle that produces a minimum value of energy deposited in any counter. These data were then used to generate a pulse height distribution, which is integrated (with dead-time and pile-up effects) to determine the total number of counts per beam luminosity (number of beam particles per beam area). Figure 2.2 shows the resulting detection efficiency of a supermonitor NM64 with BF_3-filled counters (see Table 2.1) for six different particle species in the vertical incident direction, as obtained by Clem (1999, 2004) and Clem and Dorman (2000). As shown, the detector response is optimized to measure the hadronic component of ground-level secondary particles. In particular, the NM response for atmospheric muons above 1 GeV is roughly 3.5 orders of magnitude below that for hadrons. In this energy region, the primary mechanisms for muon-induced counts are neutron production in photonuclear interactions and electromagnetic showers resulting in multiple ionization tracks in a counter. Below 1 GeV, stopped negatively charged muons are captured by a lead nucleus into a mesic orbit and absorbed by the nucleus. The de-excitation of the nucleus occurs with the emission of neutrons, which is reflected in the rise in detection efficiency with decreasing energy (Clem and Dorman 2000).

As expected (see Chapter 5), there is practically no difference in the response between neutrons and protons in the high-energy region, while at lower energies the

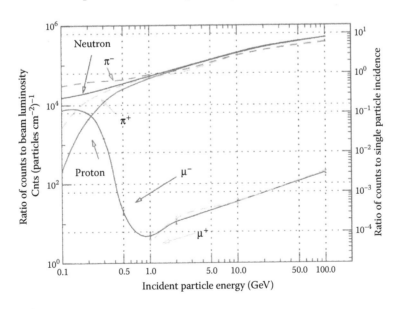

FIGURE 2.2
Standard NM64 (defined in Table 2.1) calculated detection efficiency for secondary atmospheric particles arriving in the vertical direction. (Courtesy of J.M. Clem.)

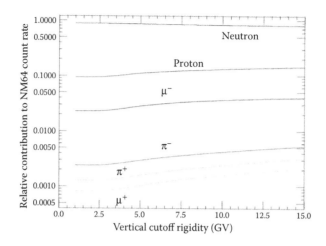

FIGURE 2.3

Relative contribution of different secondary particle species to the NM64 counting rate for a sea-level latitude survey. (Courtesy of J.M. Clem.)

ionization energy loss of protons becomes significant, greatly reducing the probability of an interaction, which is reflected in the decreasing detection efficiency (Clem and Dorman 2000).

Clem (2004) reported a comparison of this calculated detection efficiency with previous calculation results obtained by Hatton (1971) for neutrons and protons and with experimental data. These latter were obtained using laboratory monoenergetic neutron beams incident on an NM64 (Shibata et al. 1997, 1999). Within the energy range of the measurements (90–400 MeV), the data is found in fair agreement with both models. This is coincidentally the only region of agreement of the two calculations as well as the region of peak response when folded with sea-level particle spectra.

To conclude this section, Figure 2.3 shows the resulting relative contributions of the different secondary particle species to the NM64 counting rate for a sea-level latitude survey, also obtained by Clem (2004). This figure confirms that the NM response at sea level is largely dominated (around 85% of the signal) by the contribution of atmospheric neutrons, followed by the contributions of protons (around 10%) and negatively charged muons (a few percent). This instrument is thus ideal to be used as a sensitive monitor of the neutron radiation background, precisely in an energy range (typically above 10 MeV) that is very pertinent with respect to the possible neutron-induced fail mechanisms on electronics at ground level (see Chapter 5). In the perspective of microelectronics real-time soft-error rate (RTSER) tests, described in Chapter 7, it can be considered a powerful metrology tool to reevaluate the atmospheric radiation constraint in real time, as illustrated with a real case in Section 2.2.

2.2 Plateau de Bure Neutron Monitor

This section presents in detail the Plateau de Bure Neutron Monitor (PdBNM), an instrument constructed and operated by the authors for the monitoring of the atmospheric radiation flux in the particular framework of RTSER experiments (Semikh et al. 2012). This instrument was installed in 2007 on a permanent test platform at mountain altitude (2552 m), the Altitude SEE Test European Platform (ASTEP), described in Chapter 7 (see Section 7.4.1 and Figure 7.12). From a geomagnetic point of view, the ASTEP site is characterized by a cutoff rigidity of 5 GV; the natural neutron flux is approximately six times higher than the reference flux measured in New York City. In other words, the so-called acceleration factor (AF) of the platform (a term introduced in Section 1.3.1.2) is approximately equal to 6. In 2006, suspecting the importance of fluctuations in the natural radiation (neutrons) background in the interpretation and fine analysis of our RTSER experiments, we initiated the construction of an NM for the ASTEP platform, precisely to survey on site and in real time (typically minute per minute) the time variations of the natural atmospheric neutron flux incident on the ASTEP platform. The integration of the instrument was finalized in June 2007; its installation on site was performed in July 2008 after approximately one year of operation and testing in Marseille. The instrument has been fully operational on ASTEP since July 23, 2008. This section briefly illustrates the design and construction of the PdBNM and reports on its operation since July 2008 and the observation of fluctuations in detected particle flux due to atmospheric pressure variations and solar events. In Section 2.2.4, we present the modeling and numerical simulation of the PdBNM using both Geant4 and MCNPX MC codes. We successively detail the physics models involved in simulations, the comparison methodology between Geant4 and MCNPX results, the input data for realistic particle sources, and, finally, the PdBNM detection responses and its sensitivity to atmospheric particles other than neutrons.

2.2.1 PdBNM Design

The PdBNM, shown in Figure 2.4, is very similar to a standard 3-NM64 NM, defined in Section 2.1.2. The detector ensemble is based on three high-pressure (2280 Torr) cylindrical ^{3}He detectors, model LND 253109. These detectors are long tubes (effective length 1828.8 mm) offering a large effective detection volume (3558 cm^3) and a very high thermal neutron sensitivity of 1267 counts nv^{-1} (LND specifications). Each tube is surrounded by a 25 mm thick coaxial polyethylene (PE) tube, which plays the role of neutron moderator, and by 20 coaxial thick (50 mm) lead rings serving as secondary neutron producers. All these elements are placed inside an 80 mm thick PE reflector box to reject low-energy (thermal) neutrons produced in the close vicinity of the instrument. The only difference between a standard 3-NM64 design and this instrument is in the geometrical shape of the lead rings: they have flat bases and are placed directly on the bottom of the polyethylene box (Stoker et al. 2000). A Canberra electronic detection chain, composed of three charge amplifiers (model ACHNP97) and a high-voltage source (3200D), was chosen to complement a Keithley KUSB3116 acquisition module for interfacing the NM with the control PC. We developed dedicated software using Visual Basic 2008 to control the PdBNM data acquisition as well as to manage and time stamp data using a GPS time-acquisition card installed on the same PC. All these operations can be remotely controlled via the Internet using a secure connection. Every minute (in real time), the PdBNM provides the uncorrected counting rates for each detection tube, plus temperature, pressure, and hygrometry values taken at the beginning of the measurement interval. These data are postprocessed to provide hourly and monthly averaged values, posted on the ASTEP website (www.astep.eu) and available for download.

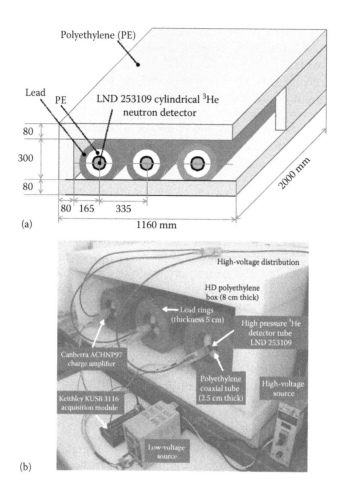

FIGURE 2.4

3D schematics (a) and detailed view (b) of the Plateau de Bure Neutron Monitor (PdBNM). The instrument is placed on a thick (40 cm) concrete floor. Dimensions in the figure are in millimeters. (Reprinted from Semikh, S. et al., The Plateau de Bure neutron monitor: Design, operation and Monte Carlo simulation, *IEEE Trans. Nucl. Sci.* 59, 303–313. © (2012) IEEE. With permission.)

2.2.2 PdBNM Installation and Operation

Assembled and first operated in Marseille during 2007–2008, the PdBNM was transported and permanently installed on the Plateau de Bure in July 2008. Figure 7.12 shows the ASTEP building, where the instrument is installed on the first floor. This building was specially constructed in 2007–2008 to host the NM; its metallic walls (including a 10 cm thick sheet of rockwool) are quasi-transparent to high-energy neutrons. The instrument is centered inside the building with respect to the circular concrete slab (thickness 40 cm) forming the floor of the room.

Figure 2.5 shows the PdBNM averaged response (one point per hour) from August 1, 2008 to October 31, 2013 (Autran et al. 2014). This uncorrected response at atmospheric pressure gives a direct image of the neutron flux variation at the ASTEP location, evidencing ~30% variations of this averaged flux at ground level, essentially due to atmospheric pressure variations (see Chapter 1). This figure also shows the importance of daily variations in the observed neutron flux, with several high-amplitude peaks corresponding to

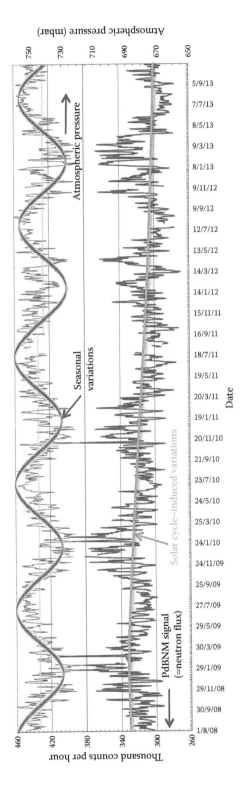

FIGURE 2.5
Plateau de Bure Neutron Monitor (PdBNM) response recorded from August 1, 2008 to October 31, 2013. Data are uncorrected from atmospheric pressure and averaged over 1 h. Approximately 30% variations in neutron flux are evidenced, primarily due to atmospheric pressure variations (with clear seasonal periodic cycles) and to long-term modulations induced by solar activity. The neutron flux is perfectly anticorrelated with both atmospheric pressure and solar cycle variations. (Reprinted from *Microelectronics Reliability*, 54, Autran, J.L., Munteanu, D., Roche, P., and Gasiot, G., Real-time soft-error rate measurements: A review, 1455–1476, Copyright (2014), with permission from Elsevier.)

the passage of severe atmospheric depressions (the largest peak corresponds to the Klaus storm on January 2009). As shown, the lowest pressure values are generally observed during winter periods, whereas the highest pressures characterize summer periods. Such seasonal variations of atmospheric pressure are, surprisingly, very regular, as illustrated by the upper curve of Figure 2.5, which appears to be very well fitted to a sinusoid curve. For the ASTEP location, the averaged level of neutron flux at ground level is consequently higher during winter than during summer. Translated into terms of AF, this point will be discussed in Chapter 7 in the context of RTSER experiments.

Finally, in addition to the impact of atmospheric pressure on the terrestrial neutron flux, the data shown in Figure 2.5 evidence a long-term modulation induced by solar activity, attested by an inverse correlation between averaged level of neutron flux and solar cycle variations, as discussed in Chapter 1. For the 2008–2013 period, which corresponds to solar cycle #24, solar activity remained extremely low throughout 2009 and then progressively increased: the resulting baseline for the neutron flux decreased slightly, as evidenced in Figure 2.5.

During its installation, the PdBNM was used to experimentally determine the AF of the ASTEP location with respect to sea level (JEDEC 2006). With exactly the same setup, two series of data, shown in Figure 2.6, were thus recorded in Marseille and on the Plateau de Bure: the difference between the counting rates and barometric coefficients for the two locations allowed us to directly evaluate the AF of ASTEP with respect to Marseille (sea level), here estimated as 6.7. Taking into account latitude, longitude, and altitude corrections for Marseille with respect to New York City (the world reference location for standardization purposes, as introduced in Chapter 1), the final value of AF is $6.7 \times 0.94 \approx 6.3$. This value is in very good agreement with the average AF (6.2) reported in Annex A of the JEDEC standard JESD89A (JEDEC 2006) and deduced from the analytical model of Gordon et al. (Equations 1.8 and following) for the Plateau de Bure research location. Such

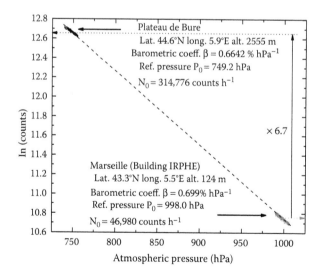

FIGURE 2.6

Experimental determination of the ASTEP acceleration factor (AF) from the barometric response of the neutron monitor successively installed in Marseille (2007–2008) and on the Plateau de Bure since July 2008. Experimental clouds correspond to one month of recording (one point per hour). (Reprinted from Semikh, S. et al., The Plateau de Bure neutron monitor: Design, operation and Monte Carlo simulation, *IEEE Trans. Nucl. Sci.* 59, 303–313. © (2012) IEEE. With permission.)

an experimental value is also in fair agreement with the value (5.9) given by the QARM model (Section 1.4.3) for quiet sun activity.

Figure 2.6 also shows the value of the PdBNM barometric coefficient β used to correct the NM counting rate for the effect of atmospheric pressure (Dorman 2004). Data for calculation of the barometric coefficient were selected during the period August 2008–December 2010, for which no disturbance of the interplanetary magnetic field and magnetosphere was reported. The least-squares method was used for the data of Figure 2.6 to obtain the regression coefficient β in the semilogarithmic representation $\ln(N)$ versus $\exp[-\beta\,(P_0-P)]$, where N is the hourly neutron monitor counting rate at atmospheric pressure P, β is the so-called barometric coefficient, and P_0 is the reference atmospheric pressure (Dorman 2004). Averaged values of $\beta = (0.6642 \pm 0.0005)\%$ hPa^{-1} have been obtained for the PdBNM with a reference pressure equal to $P_0 = 749.2$ hPa at the ASTEP location. At this reference pressure, a counting rate of $N_0 = 314{,}776$ counts.h^{-1} is measured, corresponding to the PdBNM reference counting rate. From these values, PdBNM data can be easily corrected for atmospheric pressure using the following well-known transformation:

$$N_{\text{Corrected}} = N_{\text{Uncorrected}} \exp\left[-\beta\left(P_0 - P\right)\right] \tag{2.1}$$

Of course, such a correction must be applied to raw measurements to deduce variations of the flux of primary CRs, which is generally the case in the NM community. But, in the framework of a strict neutron flux monitoring at ground level, uncorrected data and their variations in time must be considered.

2.2.3 Connection to the Neutron Monitor Database

Since May 2011, the PdBNM has been connected to the Neutron Monitor Database (NMDB), an e-infrastructures project launched in 2008 and initially supported by the European Commission in the Seventh Framework Programme. NMDB offers a real-time database for high-resolution NM measurements, including data from as many NMs as possible. This digital repository with CR data is available via the Internet (www.nmdb.eu), offering a direct access to the database through standardized web interfaces. The Forbush effect illustrated in Figure 1.8 is an example of pressure-corrected data signals from five NMs directly retrieved from the NMDB Event Research Tool (NEST) web interface (www.nmdb.eu/nest/search.php).

2.2.4 PdBNM Monte Carlo Simulation

This section reports in detail the modeling and MC simulation of the PdBNM detector to study the response of the monitor to the total flux of particles in atmospheric showers coming from the primary CRs. The initial objective was to estimate the overall counting rate (and the detection efficiency) of the PdBNM with respect to the total incoming flux and to extract partial contributions from each type of particle (in particular neutrons) to the overall counting rate. For this purpose, the atmospheric fluxes of all the basic primary particles should be taken into account, with corresponding energy spectra and angular dependencies. Following the reference work performed by Clem in this field and summarized in Section 2.1.3, we performed simulations for neutrons, muons (μ^-, μ^+), protons, charged pions (π^+, π^-), and, in addition, photons, which are also considered primary particles for PdBNM. The impact of surroundings (concrete floor, building, etc.) on the functioning of

PdBNM should also be estimated. As an important intermediate step, evaluation of the PdBNM detection response functions for each type of the incoming particles is required.

Like any other detector system, PdBNM requires an explicit calibration procedure, either with a neutron source or with another (calibrated) detector. MC simulation is an important step in completing the experimental calibration. In particular, the calibration procedure (measurements of the counters' spectra) also serves for the correct definition of the detection event in terms of region of interest (ROI) in the energy range. For PdBNM, explicit calibration is planned for the future, but presents serious practical difficulties due to its specific location, which increases the role of MC simulation.

Every single instance of neutron capture on the ^{3}He nucleus in the neutron counter is considered a neutron detection event (Clem and Dorman 2000). Generally, the neutron detection event should be defined in terms of response signal level from the proportional counter for a certain ROI, because not all the events of neutron capture produce the same response, due to wall effects and so on. This also allows one to reject possible background and noise. Our experience shows that, if the ROI is correctly defined, the difference between the adopted definition of a detection event and the ROI definition is not huge in terms of the total counting rate; at least, it is enough for the estimation of the relative contributions from each type of particle.

Besides, due to the processes of multiple secondary neutron production in the elements of the PdBNM (basically, in the lead producer), in this study we have to distinguish between the following quantities: overall counting rate of the PdBNM, its total counts, and detection efficiencies for given particle species. By definition, the efficiency of detecting the given primary particle cannot exceed 1; for the total counts (which always refer to the given particle type) this is not the case, because, for any single primary incoming particle having high enough kinetic energy, there is a certain probability of multiple neutron production and detection (see the exact definition of the "total counts" quantity below). By overall counting rate we always mean the sum of total counting rates, corresponding to all the components of natural background. The multiplicity of secondary neutron production in the parts of the PdBNM and their detection in counters will be discussed below.

2.2.4.1 Geant4 Physics Models Involved in Simulations

MC simulations of the PdBNM detector have been performed using the Geant4.9.1 toolkit, considering a list of physical processes composed from the reference physics list termed QGSP_BIC_HP (see the details on the Geant4 reference physics list webpage at http://geant4.web.cern.ch/). The QGSP_BIC_HP list includes binary cascade for primary protons and neutrons with energies below ~10 GeV, and also uses the binary light ion cascade for inelastic interaction of ions up to a few gigaelectronvolts per nucleon with matter. In addition, this package includes the High Precision neutron package (NeutronHP) to transport neutrons below 20 MeV down to thermal energies. Because the QGSP_BIC_HP list in Geant4.9.1 does not contain thermal scattering, this process was introduced separately, in view of the fact that the moderation of neutrons with kinetic energies below 4 eV in PE has to be considered in a special way (Garny et al. 2009). Indeed, in this low-energy region, the scattering of neutrons on the hydrogen nuclei in PE cannot be treated as scattering on free protons due to the possible excitation of vibrational modes in PE molecules. Such collective motion of molecules significantly changes the thermal neutron-scattering characteristics in PE, so a dedicated thermal scattering dataset and model should be included for neutron energies less than 4 eV to allow the correct treatment of neutron moderation and capture processes in the elements of PdBNM. As a result, the list of neutron interactions

TABLE 2.2

List of Considered GEANT4 Classes in Simulation Flow for Description of Neutron Interactions

Neutron Process	Energy	GEANT4 Model	Dataset
Elastic	<4 eV	G4NeutronHPThermalScattering	G4NeutronHPThermalScatteringData
	<20 MeV	G4NeutronHPElastic	G4NeutronHPElasticData
	>20 MeV	G4LElastic	—
Inelastic	<20 MeV	G4NeutronHPInelastic	G4NeutronHPInelasticData
	(20 MeV, 10 GeV)	G4BinaryCascade	—
	(10 GeV, 25 GeV)	G4LENeutronInelastic	—
	(12 GeV, 100 TeV)	QGSP	—
Fission	<20 MeV	G4NeutronHPFission	G4NeutronHPFissionData
	>20 MeV	G4LFission	—
Caption	<20 MeV	G4NeutronHPCapture	G4NeutronHPCaptureData
	>20 MeV	G4LCapture	—

was revised, as shown in Table 2.2 (models and datasets are available from Geant4.8.2). Another modification of the QGSP_BIC_HP list in the current study consists in adding (enabling) the process of muon–nuclear interaction for high-energy muons from atmospheric showers, which is disabled by default. However, it was found that the contribution of such processes is not very important here. No other changes to the QGSP_BIC_HP list were made.

2.2.4.2 Comparison with MCNPX Simulation Results

To check the adequacy of the adopted physics in Geant4, we considered a series of simple detectors (^{3}He neutron counter LND 253109 inserted into PE tube with variable external diameter; see Figure 2.7), and we compared the Geant4 simulation results obtained for such simplified architectures with those obtained using a second Monte Carlo simulation program, MCNPX.

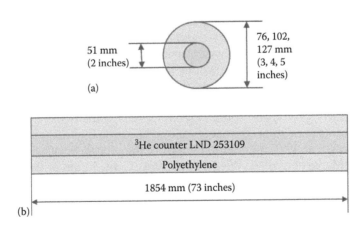

FIGURE 2.7

Front (a) and side (b) view of the test devices for comparative modeling between Geant4 and MCNPX codes. The ^{3}He neutron counter model LND 253109, 2 inches in diameter, is inserted into a polyethylene tube with variable external diameter. (Reprinted from Semikh, S. et al., The Plateau de Bure neutron monitor: Design, operation and Monte Carlo simulation, *IEEE Trans. Nucl. Sci.* 59, 303–313. © (2012) IEEE. With permission.)

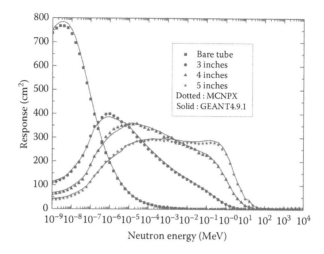

FIGURE 2.8
Response functions of the system defined in Figure 2.7 for a bare counter and different external polyethylene tube diameters. (Reprinted from Semikh, S. et al., The Plateau de Bure neutron monitor: Design, operation and Monte Carlo simulation, *IEEE Trans. Nucl. Sci.* 59, 303–313. © (2012) IEEE. With permission.)

This very popular code is a software package developed by Los Alamos National Laboratory for simulating nuclear processes in a wide area of applications in nuclear physics and nuclear engineering (Pelowitz 2008). Using both Geant4.9.1 and MCNPX 2.6.0, we studied the dependence of this series of detector responses on the energy of monoenergetic neutron flux, taking the neutron source for the simulation as a homogeneous and parallel neutron beam fully illuminating the lateral surface of the system perpendicular to the counter's axis. Figure 2.8 shows the results obtained for the bare counter and for different thicknesses of surrounding PE. Symbols correspond to MCNPX simulation results, and solid lines to Geant4 results. Statistical uncertainty for MCNPX results is less than 1% and is not shown. For Geant4, the number of primary neutrons for each energy value is 6×10^5, so for most points the statistical uncertainty is much less than 1% and is also not shown. Although the divergence between Geant4 and MCNPX at the lowest and highest energies is still significant, we clearly observe a general agreement in the shape of curves and reasonable numerical agreement, which confirms the correctness of the adopted physics list in Geant4.

The evaluation of these response functions in Geant4 simulation in practice consists in directly counting neutron detection events, that is, events of neutron capture on ^{3}He nuclei in the active volume of the counter. During the simulation, such events are identified by the reaction products (proton and triton in the final state of the reaction). But it is also very helpful to use Geant4 to calculate the energy-binned neutron fluence in the active volume of the counter (Mares et al. 1991) in order to reproduce the procedure of calculating the "Tally F4" in MCNPX (Mares et al. 1991; Wiegel et al. 1994). Convolution of this fluence with the neutron capture cross section on ^{3}He is proportional to the value of the response function. When this operation was performed in the current study, it was explicitly checked that both methods of test system response evaluation led to the same results (Semikh et al. 2012).

2.2.4.3 Input Data for Realistic Particle Sources

To study the response of the PdBNM to atmospheric radiation, realistic energy spectra (above 1 MeV) for each component of the natural background (n, μ⁻, μ⁺, p, π⁺, π⁻, γ) are needed as simulation inputs. As introduced in Chapter 1 (Section 1.3), for the neutron spectrum we used the tabulated values measured by Gordon and Goldhagen in New York City (JEDEC 2006) and, considering that the shape of the neutron spectrum above 5 MeV does not change significantly with altitude, cutoff, or solar modulation (Section 1.3.2.2), we applied Equation 1.8 to scale this reference spectrum to the ASTEP location. For the other particles, we considered energy distributions given by the EXPACS/PARMA model (see Section 1.4.2) for the ASTEP location.

Another important issue in MC simulation is the strong zenith angular dependence of atmospheric showers. To make Geant4 primary particle sources more realistic, we introduced in simulations the angular dependence of the primary flux intensity in the form of a $\cos^n(\theta)$ law, where θ is the zenith angle, as discussed in Section 1.3.1. The built-in general particle source (GPS) of Geant4 (http://reat.space.qinetiq.com/gps/) was employed to generate primary particles with such a given angular distribution. For neutrons we adopt $n = 3.5$ and for muons $n = 2$ (see Section 1.3.1).

2.2.4.4 Simulation Results

2.2.4.4.1 Detection Response Functions

In Clem and Dorman (2000), the detection response functions of NM-64 were evaluated for different particle species as "total counts" versus the energy of monoenergetic particle fluxes arriving at the upper edge of the monitor. For this purpose, the FLUKA simulation toolkit was employed (see Section 2.1.3). To perform an analysis as close as possible to that performed by Clem, but in this case for the PdBNM using Geant4, we considered incident monoenergetic particles to be uniformly distributed upon the upper edge of the PdBNM and arriving in the vertical direction. The obtained dependences of the total counts on the energy for different particle species are shown in Figure 2.9. Note that, for the given simulation run, by definition one has

$$\text{Total counts} = \frac{\text{Number of captured neutrons}}{\text{Number of primary particles}} \qquad (2.2)$$

and for primary particles with high energy this value exceeds 1 due to the massive capture of neutrons coming from the processes of secondary neutron production with high multiplicity (see detailed analysis in Section 2.2.4.4.2). In the case of primary neutrons, if no secondary neutrons are produced or captured, the total count is exactly equal to the detection efficiency.

As previously mentioned, the design of the PdBNM monitor is very similar to that of NM64 considered in (Clem and Dorman 2000): the difference is basically in geometrical dimensions and in the global shape of the lead rings, but they are still close to each other, so the detection response functions of these monitors can be qualitatively compared. Although the corresponding curves from Clem (Figure 2.2) cannot be superimposed on Figure 2.9, their behavior is in convincing agreement. In particular, all the relative variations are very similar, although for NM64 the functions are not normalized to the surface area of the monitor. Additionally, the same modeling of the PdBNM was performed within MCNPX 2.6.0 for several values of initial energies of neutrons, negative muons, and protons (these particle types make a dominant contribution to the overall counting rate). MCNPX data points are represented by symbols, as explained in the caption of Figure 2.9. The models employed are CEM03.01 and LAQGSM03.01 (Gudima et al. 2001; Mashnik

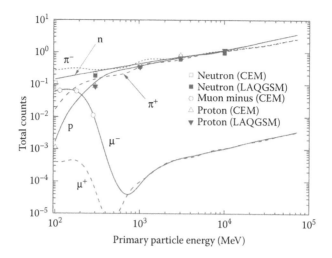

FIGURE 2.9
Plateau de Bure neutron monitor detection responses for neutrons, protons, muons, and pions. Filled symbols and lines correspond to Geant4; large open symbols correspond to MCNPX (CEM and LAQGSM models). For any given kinetic energy, the simulation run consists of 10^6 primary particles (events). (Reprinted from Semikh, S. et al., The Plateau de Bure neutron monitor: Design, operation and Monte Carlo simulation, *IEEE Trans. Nucl. Sci.* 59, 303–313. © (2012) IEEE. With permission.)

et al. 2005, 2006, 2008). The results obtained demonstrate good agreement, and we can also conclude again that the physical picture of the neutron interactions with matter adopted in Geant4.9.1 is adequate and close enough to that employed in FLUKA and in MCNPX.

2.2.4.4.2 Secondary Neutron Multiplicity

Like any NM, the PdBNM contains a large quantity of lead (about 2 tons), so the contribution of the secondary neutron production to the total counting rate deserves separate investigation. First of all, MC simulations have been used to investigate the distribution of secondary neutron production vertices within the volume of the monitor. An example of such a distribution is given in Figure 2.10 for the case of incident muons with energy distribution following the EXPACS/PARMA spectrum for ASTEP conditions and arriving in the vertical direction (z axis). For simplicity, we do not reproduce the realistic angular distribution of atmospheric muons in this investigation. As expected and illustrated in Figure 2.10, the lead tubes produce most such vertices, although the PE walls also contribute. To analyze the role of secondary neutrons in the detection of primary particles, Geant4 simulations were performed for both monoenergetic primary particles and a realistic spectrum. Table 2.3 shows simulation results for five values of primary neutron kinetic energy E_{prim}: the detection efficiency, the total counts, and the ratio of the total number of secondary neutrons produced in the run N_{sec} to the number of primary neutrons $N_{prim} = 10^6$. Any primary particle (neutron, muon, etc.) is considered as being detected in PdBNM if at least one neutron (either primary or secondary) is captured in the ^{3}He tube in the current event. Neutrons with energies below 1 MeV are almost completely reflected by 80 mm PE walls of monitor. Starting from a few megaelectronvolts, N_{sec} increases very rapidly, finally exceeding N_{prim} at energies of a few tens of megaelectronvolts. The observed difference between efficiency (which is proportional to the number of detected primary particles) and total counts is explained by the capture of these additional neutrons in ^{3}He tubes within the given event.

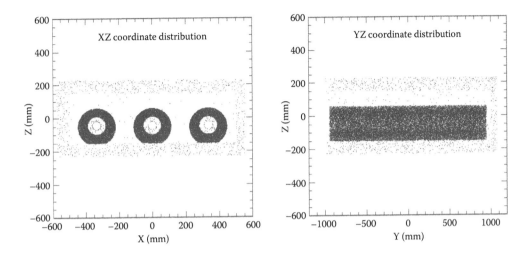

FIGURE 2.10

Two-dimensional histograms corresponding to the projections of PdBNM to coordinate planes (front and side views) and describing distribution of the secondary neutron production vertices from incoming muon flux. (Reprinted from Semikh, S. et al., The Plateau de Bure neutron monitor: Design, operation and Monte Carlo simulation, *IEEE Trans. Nucl. Sci.* 59, 303–313. © (2012) IEEE. With permission.)

In principle, to estimate the total counting rate of PdBNM from the atmospheric neutron flux it is now sufficient to convolute the neutron curve from Figure 2.9 with the spectrum from Figure 1.9. Another method, chosen in the following, is to perform direct MC simulation of PdBNM with the use of the Geant4 GPS source mimicking the atmospheric neutron spectrum, thus avoiding, for instance, the errors of numerical interpolation and integration of strongly varying functions over the huge interval of energy.

The different histograms obtained from this direct simulation for atmospheric neutrons are shown in Figure 2.11. These curves connect the number of secondary neutrons per event and the number of neutron captures in ^{3}He counters. The neutron production multiplicity here can achieve very high values (up to 120 secondary neutrons per event), and the same is true of the neutron capture multiplicity (up to 14 captured neutrons per event). From these data, the contribution of the secondary neutrons to the total counting rate can be analyzed. In fact, histogram 1 in Figure 2.11 corresponds to the events characterized by a single captured neutron, primary or secondary. Its first bin contains the events with no secondary neutrons produced, thus describing the capture of only (moderated) primary

TABLE 2.3

Dependence of Efficiency, Total Counts, and Secondary Neutron Production Ratio on Primary Neutron Energy

E_{prim} (MeV)	Detection Efficiency	Total Counts	N_{sec}/N_{prim}
100	0.112	0.143	1.786
300	0.157	0.251	3.589
1,000	0.206	0.446	6.700
3,000	0.236	0.721	10.950
10,000	0.275	1.164	17.680

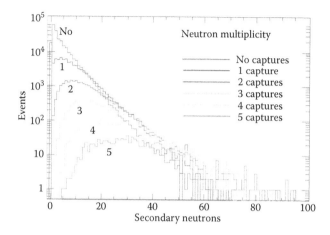

FIGURE 2.11
Set of histograms connecting the number of secondary neutrons per event and the number of neutron captures in ^{3}He counters for the JEDEC primary neutron spectrum. Each curve corresponds to the fixed number of neutron captures in the event. (Reprinted from Semikh, S. et al., The Plateau de Bure neutron monitor: Design, operation and Monte Carlo simulation, *IEEE Trans. Nucl. Sci.* 59, 303–313. © (2012) IEEE. With permission.)

neutrons, which corresponds to a (limited) percentage of 9.7% of all the captured neutrons in the simulation run. In total, this histogram accumulates 51.1% of the captured neutrons. The other 48.9% of neutrons are involved in multiple captures (double, triple, etc.), for which the distribution is summarized in Table 2.4 in the following way: the integral from histogram 1 in Figure 2.11 divided by the total number of captured neutrons in the run is equal to 0.511 (51.1%); the integral from histogram 2 (events with double neutron captures) multiplied by 2 and divided by the total number of captured neutrons is 0.26 (26%); and so on.

The same analysis can be performed for primary particles of any type. It is clear that, in detection of particles other than neutrons, the production of secondary neutrons plays a principal role (they are detected in so far as they are able to produce neutrons). Considering primary protons and negatively charged muons for the corresponding curves in Figure 2.9, we obtain the data reported in Table 2.5. It is clear that for protons the secondary neutron production multiplicity is even higher than for primary neutrons, and the corresponding multiple neutron capture processes in ^{3}He are more intensive, which follows from Table 2.4. For negative muons, there is a primary energy interval with a high secondary neutron production ratio, where the processes of muon moderation and capture in the

TABLE 2.4

Contributions to Total Counting Rate for Given Particle Type from Events with Different Multiplicities of Neutron Captures in ^{3}He Counters

Multiplicity	Contributions from Different Multiplicities of Neutron Captures (%)								
	1	2	3	4	5	6	7	8	9
Neutron	51.1	26.0	12.5	5.5	2.6	1.2	0.6	0.3	0.1
Proton	24.2	23.7	17.2	11.5	7.3	5.0	3.2	2.1	1.4
Negative muon	88.2	10.1	1.1	0.2	0.2	0.1	0	0	0

TABLE 2.5

Dependence of Efficiency, Total Counts, and Secondary Neutron Production Ratio on Primary Proton and Negative Muon Energy

E_{prim} (MeV)	Protons			Negative Muons		
	Detection Efficiency	Total Counts	N_{sec}/N_{prim}	Detection Efficiency	Total Counts	N_{sec}/N_{prim}
113	0.0038	0.0040	0.093	0.0614	0.0657	0.739
179	0.0248	0.0301	0.549	0.0556	0.0594	0.661
284	0.0743	0.1031	1.642	0.0108	0.0115	0.153
450	0.1424	0.2252	3.447	0.00024	0.00026	0.00285
713	0.1714	0.3088	4.829	3.9×10^{-5}	3.9×10^{-5}	0.00068
1,130	0.1941	0.4028	6.338	0.00010	0.00011	0.00194
...	...	...	...	...	...	...
45,000	0.3005	2.787	42.18	0.0019	0.0026	0.0374
71,300	0.3075	3.633	54.95	0.0025	0.0035	0.0474

producer are effective. But, for any energy, the difference between detection efficiency and total counts does not become as large as it is for primary neutrons and protons.

2.2.4.4.3 *Impact of Surroundings on PdBNM Counting Rate*

Another important issue for simulation is the impact of the surroundings on the PdBNM overall counting rate. As described in Section 2.2.2, the monitor is situated inside the first floor of the ASTEP building (see Figure 7.12), which possibly distorts the original radiation background. Although the building is relatively light, it contains significant amounts of steel and concrete, which could lead to secondary neutron production (in the steel walls) and neutron reflection (from the concrete floor), so its effect on the counting rate should be estimated. The building was included in the total geometry of modeling, as shown in Figure 2.12a. In this simulation, we found that the number of secondary neutrons produced in the system "building + monitor" is approximately three times higher than in the stand-alone monitor for the primary protons and neutrons, and about five times higher for the primary negative muons. But the resulting impact of this additional neutron flux on the total counts of PdBNM is not so significant: it increases the total counts only by 4.1% for primary neutrons, 4.5% for protons, and 1.7% for negative muons. Thus, these extra neutrons are mostly not detected by PdBNM. To conclude this section, Figures 2.12b and c illustrate different views of a simulated event for five incoming atmospheric neutrons (with energy 100 MeV) interacting with the matter of the PdBNM.

2.2.4.4.4 *Contributions of Different Particle Species to PdBNM Counting Rate*

It is clear that, until the experimental calibration of the PdBNM is done, only the relative contributions from the different particle species of the natural radiation environment to the overall counting rate of the PdBNM monitor can be reliably estimated by MC simulation. Such contributions are directly proportional to the following factors: (i) the partial detection efficiency (or total counts; see Figure 2.9) for the given particle type and (ii) its partial flux intensity in natural conditions, depending on the latitude, longitude, altitude, atmospheric pressure, solar cycle, and so on. To obtain realistic partial fluxes for the given conditions, we employed the EXPACS/PARMA model and constructed a Geant4 particle source as described in Section 2.2.4.3. Calculated relative contributions to the PdBNM overall counting rate from different types of particles are listed in Table 2.6 for two completely

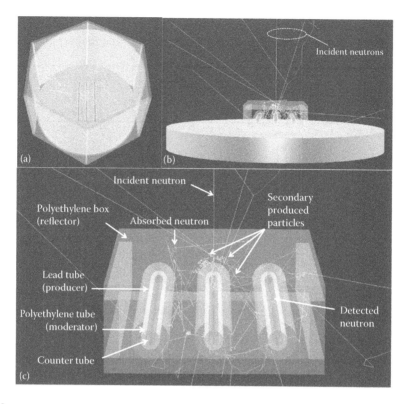

FIGURE 2.12
Root screenshots of Geant4 simulations showing the tracks of primary and secondary particles for different views of the instrument: (a) top view including the ASTEP surrounding building (first floor); (b) face view of the instrument placed on the concrete floor of the building (first floor); (c) detailed view at the level of the polyethylene box of the neutron monitor.

different locations—Marseille and Plateau de Bure. The upper and lower limits for the percentages are due to the difference in efficiency and total counts between the detection species, which, in turn, is caused by the detection of secondary neutrons, as described above. It is also worth noting that to evaluate total counts for primary neutrons it is necessary to consider not only the high-energy part of the neutron spectrum (>1 MeV) but also

TABLE 2.6

Relative Contributions of Different Particle Species to PdBNM Overall Counting Rate

Particles	Marseille (Sea Level) (%)	ASTEP (Elevation 2555 m) (%)
Neutrons	83.2–83.4	83.4–86.6
Protons	6.2–8.6	9.0–13.2
Negative muons	6.0–7.9	2.3–3.0
Gamma rays	2.0–2.7	1.1–1.4

Note: For the simulation of the instrument response at the ASTEP location, the impact of the building is taken into account.

the lower part, typically from 1 eV to 1 MeV (see Figure 1.11), which makes a significant contribution due to the nonnegligible flux in this low-energy region.

The values reported in Table 2.6 demonstrate that the PdBMN is highly sensitive to atmospheric neutrons, but, obviously, it is also quite sensitive to atmospheric protons and negatively charged muons. The latter, when moderated in its materials and captured in nuclei, produce many secondary neutrons (also shown in Table 2.4). The sensitivity of PdBNM to other particle species (like electrons and pions) is estimated to be very low, due to either the small corresponding value of partial detection efficiency (for electrons) or low atmospheric partial flux intensity (for charged pions). The numerical results of Table 2.6 can be compared with the relative contributions of the different secondary particle species determined by Clem (2004) for the NM64 at sea level and shown in Figure 2.8. Good agreement is found between the two series of data obtained with similar setups using two different chains of MC simulations. The main conclusion is that atmospheric neutrons dominate the response of these different instruments and contribute around 80%–85% of the NM counting rate at sea level.

2.2.5 Concluding Remarks

In conclusion, Section 2.2 summarized a seven-year effort to completely develop, install, and characterize by simulation a new NM permanently installed on the Plateau de Bure in the French South Alps. The primary purpose of the instrument was real-time measurement of the atmospheric neutron flux impacting microelectronics experiments deployed at altitude to precisely investigate the impact of natural radiation on electronics. Nevertheless, data obtained with this instrument can be also considered for CR investigations, along with other monitors installed around the world. Almost five years of continuous operation have demonstrated the high stability and reliability of the instrument, which will soon be integrated into the NMDB network for real-time data accessibility on the web.

Our modeling and numerical simulation work, performed with two different MC codes, Geant4 and MCNPX, allowed us to obtain the NM detection response functions for neutrons, muons, protons, and pions and their respective contribution to the overall counting rate of the instrument, considering realistic atmospheric particle sources. We highlighted the relative importance of the contribution of protons and negative muons to the monitor response (for muons especially at sea level) and the negligible impact of the surrounding ASTEP building on the counting rate. Finally, we carefully characterized the secondary neutron multiplicity processes, essential to understand the physics of NMs.

References

Autran, J.L., Munteanu, D., Roche, P., and Gasiot, G. 2014. Real-time soft-error rate measurements: A review. *Microelectronics Reliability* 54(8):1455–1476.

Carmichael, H. 1964. Cosmic rays. In *IQSY Instruction Manual No. 7*. London: IQSY Secretariat. Reprinted as Chapter 13, *Annals of the IQSY*, 1969, vol. 1. Cambridge, MA: MIT Press, p. 178.

Clem, J.M. 1999. Atmospheric yield functions and the response to secondary particles of neutron monitors. In *Proceedings of the 26th International Cosmic Ray Conference*, vol. 7, p. 317, August 17–25, Salt Lake City.

Clem, J.M. 2004. Neutron monitor detection efficiency. Annual CRONUS Collaboration Meeting. Available at: www.physics.purdue.edu/cronus/files/cronus04.ppt.

Clem, J.M. and Dorman, L.I. 2000. Neutron monitor response functions. *Space Science Review* 93:335–359.

Dorman, L.I. 2004. *Cosmic Rays in the Earth's Atmosphere and Underground*. Dordrecht: Kluwer Academic.

Ferrari, A., Sala, P.R., Fasso, A., and Ranft, J. 2005. FLUKA: A multi-particle transport code. CERN-2005-10, INFN/TC_05/11, SLAC-R-773.

Fuller, N. 2009. Neutron monitors. Neutron Monitor DataBase website, http://www.nmdb.eu.

Garny, S., Leuthold, G., Mares, V., Paretzke, H.G., and Ruhm, W. 2009. GEANT4 transport calculations for neutrons and photons below 15 MeV. *IEEE Transactions on Nuclear Science* 56:2392–2396.

Gudima, K.K., Mashnik, S.G., and Sierk, A.J. 2001. User manual for the code LAQGSM. Los Alamos National Laboratory Report LA-UR-01-6804.

Hatton, C.J. 1971. The neutron monitor. In *Progress in Elementary Particle and Cosmic Ray Physics*, vol. X. New York: American Elsevier.

JEDEC. 2006. Measurement and reporting of alpha particle and terrestrial cosmic ray-induced soft errors in semiconductor devices, JEDEC Standard JESD89A, October 2006. Available at: http://www.jedec.org/download/search/jesd89a.pdf.

Mares, V., Schraube, G., and Schraube, H. 1991. Calculated neutron response of a Bonner sphere spectrometer with ^{3}He counter. *Nuclear Instruments and Methods in Physics Research* A307:398–412.

Mashnik, S.G., Gudima, K.K., Prael, R.E., Sierk, A.J., Baznat, M.I., and Mokhov, N.V. 2008. CEM03.03 and LAQGSM03.03 event generators for the MCNP6, MCNPX, and MARS15 transport codes. In invited lectures presented at the Joint ICTP-IAEA Advanced Workshop on Model Codes for Spallation Reactions, LANL Report LA-UR-08-2931.

Mashnik, S.G., Gudima, K.K., Sierk, A.J., Baznat, M.I., and Mokhov, N.V. 2005. CEM03.01 user manual. Los Alamos National Laboratory Report LA-UR-05-7321.

Mashnik, S.G., Sierk, A.J., Gudima, K.K., and Baznat, M.I. 2006. CEM03 and LAQGSM03: New modeling tools for nuclear applications. *Journal of Physics: Conference Series—EPS Euroconference XIX Nuclear Physics Divisional Conference* 41:340–351.

Pelowitz, D.B. 2008. MCNPX™ User's Manual Version 2.6.0. Los Alamos National Laboratory, LA-CP-07-1473.

Semikh, S., Serre, S., Autran, J.L., et al. 2012. The Plateau de Bure neutron monitor: Design, operation and Monte Carlo simulation. *IEEE Transactions on Nuclear Science* 59:303–313.

Shea, D.A. and Morgan, D. 2010. The helium-3 shortage: Supply, demand, and options for Congress. Congressional Research Service, Report R41419.

Shea, M.A. and Smart, D.F. 2000. Fifty years of cosmic radiation data. *Space Science Review* 93:229–262.

Shibata, S., Munakata, Y., Tatsuoka, R., et al. 1997. Calibration of neutron monitor using an accelerator. *Proceedings of the International Cosmic Ray Conference* 1:45.

Shibata, S., Munakata, Y., Tatsuoka, R., et al. 1999. Calibration of neutron monitor using accelerator neutron beam. *Proceedings of the International Cosmic Ray Conference* 7:313.

Simpson, J.A. 1958. Cosmic-radiation neutron intensity monitor. *Annals of the International Geophysical Year IV*, Part VII. London: Pergamon Press, p. 351.

Simpson, J.A. 2000. The cosmic ray nucleonic component: The invention and scientific uses of the neutron monitor. *Space Science Reviews* 93:12–32.

Stoker, P.H., Dorman, L.I., and Clem, J.M. 2000. Neutron monitor design improvements. *Space Science Reviews* 93:361–380.

Wiegel, B., Alevra, A.V., and Siebert, B.R.L. 1994. Calculations of the response functions of Bonner spheres with a spherical ^{3}He proportional counter using a realistic detector model. PTB-Bericht N-21, ISBN 3-89429-563-5.

3

Natural Radioactivity of Electronic Materials

3.1 Radioactivity

Radioactivity is a natural physical phenomenon in which unstable radioactive nuclei (radionuclides) spontaneously transform to more stable nuclei, releasing energy in the form of various radiations. Radioactivity was discovered in 1896 by Henri Becquerel (1852–1908) in uranium and was quickly confirmed by Marie Skłodowska-Curie (1867–1934) for radium. In 1903, Becquerel shared the Nobel Prize in Physics with Pierre and Marie Curie "in recognition of the extraordinary services he has rendered by his discovery of spontaneous radioactivity."

Radioactive nuclei are called radioisotopes, and their spontaneous nuclear transformation is a physical phenomenon called radioactive decay. The emitted radiations are α-rays, β-rays, and γ radiation.

The decay rate of a radioactive substance is described by the nuclear activity, defined as the number of atoms of the radioactive substance disintegrating per unit time. The unit of the activity is the Becquerel (Bq). An activity of 1 Bq corresponds to one disintegration per second. Another unit of activity is the Curie (Ci), which is defined as the activity of 1 g of ^{226}Ra. 1 Ci is equal to 3.7×10^{10} Bq.

3.1.1 Radioactive Decay

As radioactivity is by nature a random process, one cannot know when a given radioactive atom will decay. However, it is possible to define a probability that a nucleus will disintegrate per unit time, as explained in the following. The law of radioactive decay explains that the activity of radionuclides in a radioactive substance decays exponentially with time, as follows:

$$dN = -\lambda N \, dt \qquad (3.1)$$

where:
- N = number of radioactive atoms of the substance
- dN = number of atoms disintegrated during an infinitesimal time, dt
- λ = a proportionality constant called the radioactive decay constant

By definition, this constant is the probability per unit time that a radioactive element will decay. The radioactive decay constant is a characteristic of the nucleus and is independent of space and time. In Equation 3.1, the negative sign means that the number of radioactive nuclei decreases.

Equation 3.1 shows that the number of atoms of the radioactive substance disintegrated per unit time, dN/dt, is proportional to the total number N of radioactive atoms present at time t, the constant of proportionality being λ. dN/dt is called the activity of the radioactive substance and is given by

$$\frac{dN}{dt} = -\lambda N \tag{3.2}$$

Rewriting Equation 3.2 gives

$$\frac{dN}{N} = -\lambda\, dt \tag{3.3}$$

By integrating Equation 3.3, one obtains

$$N = N_0\, e^{-\lambda t} \tag{3.4}$$

where N_0 is the number of radioactive atoms present at $t=0$. Using Equations 3.2 and 3.4, the activity of the radioactive substance is given by

$$A = \left|\frac{dN}{dt}\right| = \lambda N = \lambda N_0\, e^{-\lambda t} \tag{3.5}$$

Equations 3.4 and 3.5 describe the exponential law of radioactive decay. They show that the activity (disintegration rate) and the number of radioactive nuclei decrease exponentially with time.

Another characteristic constant of a radioisotope is the half-life, denoted $t_{1/2}$, which represents the time required for half of the radioactive nuclei initially present to disintegrate. This period can be calculated by replacing $N=N_0/2$ and $t=t_{1/2}$ in Equation 3.4:

$$t_{1/2} = \frac{\ln 2}{\lambda} = \frac{0.693}{\lambda} \tag{3.6}$$

From Equations 3.4 and 3.5, one can easily deduce that the number of radioactive atoms and the disintegration rate decrease to one-half in one half-life, to one quarter in two half-lives, to one eighth in three half-lives, and so on (Valkovic 2000).

The amount of energy emitted during the radioactive decay and the nature of radiation vary considerably from one radioactive element to another. Highly radioactive substances such as cesium 137 disintegrate very rapidly, many times per second, and therefore have a relatively short half-life. Other isotopes, such as uranium 235 or uranium 238, manifest only a few disintegrations per second and are characterized by a half-life of several million years.

Another interesting quantity is the specific activity of a radioactive substance, which is defined as the activity per unit mass. The specific activity is measured in Becquerels per gram or Curies per gram and is given by

$$A = \frac{N_A}{M}\lambda \tag{3.7}$$

where:

M = molar mass of the radioactive substance

N_A = Avogadro's number

A radionuclide, during its disintegration, forms another nuclide. The original radionuclide is called the parent and the decay product is called the daughter. The daughter may also be a radionuclide. This new radionuclide also undergoes decay, yielding a new nuclide, which may or may not be stable. A succession of nuclides, each of which transforms by radioactive disintegration into the next until a stable nuclide results, is called a radioactive series (Valkovic 2000). There are three naturally occurring decay series:

1. The uranium decay series, in which ^{238}U decays through 14 daughter radioactive nuclei to a stable nucleus ^{206}Pb. This series is also called the 4n+2 series (n is an integer), because all nuclides have a mass number that, when divided by 4, has a remainder of 2.

2. The thorium decay series (also called the 4n series), in which ^{232}Th decays through 10 intermediate daughter radionuclides to a stable nucleus ^{208}Pb.

3. The actinium or ^{235}U decay series (also called the 4n+3 series), in which ^{235}U decays through 11 intermediate radionuclides to a stable nucleus ^{207}Pb.

These series will be presented in detail in Section 3.3.

Radioactive equilibrium refers to that state in which the ratios between the amounts of successive members of the series remain constant (Valkovic 2000). Under these conditions, the disintegration rates of the parent and all the subsequent radioactive daughters will be the same. Three situations are possible:

1. When the radionuclide and decay-product half-lives are similar, the total activity (i.e., the combined decay of both radionuclides) decays at about the same rate as the original radionuclide. This is known as *transient equilibrium.*

2. When the half-life of the original radionuclide is much longer than the half-life of the decay product, the decay product decays at the same rate as it is produced. This state is called *secular equilibrium.*

3. When the half-life of the decay product is much longer than that of the original radionuclide, equilibrium cannot occur.

There are several modes of radioactive decay, which can be divided into six main groups:

1. Alpha decay (α)

2. Beta decay (β)

3. Gamma decay (γ)

4. Spontaneous fission

5. Proton emission

6. Neutron emission

Among all these emissions, only alpha emission and spontaneous fission may produce ions capable of ionizing matter and inducing a malfunction in microelectronic devices.

Since spontaneous fission is a much rarer process than alpha emission (Wrobel 2008), we describe only alpha emission.

3.1.2 Alpha-Particle Emission

An alpha particle corresponds to a nucleus of an atom of helium (He). Unlike neutrons, alpha particles directly ionize matter and are capable of triggering a single-event upset (SEU) in the circuit. The phenomenon of ionization (electron–hole creation) and its implications for devices and circuits will be explained in detail in Sections 5.2.1 and 5.6. In contrast to neutrons, which represent external radiation constraints (since they come from the atmosphere), alpha particles are an internal radiation constraint of electronic circuits, as they result from radioactive decay of unstable radioactive nuclei present in the different materials constituting the circuit, the chip connections, and the surrounding package. These radionuclides are called alpha emitters. There are 815 alpha emitters, natural or artificial (Magill and Galy 2005). Among these alpha emitters, heavy elements ($Z > 83$) are natural alpha emitters. The spontaneous disintegration of a nucleus X by alpha emission can be written as follows:

$$\,^{A}_{Z}X \rightarrow \,^{A-4}_{Z-2}Y + \,^{4}_{2}He$$

$$(3.8)$$

where A and Z are, respectively, the atomic mass and the atomic number of the nucleus X. Alpha radiation reduces the ratio of protons to neutrons, resulting in a more stable configuration. As shown in Equation 3.8, in alpha decay the atomic number of the decay product changes, so the parent and the daughter atoms are different elements, and therefore they will have different chemical properties.

Alpha emitters may be classified into two categories:

1. Radioactive isotopes present in the form of trace amounts of impurities in common materials used in the device fabrication; the most common radioactive impurities are ^{238}U, ^{232}Th, and ^{235}U. These isotopes have long half-lives, comparable to the earth's age, and are found naturally in trace amounts in commonly used materials and in the environment. The materials used in the microelectronics industry contain these radioactive isotopes (in an amount that is of the order of parts per billion for ^{238}U and ^{232}Th), although very intensive purification processes are used (Gedion 2012). Moreover, ^{238}U and ^{232}Th isotopes produce other alpha emitters in their decay chain. The decay chain of ^{238}U has eight alpha emitters and that of ^{232}Th has six alpha emitters, as will be shown in Sections 3.3.1 and 3.3.2.

2. Radioactive isotopes of chemical elements introduced recently in the fabrication of microelectronic devices. One example is hafnium oxide, a high-permittivity insulator introduced in order to replace the conventional SiO_2 in the gate stack in ultraminiaturized CMOS technologies (for more detail see Chapter 12). Hafnium has a natural isotope (^{174}Hf, with a natural abundance of 0.162%) that decays by emitting alpha particles.

These types of alpha emitters and related issues concerning their impact on radiation sensitivity and SEU occurrence in integrated circuits will be addressed in detail in Sections 3.3 and 3.6.

3.2 Radioactive Nuclides in Nature

As explained in Valkovic (2000), radionuclides that can be found in the environment can be divided into three general categories:

1. Primordial radionuclides: naturally occurring nuclides of very long half-life, which have persisted since the formation of the earth, and their shorter-lived daughter nuclides
2. Cosmogenic radionuclides: naturally occurring nuclides, which have short half-lives on the geological timescale, but which are being continuously produced by cosmic-ray radiation
3. Human-produced (or anthropogenic) radionuclides: radionuclides released into the environment due to human activity and by accident

The first two types of radionuclides constitute natural radionuclides, while radionuclides of the third type are artificial radionuclides. Natural radionuclides (primordial and cosmogenic) will be described in detail in Sections 3.3 and 3.4.

Concerning artificial radionuclides, humans have used radioactivity for one century, but the amounts of artificial radionuclides are small compared with those of natural radionuclides. Historically, the first experiment in artificially initiating nuclear transformation was carried out by the British physicist Ernest Rutherford (1871–1937) in 1919 (Valkovic 2000). It was shown that, when a particle emitted from ^{214}Po was absorbed on striking a nitrogen atom, oxygen and a proton were produced by the reaction:

$$^{14}N + {}^4He \rightarrow {}^{17}O + {}^1H \tag{3.9}$$

also noted by $^{14}N(\alpha, p)^{17}O$ (see Chapter 5, Section 5.1.2 for the notation of nuclear reactions). All the nuclides obtained by Rutherford were stable nuclides (Valkovic 2000). In 1934, the French physicists Irene Curie (1897–1956) and Frederic Joliot (1900–1958) discovered artificial radioactivity, taking a great step toward the use and control of radioactivity. In 1935 they were awarded the Nobel Prize in Chemistry for this discovery. In particular, they succeeded in artificially producing phosphorus-30 by bombarding aluminum with α-particles, following the reaction $^{27}Al(\alpha, n)^{30}P$:

$$^{27}Al + {}^4He \rightarrow {}^{30}P + {}^1n \tag{3.10}$$

This was the first man-made radionuclide. Since then, many artificial radioactive nuclei have been synthesized and studied. An artificial radionuclide is created by the bombardment of stable nuclei with neutrons, protons, and x-rays. Man-made radioactivity has found many useful applications in numerous domains, such as the following:

1. In nuclear medicine, for diagnosis, treatment, and research. The use of radionuclides and radioactivity in diagnosis and treatment of diseases is well established in practice (Valkovic 2000).
2. In food preservation, to reduce the risk of food-borne illness, to kill parasites, and to stop the sprouting and control the ripening of stored fruit and vegetables.

TABLE 3.1

Some Artificial (Man-Made or
Anthropogenic) Radionuclides

Artificial Radionuclide	Half-Life (years)	Decay Mode
^{99}T	2.1×10^5	β
^{137}Cs	30	β, γ
^{134}Cs	2.1	β, γ
^{90}Sr	28	β
^{129}I	1.6×10^4	β, γ
^{238}Pu	87	α
^{239}Pu	2.4×10^{14}	α
^{240}Pu	6.56×10^3	α
^{241}Pu	14	γ
^{242}Pu	3.7×10^5	α

3. In biochemistry and genetics, to label molecules and to trace chemical and physiological processes occurring in living organisms.

4. In industry and mining, to examine welds, to detect leaks, and to study the rate of wear, erosion, and corrosion of metals.

5. In geology, archaeology, and paleontology, to measure ages of rocks, minerals, and fossil materials.

However, accidental release of man-made radionuclides into the environment has occurred, especially during nuclear accidents (e.g., Chernobyl, Fukushima), discharges from nuclear installations, or atmospheric nuclear weapons testing. These radionuclides are dispersed in nature through the vegetation, oceans, and air and contaminate the natural environment (Gedion 2012). Some artificial radioactive contaminants and their characteristics are listed in Table 3.1.

3.3 Primordial Radionuclides

Primordial radionuclides are naturally occurring radionuclides with typically very long half-times, often on the order of hundreds of millions of years. Primordial radionuclides (also called terrestrial background radiation) are formed by the process of nucleosynthesis in stars. Only those radionuclides with half-lives comparable to the age of the earth, and their decay products, can still be found today on earth, for example, ^{40}K and the radionuclides from the ^{238}U and ^{232}Th series. Some of the long-lived naturally occurring radionuclides are shown in Table 3.2. In this table, nuclides that are of particular interest to microelectronics are uranium and thorium isotopes and their series, presented in Sections 3.3.1 and 3.3.2.

3.3.1 Uranium Decay Chain

Natural uranium consists of three isotopes, ^{238}U, ^{235}U, and ^{234}U, having isotopic abundances of 99.2745%, 0.72%, and 0.0055%, respectively. The most abundant isotope, ^{238}U, is present

TABLE 3.2

Some Primordial Radionuclides with Long Half-Lives Present on the Earth

Primordial Radionuclide	Half-Life ($\times 10^9$ years)	Isotopic Abundance (%)	Decay Mode
^{40}K	1.27	0.0117	β
^{50}V	6×10^5	0.25	β
^{87}Rb	47	27.835	β
^{115}In	6×10^5	95.72	β
^{138}La	110	0.0902	β
^{142}Ce	6×10^6	11.08	α
^{147}Sm	110	15	α
^{148}Sm	1.2×10^4	11.23	α
^{149}Sm	4×10^5	13.8	α
^{152}Gd	1.1×10^5	0.2	α
^{174}Hf	4.3×10^6	0.18	α
^{144}Nd	5×10^6	23.8	α
^{190}Pt	700	0.01	α
^{192}Pt	10^6	0.79	α
^{204}Pb	1.4×10^8	1.4	α
^{232}Th	14	100	α
^{235}U	0.71	0.72	α
^{238}U	4.5	99.3	α

in most terrestrial materials; its presence in ultratrace amounts (at sub-ppb or ppt levels) is attested even in purified and ultrapurified materials used in microelectronics. The ^{238}U decay chain (4n + 2 series), illustrated in Figure 3.1, comprises 14 radioactive isotopes, ending in stable lead 206. The parent nucleus ^{238}U undergoes alpha decay with a half-life of 4.5×10^9 years and has a specific activity of 12.44×10^4 MBq kg^{-1}. The majority of nuclides of the uranium series have short half-lives, only five nuclides having half-lives exceeding 1 year: ^{238}U, ^{234}U, ^{230}Th, ^{226}Ra, and ^{210}Pb (Valkovic 2000).

During decay, each nuclide of the uranium decay chain will emit characteristic radiation. The decay will result in the emission of alpha particles or beta particles with characteristic energies and probabilities of emission (Valkovic 2000). In the ^{238}U series, there are eight alpha-decay and six beta-decay steps. Table 3.3 presents the different isotopes of the ^{238}U decay chain, their half-lives, and their disintegration modes. For isotopes undergoing alpha decay, different characteristics of the emitted alpha particles are summarized in Table 3.3, such as energy, range in silicon, and linear energy transfer (LET). As mentioned above, among its 14 radioactive isotopes, the uranium decay chain has eight daughter nuclides that decay primarily by alpha emission: ^{238}U, ^{234}U, ^{230}Th, ^{226}Ra, ^{222}Rn, ^{218}Po, ^{214}Po, and ^{210}Po. For these alpha-emitter nuclides, the emitted alpha particles have energy ranging between 4.19 and 7.68 MeV (Table 3.3). Their corresponding ranges in silicon vary from 19 to 46 μm, and their initial LET from 0.47 to 0.68 MeV/(mg cm^{-2}).

The uranium series contains two radionuclides of particular interest: ^{226}Ra and its daughter ^{222}Rn. Radium is present in rock and soils and has a half-life of 1602 years and high specific activity. It decays through alpha-emission to ^{222}Rn, which has a half-life of 38 days and is a gaseous decay product. ^{222}Rn is the main isotope responsible for radiation

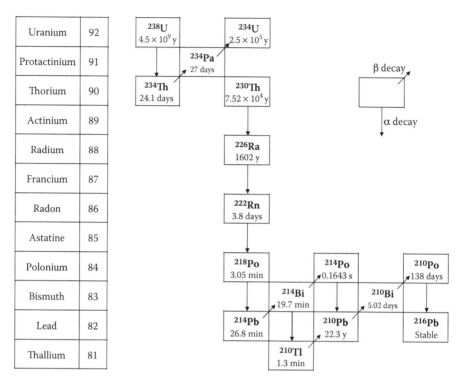

FIGURE 3.1
Uranium 238 radioactive decay chain. The half-life is also indicated for all elements (y = years, min = minutes, s = seconds).

exposure from indoor radon. The particular implication of this gas for microelectronics will be discussed in Section 3.5.

Another isotope of natural uranium, ^{235}U, is a naturally occurring radionuclide with long half-life (0.7×10^9 years) and a specific activity of 8×10^4 MBq kg^{-1}. As explained in Section 3.1.1, the parent nucleus ^{235}U decays to stable lead 207 through 11 daughter radionuclides; this is the actinium series (4n + 3). Among the daughters of the ^{235}U decay chain, there are seven nuclides that primarily decay by alpha-particle emission. Due to the very low natural abundance of ^{235}U (0.72%) compared with the abundance of ^{238}U (99.2745%), the number of alpha disintegrations occurring in a microelectronic device due to radionuclides of the ^{235}U decay series represents only 4% of the number of alpha disintegrations due to isotopes of the ^{238}U decay chain. Therefore, the impact on microelectronics devices of ^{235}U and its daughter radionuclides is usually neglected.

3.3.2 Thorium Decay Chain

Natural thorium consists of 100% ^{232}Th and has a specific activity of 4.06×10^4 MBq kg^{-1}, lower than that of uranium. ^{232}Th is the parent of the thorium decay chain (4n series), shown in Figure 3.2, which includes 10 radioactive isotopes. The ^{232}Th parent nucleus decays through six alpha and four beta emissions to the terminal nuclide, which is the stable lead 208. Except for the ^{232}Th parent, which has a half-life of 1.4×10^{10} years, the half-lives of the daughter nuclides are all less than 7 years (Valkovic 2000). Table 3.4 shows the main isotopes of the thorium series, with their main characteristics, such as half-lives and

TABLE 3.3

Isotopes of the ^{238}U Decay Chain and Their Characteristics (Half-Life and Decay Mode)

Isotope	Half-Life	Decay Mode	Characteristics of the Emitted Alpha Particle		
			Energy (MeV)	Range in Silicon (μm)	LET (MeV cm² mg⁻¹)
^{238}U	1.41×10^{17}	α	4.19	18.95	0.677
^{234}Th	2.08×10^6	β	–	–	–
^{234}Pa	2.41×10^4	β	–	–	–
^{234}U	7.76×10^{12}	α	4.68	22.17	0.634
^{230}Th	2.38×10^{12}	α	4.58	21.49	0.642
^{226}Ra	5.05×10^{10}	α	4.77	22.78	0.627
^{222}Rn	3.30×10^5	α	5.49	27.94	0.575
^{218}Po	1.86×10^2	α	6.00	31.86	0.545
^{214}Pb	1.61×10^3	β	–	–	–
^{214}Bi	1.19×10^3	β	–	–	–
^{214}Po	1.64×10^{-4}	α	7.68	46.22	0.468
^{210}Pb	7.04×10^8	β	–	–	–
^{210}Bi	4.33×10^5	β	–	–	–
^{210}Po	1.20×10^7	α	5.31	26.61	0.588

Note: For isotopes undergoing alpha decay, some characteristics of the emitted alpha particles are indicated: energy, range in silicon, and initial linear energy transfer (LET).

decay mode. In the thorium decay chain, there are six nuclides that disintegrate primarily by alpha emission. For these alpha emitters, Table 3.4 also presents the energies of the emitted alpha particles. As shown in Figure 3.2 and Table 3.4, ^{212}Bi can undergo two types of disintegration (α and β) with significant probabilities of occurrence. In the first case, ^{212}Bi undergoes beta decay at a percentage of 64% and generates the ^{212}Po nucleus; this latter will emit an alpha particle. In the second case, ^{212}Bi undergoes alpha disintegration at a percentage of 36% and generates ^{208}Tl, which will undergo beta decay. Particles emitted by the thorium decay chain have a wider range and higher maximum than the ^{238}U decay chain, with energies between 3.95 and 8.78 MeV (Valkovic 2000).

3.4 Cosmic-Ray-Produced Radionuclides

Cosmogenic radionuclides, such as ^{3}H, ^{7}Be, ^{10}Be, ^{14}C, and ^{22}Na, are produced by the interaction of cosmic-ray particles in the earth's atmosphere. As explained in Chapter 1, cosmic rays are divided into two categories: *primary* and *secondary*. Primary cosmic rays are high-energy particles that come from outer space and continually bombard the earth's atmosphere; the total number of primary cosmic rays striking the atmosphere is roughly 10^4 m^{-2} s^{-1}. The nuclear component of the primary cosmic rays comprises (at the top of the atmosphere) about 90% protons, 9% He, and 1% heavier nuclei (Valkovic 2000). The secondary cosmic rays are the result of the interaction of primary cosmic rays with the

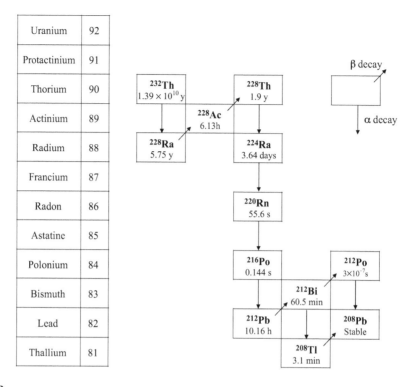

FIGURE 3.2
Thorium 232 radioactive decay chain. The half-life is also indicated for all elements (y = years, h = hours, min = minutes, s = seconds).

nuclei in the atmosphere, which produces a cascade of interactions and secondary reaction products. During passage through the atmosphere, the cosmic ray interactions also produce a wide variety of radioactive nuclei called cosmogenic radionuclides, such as those presented in Table 3.5.

Some cosmogenic radionuclides are formed *in situ* in soil and rock exposed to cosmic rays. Among all these radionuclides, some provide interesting markers in different domains. For example, the production rate in the atmosphere of the two isotopes of beryllium, ^{7}Be and ^{10}Be, is directly linked to solar activity, since this latter affects the magnetic field that shields the earth from cosmic rays (see Chapter 1). In the field of geology and archaeology, the presence of ^{14}C in organic materials is exploited to estimate the date of samples: this method is called radiocarbon dating. Note that the atmospheric nuclear tests that occurred between 1955 and 1980 dramatically increased the amount of ^{14}C in the atmosphere and subsequently in the biosphere during this period; this created an important anomaly in the concentration of ^{14}C, which doubled in the northern hemisphere.

3.5 Radon

Radon (^{222}Rn) and *thoron* (^{220}Rn) are radioactive noble gases. They are formed in rocks and soil as progeny of uranium and thorium, respectively. Both have 86 protons, but they have

TABLE 3.4

Isotopes of the ^{232}Th Decay Chain and Their Characteristics
(Half-Life and Decay Mode)

| Isotope | Half-Life | Decay Mode | Characteristics of the Emitted Alpha Particle | | |
			Energy (MeV)	Range in Silicon (μm)	LET (MeV cm^2 mg^{-1})
^{232}Th	4.39×10^{17}	α	4.00	17.76	0.695
^{228}Ra	1.80×10^{8}	β	–	–	–
^{228}Ac	2.20×10^{4}	β	–	–	–
^{228}Th	6.0×10^{7}	α	5.40	27.27	0.582
^{224}Ra	3.11×10^{5}	α	5.67	29.30	0.565
^{220}Rn	55.6	α	6.29	34.18	0.530
^{216}Po	0.144	α	6.78	38.25	0.505
^{212}Pb	3.82×10^{4}	β	–	–	–

Two possibilities of radioactive decay:

(1) First case:

^{212}Bi (64%)	3.66×10^{3}	β	–	–	–
^{212}Po	3.0×10^{7}	α	8.78	56.82	0.428

(2) Second case:

^{212}Bi (36%)	9.84×10^{3}	α	6.32	34.42	0.528
^{208}Tl	1.86×10^{2}	β	–	–	–

Note: For isotopes undergoing alpha decay, some characteristics of the emitted alpha particles are indicated: energy, range in silicon, and initial linear energy transfer (LET). ^{212}Bi can undergo two types of radioactive decay at comparable levels.

TABLE 3.5

Some Cosmogenic Radionuclides and Their Decay Characteristics (Half-Life and Decay Mode)

Cosmogenic Radionuclides	Half-Life (years)	Decay Mode
^{10}Be	1.6×10^{6}	β
^{26}Al	7.2×10^{5}	β, γ
^{36}Cl	3.0×10^{5}	β
^{80}Kr	2.13×10^{5}	K X-rays
^{14}C	5730	β
^{32}Si	650	β
^{39}Ar	269	β
^{3}H	12.33	β
^{22}Na	2.6	β, γ

Source: Numerical data from Valkovic, V., *Radioactivity in the Environment*, Elsevier, Amsterdam, 2000.

different numbers of neutrons and therefore different atomic mass numbers. As previously shown, ^{222}Rn is a decay product of ^{226}Ra, which is itself a decay product of the primordial radionuclide ^{238}U. The parent isotope for thoron is ^{232}Th, a primordial element (see Section 3.3.2), which is widespread in soils and rocks. Both radon and thoron are emitted from the ground into the atmosphere, where they decay and form daughter products, isotopes of polonium, bismuth, and lead (see Figures 3.1 and 3.2), which either remain airborne until they decay or are deposited in rain and by diffusion to the ground (Valkovic 2000).

Typical concentrations of ^{222}Rn in the environment range from less than 1–10 Bq m^{-3} outdoors to 100–1000 Bq m^{-3} indoors, as a function of location on the earth. As noble gases, radon and thoron are not very chemically reactive and can diffuse into and through materials prior to decaying, posing a potential reliability problem for semiconductor packages, as radon and several of its progenies decay via alpha emission (Wong et al. 2010). Radon (thoron) diffusion and transport through packaging materials (ceramic, resins, loaded polymers) is a complex process and is affected by several factors. For any material medium, the porosity, permeability, and diffusion coefficient are the parameters that can quantify its capability to hinder the flow of radon (Narula et al. 2010). An increase in porosity will provide more air space within the material for radon to travel, thus reducing resistance to radon transport. The permeability of a material describes its ability to act as a barrier to gas movement when a pressure gradient exists across it, and is closely related to the porosity of the material. The radon diffusion coefficient of a material quantifies the ability of radon gas to move through it when a concentration gradient is the driving force. This parameter is proportional to the porosity and permeability of the medium (Narula et al. 2010).

Only a few published studies have investigated this problematic of radon diffusion in packaged electronics and its impact on the soft-error rate (SER) of circuits.

In Wilkinson and Hareland (2005), the authors investigated a novel packaging scheme for an SRAM component for possible use in an implantable medical device. Standard testing revealed a high incidence of soft errors, which was quickly traced to alpha-particle emissions, primarily (about 98%) from the solder balls used for the direct bonding of the silicon onto the circuit board, which can help reduce the total volume of an implantable product. The remaining 2% of the observed errors were too far from the solder spheres to be explained by that mechanism. The possibility that the observed error rate was caused by alpha emission from ^{222}Rn decay was considered. Based on the observed sensitivity to alpha-particle upset and the gas volume exposed to the die face, a value of 10 pCi L^{-1} would be required to cause the observed error rate. Purging the shielding chamber with pure N$_2$ did not reduce the residual error rate. The laboratory air was measured at 1 pCi L^{-1}, confirming that the alpha particles responsible for these errors were from another material. The use of an underfill layer quickly removed the excess 2% alpha particles from the error rate, providing a robust fix for the nonsolder-ball alpha-particle sources (Wilkinson and Hareland 2005).

In Wong et al. (2010), the authors conducted an experimental study to determine whether radon can diffuse into the semiconductor-molding compound, decay near the silicon surface, and increase the SER of the device. They used wire-bonded SRAM memories packaged with ultralow-alpha (ULA)-emission molding compound with a nominal emissivity below 0.002 α h^{-1} cm^{-2}. The devices were loaded into a real-time test setup and monitored for 100 days to assess the failure rate of the devices in typical operating conditions under ambient radon level. A second run was completed in the same conditions, except that a gas-tight metal tank was constructed to enclose the test setup together with a radon source. Typical radon concentration during this second run was about 16,000 Bq m^{-3}, which is about 3000 times higher than ambient radon. Despite this extreme elevated concentration of radon and the duration of the experiment (100 days' exposure), the SER of the wire-bonded memories was not affected by the 3000-fold increase in radon concentration. The packaging material and thickness of the molding compound appear, in this case, to provide a sufficient barrier to minimize radon diffusion. Follow-up experiments (not yet published) may be completed on a flip-chip device to determine whether radon would diffuse into the flip-chip package.

3.6 Radionuclides and Radioactive Contamination in Advanced CMOS Technologies

With terrestrial cosmic rays presented in Chapter 1, alpha emitters present in circuit or packaging materials represent the second radiation source of reliability degradation for electronics. We examine in this section the important issue related to the natural alpha radioactivity of materials used in microelectronics. As already introduced in Section 3.1.2, alpha radioactive nuclei can be categorized into two families: radioactive isotopes present in the form of ultratrace amounts of impurities in materials and radioactive isotopes of chemical elements introduced recently into microelectronic device fabrication. In Wrobel et al. (2008b), the authors term the first category *radioactive impurities* and the second *radioactive materials*. Hafnium, already used in gate-oxide composition, and platinum, which may be used in platinum-base silicide contacts, are two examples of radioactive materials; uranium and thorium, naturally present in all terrestrial materials and compounds, are radioactive impurities.

In order to roughly evaluate the contribution of both radioactive materials and impurities to the SER of circuits, Wrobel et al. introduced a quantity or metric that corresponds to the disintegration rate of a sheet of material of 1 mm²×1 μm during 10^9 h. This particular choice is guided by the fact that: (i) the SER is generally expressed as failure in time (FIT), which expresses the number of errors during 10^9 h; (ii) a volume of 1 mm²×1 μm is representative of the sensitive volume of a typical circuit in current bulk technologies.

Table 3.6 shows the disintegration rate for natural radioactive materials. The number of radioactive nuclei has been calculated by considering their natural abundances (Wrobel et al. 2008b). Note that the values indicated in Table 3.6 should be enhanced for platinum (by a factor of two), gadolinium (by a factor of three), and samarium (by a factor of two), since their decay leads to several daughter nuclei that are alpha emitters (Wrobel et al. 2008a). These calculations show that the most active material is platinum, followed by neodymium, gadolinium, and samarium. Hafnium shows the lowest disintegration rate. All these materials are susceptible to being used in advanced CMOS technologies. For example, platinum can be used in different alloys and silicides for metal electrodes; it may be found close to the sensitive regions of transistors (gate electrode, source and drain contacts). Neodymium, gadolinium, and samarium can also be potentially used as high-permittivity gate insulators; this is already the case for hafnium, used in commercial CMOS processes since 2011. To evaluate their impact on the SER in circuits, Gedion et al. performed a worst-case simulation study in Gedion et al. (2010) considering the different forms (alloys, oxides, silicides) in which these radioactive isotopes may be introduced at device level. Table 3.7 summarizes the results of this estimation. Gd, Hf, and Nd-based materials do not play a major role in triggering soft errors. This is due to their long half-life,

TABLE 3.6

Disintegration Rates of Radioactive Materials

Element	Half-Life	Natural Abundance (%)	Disintegration Rate $\times 10^4\,\alpha/$ $(mm^2 \times \mu m \times 10^9\,h)$
^{190}Pt	2.05×10^{19}	0.014	85.21
^{144}Nd	7.23×10^{22}	23.8	41.07
^{152}Gd	3.41×10^{21}	0.2	7.32
^{148}Sm	2.21×10^{23}	11.24	6.35
^{187}Re	1.37×10^{18}	62.6	5.70
^{186}Os	6.31×10^{22}	1.59	3.14
^{174}Hf	6.31×10^{22}	0.16	0.32

Source: Data from Wrobel, F. et al., *IEEE Trans. Nucl. Sci.* 55, 3141–3145, 2008; Wrobel, F. et al., *Appl. Phys. Lett.* 93, 064105, 2008.

TABLE 3.7

Worst-Case Estimation of Soft-Error Rate for SRAM Memories with Different Compounds Incorporating Radioactive Isotopes

Compound	Used as	SER (FIT/Mbit)
HfO_2	Gate oxide	0.08
$GdScO_3$	Gate oxide	3.3
Gd_2O_3	Gate oxide	4.8
Nd_2O_3	Gate oxide	9.6
$SmScO_3$	Gate oxide	7.9×10^3
PtSi	P-MOSFET metal electrode	34.2
Pt_3Ge_2	P-MOSFET metal electrode	38.0
$Ni_{70}Gd_{30}$	N-MOSFET metal electrode	6.5

Source: Data from Gedion, M. et al., *J. Phys. D: Appl. Phys.* 43, 275501, 2010.

resulting in a low disintegration rate. The authors concluded that these elements may be used in gate oxides without taking account of their contributions to soft errors. In contrast, when Sm is used as a gate-oxide compound ($SmScO_3$), its high disintegration rate logically induces a very high alpha contribution to SER. For this reason, this material should not be employed in gate dielectrics without isotopic purification. Finally, the results of Table 3.7 show that platinum in its different forms represents a potential source of soft errors. Even if its activity is much lower than that of samarium, its contribution to soft errors must be taken into account because of its proximity to the sensitive zones.

For uranium and thorium radioactive impurities, Table 3.8 reports the values of the disintegration rate for the 1 $mm^2 \times 1\ \mu m$ volume, assuming a typical concentration of 1 ppb. For the same reason as previously mentioned for Table 3.6, the disintegration rate for uranium should be enhanced by a factor of eight, since its decay leads to eight daughter nuclei that are alpha emitters; in the case of thorium, the disintegration rate should be enhanced by a factor of six, since its decay leads to six alpha-emitting daughter nuclei (Wrobel et al. 2008a). It is seen that the emission rate from 1 ppb of these impurities is much lower than

TABLE 3.8

Disintegration Rates of Radioactive Impurities (at 1 ppb Concentration) That May Be Unintentionally Introduced into Electronic Materials

Element	Half-Life	Natural Abundance (%)	Disintegration Rate $\alpha/(\text{mm}^2 \times \mu\text{m} \times 10^9\,\text{h})$
^{238}U	1.41×10^{17}	99.2742	878
^{234}U	7.75×10^{12}	5.40×10^{-3}	869
^{232}Th	4.43×10^{17}	100	282
^{234}U (fission)	7.75×10^{12}	5.40×10^{-3}	$<7.82 \times 10^{-4}$
^{232}Th (fission)	4.43×10^{17}	100	$<2.53 \times 10^{-4}$

Source: Data from Wrobel, F. et al., *Appl. Phys. Lett.* 93, 064105, 2008.

that of the radioactive materials added to devices (Table 3.6). However, the fundamental difference is that the quantity and position of elements added to devices can be controlled, whereas those of unintentional impurities cannot. For this reason, the contribution of impurities to the SER must be carefully evaluated for a given circuit; it remains strongly dependent on the fabrication processes, the selection of materials, and the quality controls performed along the whole fabrication chain, from raw materials to packaged chips.

To conclude this section, the results of Tables 3.6 and 3.8 have been compared in Wrobel et al. (2008b) with the number of nuclear reactions induced by atmospheric neutrons (see Chapter 5) at both avionic (12 km) and ground levels. The number of reactions produced by neutrons at ground level is typically 4000 $(\text{mm}^2 \times \mu\text{m} \times 10^9\,\text{h})^{-1}$, and 300 times higher at avionic altitude. At ground level, this is lower than the number of alpha particles produced by the 1 μm layer of the alpha emitters described above, confirming that the alpha-emitter issue is crucial for ground-level applications (Wrobel et al. 2008b).

3.7 Alpha Radiation from Interconnect Metallization and Packaging Materials

Although silicon and other front-end-of-line (FEOL) materials can play a role in the alpha-particle-induced SER of circuits, in many cases the main alpha-particle sources are found in materials adjacent to the chip, in the solders, and in the packaging materials (Heidel et al. 2008). In particular, due to the increasing use of flip-chip packages and developments toward 3D packaging, the solder bumps have moved very close to the active Si devices, where even low-energy alphas with short ranges are able to induce soft error (Kumar et al. 2013). One of the major sources of alpha-particle radiation is lead (Pb) solder in the form of solder balls used for joining components in the packaging. This is illustrated in Figure 3.3. Among all the lead isotopes, the ^{210}Pb radionuclide of the ^{238}U decay chain (see Figure 3.1) has the shortest half-life (22.2 years). While it is not directly an alpha emitter, ^{210}Pb beta decays into ^{210}Bi in 22 years, then beta decays to ^{210}Po in 5 days, and finally alpha decays to ^{206}Pb in 138 days (this latter isotope is stable). During its decay, ^{210}Po emits an alpha particle of 5.407 MeV. The major problem of ^{210}Pb is that it is chemically inseparable from stable Pb isotopes. Therefore, obtaining very-low-alpha-particle-emitting Pb is a challenge for the semiconductor industry (Heidel et al. 2008). Since the half-life of ^{210}Po is 22 years,

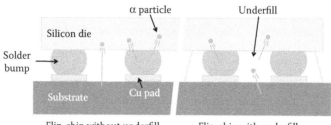

FIGURE 3.3
Schematic cross section of a flip-chip package with and without underfill.

extremely old lead would not be expected to have high activity. For example, archaeological lead recovered from ancient galleons stranded on the seabed has an activity of less than 0.001 counts cm^{-2} h^{-1}. However, there is a very limited supply of this material. The semiconductor industry is currently using solders with alpha-particle emissivity levels ranging between 0.05 and 0.01 counts cm^{-2} h^{-1}. With increasing circuit density, decreasing power-supply voltage, and further miniaturization of IC devices, the requirement for alpha-particle emission level, especially in medical, military, automotive, aerospace, and telecommunications applications, has dropped from 0.05 to <0.002 counts cm^{-2} h^{-1} over the last decade (Kumar et al. 2013).

Because of legislations such as Restriction of the use of certain Hazardous Substances (RoHS) and Waste Electrical and Electronic Equipment (WEEE), there is a global shift toward the use of Pb-free electronic products. Semiconductor packaging and electronics manufacturing companies now employ Pb-free solder for their electronic products (Kumar et al. 2013). But lead-free solder does not guarantee low alpha-particle levels, since tin (Sn), for example, can have an appreciable alpha-particle component because of the incorporation of impurities during Sn refining (Heidel et al. 2008). Samples of Sn have been observed to emit alpha particles at rates as high as or higher than those of Pb. Despite continuous efforts to achieve the lowest-alpha-emitting solder for both Pb solder and Pb-free solder, solder balls remain a significant source of alpha particles. The alpha-particle emission rates for solder are typically in the range of 5–50 counts cm^{-2} h^{-1} (Heidel et al. 2008).

The different packaging materials represent another source of alpha-particle emission. Underfill, overmolds, organic packages, and ceramic packages can all be sources of alpha particles. While mold compounds are generally in the range of 1–5 α cm^{-2} h^{-1}, this alpha-particle contribution to the SER cannot be ignored for advanced CMOS technologies. Organic and ceramic packages can also emit a significant amount of alpha particles, on the order of 80–100 counts cm^{-2} h^{-1} (Heidel et al. 2008).

Different mitigation strategies can be envisaged to reduce the alpha-particle-induced SER of circuits (Cabral et al. 2007). The first is an effort to drive the alpha-particle levels of the material to even lower emission rates. This requires the development and manufacturing of products using new material compositions, as well as better qualification and monitoring of incoming materials, which in turn requires the development of more sensitive alpha-particle detection equipment that can isolate subtle changes in the alpha-particle emission rate of materials.

The second is an effort to develop an alpha-particle barrier that prevents all alpha particles from external sources from affecting product circuits. The barrier could reside either on the chip, directly under the solder balls, or on the packaging above the thin-film wiring levels. Both these approaches add processing complexity and cost, and they still require a level of internal alpha-particle contamination control by manufacturing lines. However, the flux of alpha particles reaching the ICs is greatly reduced, effectively switching this problem from one of internally generated alpha particles to solely an issue of cosmic-neutron particle flux.

3.8 Emissivity Model

Alpha-particle surface emissivity is the most accessible and measurable physical quantity to characterize the alpha radioactivity of materials. In complement to the techniques of emissivity measurement and ultratrace detection that will be presented in Chapter 4, modeling and simulation are an indispensable approach for the estimation of the internal radioactivity of bulk or (multi)layered materials from emissivity measurements. Recently, Martinie et al. (2011) have developed a full analytical model describing the fraction of alpha particles generated in a bulk monolayer (due to the presence of a certain amount of uranium and/ or thorium radioactive impurities) and escaping from the surface of this layer. From this calculation, the layer surface emissivity, generally measured at wafer level, as well as the corresponding energy spectrum of emitted alpha particles can be easily deduced. This analytical modeling approach has been generalized and extended to the case of a multilayered system of different materials. We recount here the main results of this study.

3.8.1 Analytical Model for Monolayers

3.8.1.1 *Monte Carlo Simulation of Uranium Decay*

As explained in Section 3.3, uranium and thorium are naturally present in all terrestrial materials; energies of the corresponding emitted alpha particles range from 4 to 9 MeV (Wrobel et al. 2008a,b; Adamiec and Aitken 1998; Gedion et al. 2011), and their corresponding ranges in silicon vary from 19 to 57 μm. In their study, Martinie et al. (2011) considered both uranium and thorium decay chains at secular equilibrium; all daughter elements of the considered decay chain then have the same activity. For a material layer contaminated with 1 ppb of uranium, the total (i.e., global) activity of this material, combining the activity of all daughter elements of the decay chain, is then equal to $A_{Si\_U} = 1950$ Bq m^{-3} (or $A_{Si\_Th} = 473$ Bq m^{-3} for a layer with 1 ppb of thorium) (Wrobel et al. 2008b).

In a first step, a numerical Monte Carlo simulation code has been specifically developed in Martinie et al. (2011) to numerically evaluate the number (or the fraction) of alpha particles coming from the alpha-emission processes of the uranium (or thorium) decay chain and escaping from a material layer of given dimensions and contamination level. For each produced particle, the code allows one to know the particle's initial energy, the corresponding energy loss during its transport into the material layer, its range, its final energy, and so on.

In these simulations, the conversions between the particle range and its energy have been obtained from Stopping and Range of Ions in Matter (SRIM) data tables (http://www.srim.org). In order to maintain a full analytical model, SRIM data have been accurately fitted in Martinie et al. (2011) using a set of fitting functions based on the ratio of polynomial expression (Martinie et al. 2014) to accurately describe the evolution of range and energy (the validity of this fitting approach is limited to 1 keV–10 MeV). Numerical results have been successfully compared with data obtained from the Monte Carlo simulator ORACLE (Wrobel and Saigné 2011), initially developed to predict the SER of a given technology subjected to various radiation environments, including space ions and protons, atmospheric neutrons, and natural alpha-particle emitters (Wrobel and Saigné 2011).

Figure 3.4a shows the spectrum of alpha particles emitted from a monolayer of silicon (1 cm^2 × 50 μm) contaminated with 1 ppb of ^{238}U and its decay daughter elements (a total of eight alpha-particle emitters). The characteristic shape of each spectrum is directly determined by the evolution of the particle energy with its range in silicon. In particular,

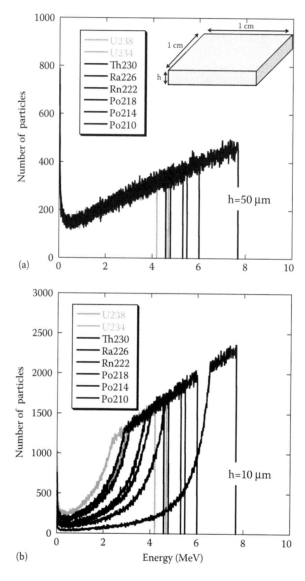

FIGURE 3.4

Numerical Monte Carlo simulations giving the number of escaping alpha particles versus the particle energy for a total of 1,000,000 alpha-particle emissions generated in the bulk of a 1 cm^2 × 50 μm (a) and 1 cm^2 × 10 μm (b) silicon layer contaminated with 1 ppb of ^{238}U and its decay daughter elements (total of eight alpha-particle emitters). The energy interval for the particle counting is equal to 10 keV. (Reprinted from Martinie, S. et al., Analytical modeling of alpha-particle emission rate at wafer-level, *IEEE Trans. Nucl. Sci.* 58, 2798–2803. © (2011) IEEE. With permission.)

TABLE 3.9

Fraction of Escaping Particles

	0.1 MeV		2 MeV	
E_{th}	50 μm (%)	10 μm (%)	50 μm (%)	10 μm (%)
^{238}U	9.2	36.3	5.9	28.7
^{234}Th	10.8	38.3	7.5	33.3
^{230}Th	10.4	38.0	7.1	32.5
^{226}Ra	11.1	38.7	7.8	33.9
^{222}Rn	13.6	40.8	10.3	37.9
^{218}Po	15.6	42.0	12.3	39.9
^{214}Po	22.7	44.4	19.4	43.5
^{210}Po	12.9	40.3	9.7	37.0
Total	13.3	38.8	10.0	35.8

Source: Data from Martinie, S. et al., *IEEE Trans. Nucl. Sci.* 58, 2798–2803, 2011.

the increase of spectra at low energy is directly linked with the great number of high-energy alpha particles, where the energy lost precisely corresponds to the maximum range (Gordon et al. 2010). For a thinner layer (1 cm^2×10 μm, see Figure 3.4b), the number of escaping particles increases and the curves reduce to peaks centered on the energy value corresponding to the maximum range of the particle track.

Table 3.9 summarizes the fraction of escaping particles considering a 0.1 and 2 MeV threshold energy (E_{th}) for each alpha emitter of the uranium decay chain. The fraction of escaping particles is obtained by the ratio of the number of particles with energy above the threshold energy to the total number of generated particles.

3.8.1.2 Analytical Modeling of Emissivity

Experimentally, the emissivity is measured in a certain energy range corresponding to the energy window of the detector used in the alpha-particle counting system. In particular, to reduce detection noise, threshold energy (E_{th}) is chosen to obtain a more believable emissivity measurement. This aspect will be introduced later in our model to take into account such an important experimental limitation. Currently, a state-of-the-art ultralow-background alpha-particle detector has a threshold energy E_{th} around 1 MeV (Gordon et al. 2009).

To evaluate the alpha-particle emissivity, we start from the following elementary formula (the unit used in the following will be α (cm^2 h^{-1})$^{-1}$):

$$\phi_\alpha = n \times D_\alpha \times h \times EP \tag{3.11}$$

where:

n   is the number of alpha emitters in the decay chain (equal to 8 for uranium and 6 for thorium)

D_α is the disintegration rate for 1 ppb of contaminant (0.878 [cm^2 μm.h]$^{-1}$ for uranium and 0.282 [cm^2 μm h]$^{-1}$ for thorium [Wrobel et al. 2008b])

h   is the bulk layer thickness

EP   corresponds to the fraction of particles escaping from the top of the material layer

As will be shown later, the number of particles escaping from the layer strongly depends on its geometric dimensions. Consequently, the main parameter in Equation 3.11 is the fraction of escaping particles, which takes into account the sample geometry and the threshold energy. previously defined.

The EP parameter is directly linked to the probability that particles escape from the bulk outside the layer. The fraction of escaping particles was first evaluated in Martinie et al. (2011) from a purely geometric point of view. Figure 3.5 illustrates two typical cases when a particle can exit (depth z_1 from the top layer) or not (depth z_2) from the silicon layer. In the case when the depth of the alpha-emission point z is less than the range of the nuclei considered (R_α), the escaping probability is the ratio of the surface area of the spherical cap to that of the complete sphere:

$$p(z, R_\alpha) = \frac{(R_\alpha - z)}{2 \times R_\alpha} \tag{3.12}$$

From Equation 3.12, the EP for alpha particles emitted from a given isotope (e.g., ^{238}U or ^{214}Po) is integrated from 0 to the minimum value between the thickness and the range of the nuclei considered:

$$P_\alpha(h, R_\alpha) = \frac{1}{h} \times \int_0^{\min(R_\alpha, h)} p(z, R_\alpha) dz \tag{3.13}$$

The integral in Equation 3.13 is analytical; then, after integration, Equation 3.13 gives

$$P_\alpha(h, R_\alpha) = \frac{2 \times R_\alpha - h}{4 \times R_\alpha} \quad \text{if } h < R_\alpha \tag{3.14}$$

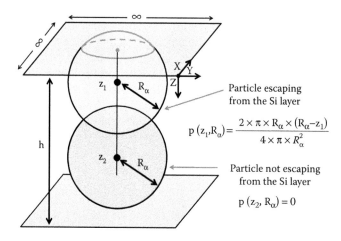

FIGURE 3.5
Schematic illustration of the geometric model considered for the exact calculation of the fraction of escaping alpha particles from a material layer with infinite dimensions and thickness h. (Reprinted from Martinie, S. et al., Analytical modeling of alpha-particle emission rate at wafer-level, *IEEE Trans. Nucl. Sci.* 58, 2798–2803. © (2011) IEEE. With permission.)

$$P_\alpha(h, R_\alpha) = \frac{R_\alpha}{4 \times h} \quad \text{if } h > R_\alpha \tag{3.15}$$

Equations 3.14 and 3.15 take into account only one alpha-emitting radioisotope without any threshold energy. To include the contribution of all elements of the disintegration chain and the effect of the threshold energy in the result, it is necessary to sum over the number of daughter elements and to introduce the effect of E_{th}. The EP parameter of 3.11 can be finally expressed as

$$EP(h, E_{th}) = \frac{1}{n} \times \sum_{i=1}^{n} P_\alpha \left(h, R_{\alpha\,(E_i)} - R_\alpha(E_{th}) \right) \tag{3.16}$$

where:

n $\quad$ = number of daughter nuclei (8 for uranium, 6 for thorium)

$R_\alpha(E_i)$ = range of the ith daughter nuclei (for uranium, for example, $R_\alpha(E_7) = 46.22\ \mu m$ corresponds to ^{214}Po daughter nuclei)

$R_\alpha(E_{th})$ = corresponding range of the threshold energy (E_{th})

The set of Equations 3.11 through 3.15 constitutes the analytical model proposed in (Martinie et al. 2011). Figure 3.6 compares the fraction of escaping particles obtained from this model with data obtained by numerical Monte Carlo simulation for different threshold energy and layer thickness. The agreement between the two series of data is perfect,

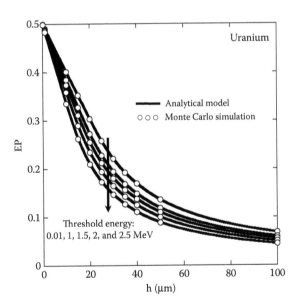

FIGURE 3.6

Comparison between the analytical model (solid lines) and Monte Carlo simulations (symbols) for the fraction of particles escaping from the top surface of a bulk silicon layer versus the layer thickness for different threshold energies (uranium decay chain). (Reprinted from Martinie, S. et al., Analytical modeling of alpha-particle emission rate at wafer-level, *IEEE Trans. Nucl. Sci.* 58, 2798–2803. © (2011) IEEE. With permission.)

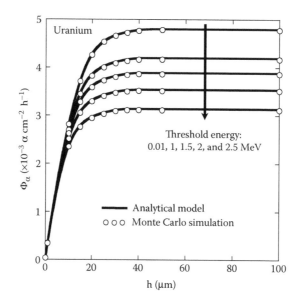

FIGURE 3.7
Comparison between the analytical model (solid lines) and Monte Carlo simulations (symbols) for the emissivity of a silicon layer (1 cm²×h μm) contaminated with 1 ppb of ^{238}U (and its decay daughter elements) as a function of silicon-layer thickness. (Reprinted from Martinie, S. et al., Analytical modeling of alpha-particle emission rate at wafer-level, *IEEE Trans. Nucl. Sci.* 58, 2798–2803. © (2011) IEEE. With permission.)

demonstrating the validity of Equation 3.15. In the case of an infinitely thin layer, the probability that a particle exits from the top surface is 0.5 (50% escapes from the top surface, 50% below); this value continuously decreases when the silicon bulk layer increases. When a given threshold energy is introduced into the calculation, all escaping particles with a final energy below E_{th} are not counted, thus reducing the probability.

Figure 3.7 compares alpha-particle emissivity values obtained from numerical simulations and using the analytical model. As explained above, when a detection threshold energy is fixed in Equation 3.16, all alpha particles escaping the layer with an energy below this threshold energy are not considered, and the emissivity level is reduced. In the same way, reduction of the layer thickness strongly decreases the emissivity value even if the fraction of escaping particles increases in the same time and tends toward 0.5, as illustrated in Figure 3.6.

3.8.1.3 Analytical Description of Emitted-Alpha-Particle Energy Distribution

From Equation 3.16, which gives the fraction of escaping particles, it is possible to analytically derive the number (or the fraction) of emitted particles in a given energy interval [E, E+ ε]:

$$N(h, E_i, E_t) = N_T \times \left[\sum_{i=1}^{n} \left[EP(h, E_i, E) - EP(h, E_i, E + \varepsilon) \right] \right] \tag{3.17}$$

where N_T is the total number of generated particles and E_i is the energy of the emitted alpha for the ith daughter nuclei.

Figure 3.8a (h = 50 µm) and 3.8b (h = 10 µm) compare numerical simulation results with data obtained with Equation 3.17 for two particular daughter nuclei of the uranium decay chain (^{238}U and ^{210}Po) in terms of number of escaping particles versus energy. These two figures illustrate the capability of the analytical model to perfectly reproduce the energy distributions of the emitted alpha particles escaping from a silicon layer of a given thickness.

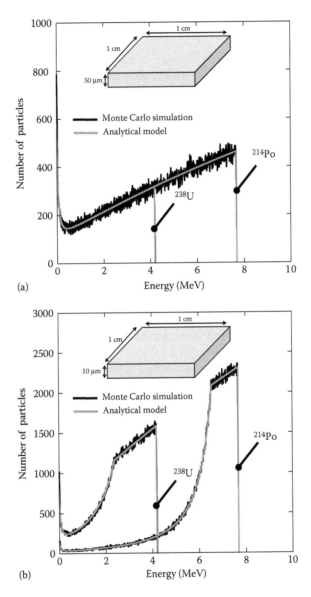

(a)

(b)

FIGURE 3.8
Comparison between results obtained from numerical simulations and from the analytical model concerning the number of escaping particles versus energy for a total of 1,000,000 alpha-particle emissions generated in the bulk of a 1 cm²×50 µm (a) and 1 cm²×10 µm (b) silicon layer contaminated with 1 ppb of ^{238}U and ^{214}Po radioisotopes. The energy interval for the particle counting is equal to 10 keV. (Reprinted from Martinie, S. et al., Analytical modeling of alpha-particle emission rate at wafer-level, *IEEE Trans. Nucl. Sci.* 58, 2798–2803. © (2011) IEEE. With permission.)

3.8.2 Analytical Model for Multilayer Stack

The principle of calculation for evaluating the global emissivity of a stack composed of several layers of different materials presented in Martinie et al. (2011) is outlined in this section. The proposed method is then applied to a typical stack representative of a back end of line (BEOL) of a silicon CMOS technology.

3.8.2.1 Modeling Approach

As explained previously, the main parameter entering into the expression of layer emissivity is the fraction of escaping particles, which takes into account the geometric dimensions of the sample as well as the threshold energy associated with the detection system. To estimate the global emissivity of a given stack, each layer of which is contaminated with X ppb of uranium (or thorium), one must evaluate the corresponding emissivity of each layer and properly add these separate contributions to obtain the global emissivity of the stack. Thereafter, one also needs to determine the fraction of escaping particles coming from a considered layer and crossing the upper layers. The resulting expression for the emissivity of a multilayer stack can be written as

$$\phi_{\alpha\_total} = n \times D_\alpha \times \sum_{m=0}^{nL-1} \left[h_m \times EP_m \times X_{ppb\_U}^m \right] \tag{3.18}$$

where:

h_m = thickness
$X_{ppb\_U}^m$ = contamination level in radioactive (uranium) impurities
EP_m = fraction of escaping particles related to the mth layer of the stack (m varying from 0 to nL, the total number of layers in the stack)

The EP_m parameter must take into account the role as a filtering barrier (with respect to the incoming alpha particles) of all layers of the stack (indexed from $m+1$ to $nL-1$) located above the considered m^{th} layer, as illustrated in Figure 3.9. Taking into account all nuclei of the considered decay chain, EP_m can be expressed as

$$EP_m = \frac{\sum_{i=1}^{n} \left[P_{\alpha 0}^{i,m} \times \prod_{j=m+1}^{nL-1} \left[2 \times p\left(h_j, R_\alpha^j (E_i) - R_\alpha^j (\xi_{j+1})\right) \right] \right]}{n} \tag{3.19}$$

$$P_{\alpha 0}^{i,m} = P_\alpha \left(h_m, R_\alpha^m (E_i) - R_\alpha^m (\xi_{m+1}) \right) \tag{3.20}$$

where:
R_α^m is the alpha-particle range-versus-energy relationship for the mth material layer
ξ is the threshold energy loss
the factor 2 comes from a normalization condition

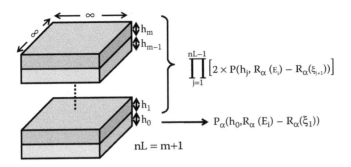

FIGURE 3.9
Schematic representation of a multilayer stack composed of nL layers of different thicknesses. The corresponding escaping probabilities for a given layer and for the entire stack are also indicated. (Reprinted from Martinie, S. et al., Analytical modeling of alpha-particle emission rate at wafer-level, *IEEE Trans. Nucl. Sci.* 58, 2798–2803. © (2011) IEEE. With permission.)

This threshold energy loss ξ represents the energy gradually lost between the considered layer and the top layer of the stack through all the intermediate layers. The corresponding formulas are

$$\xi_{nL} = E_{th} \tag{3.21}$$

$$\xi_m = E_\alpha^m \left(h_m + R_\alpha^m \left(\xi_{m+1} \right) \right) \tag{3.22}$$

where E_α^m and R_α^m are the energy-versus-range and range-versus-energy relationships of the considered layer, respectively, also described by polynomial fitting functions. In the particular case of a stack reduced to a single layer, this threshold energy-loss factor corresponds to the threshold energy E_{th} defined in Section 3.8.1.2.

3.8.2.2 Example of Multilayer

To illustrate this approach, Martinie et al. considered a multilayer stack composed of different layers Si/Al/Cu/Al/Cu (see inset of Figure 3.10b). Figure 3.10a shows the contribution to the emissivity at the top surface of the stack of the different individual layers, each layer being contaminated with 1 ppb of uranium. Figure 3.10b shows the total emissivity at the top layer of the stack corresponding to the sum of all different contributions shown in Figure 3.10a (1 ppb of uranium).

The two figures demonstrate good agreement between numerical simulations performed with MC-ORACLE (Wrobel and Saigné 2011) and data obtained with the analytical model. Finally, this analytical approach can be easily generalized to any multilayer system containing more than one type of radioactive impurity.

3.8.3 Universal Nomogram for Bulk Silicon

Finally, Martinie et al. proposed a universal nomogram for silicon, giving the alpha emissivity as a function of the level of residual impurities in uranium, thorium, or both (Martinie et al. 2011). The analytical model presented previously was used for

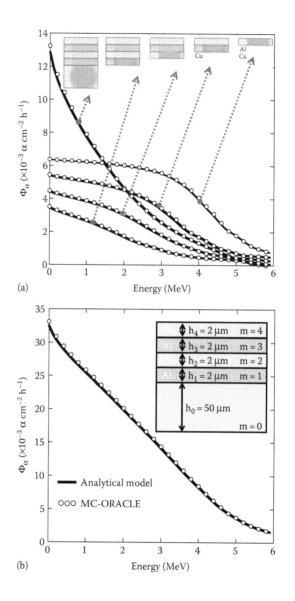

FIGURE 3.10

Alpha-particle emissivity calculated for different individual layers (a) and for the whole stack of layers (b). All these layers are subjected to the same contamination of 1 ppb of ^{238}U (and its decay daughter elements). Solid lines correspond to results obtained with the analytical model; symbols correspond to Monte Carlo simulation data obtained with MC-ORACLE. The threshold energy for the alpha-particle detection system is fixed at 2 MeV. (Reprinted from Martinie, S. et al., Analytical modeling of alpha-particle emission rate at wafer-level, *IEEE Trans. Nucl. Sci.* 58, 2798–2803. © (2011) IEEE. With permission.)

the exact estimation of these emissivity values for all the alpha emitters in the energy range 1 keV–10 MeV. The impact of the residual traces of both thorium and uranium on the emissivity of the considered silicon layer is estimated using the following relation:

$$\phi_{\alpha\_total} = \phi_{\alpha\_U} \times X_{ppb\_U} + \phi_{\alpha\_Th} \times X_{ppb\_Th} \tag{3.23}$$

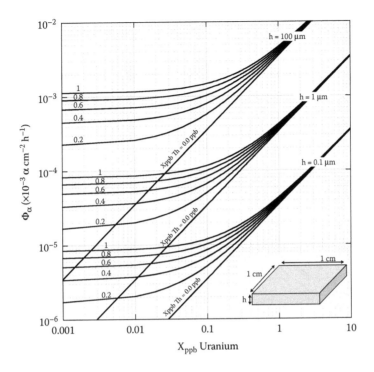

FIGURE 3.11
Universal nomogram giving alpha-particle emissivity as a function of uranium and thorium contamination level for a silicon layer of a given thickness. Emissivity versus uranium contamination for different amounts of thorium contamination is plotted. The threshold energy for the alpha-particle detection system is fixed at 2 MeV. (Reprinted from Martinie, S. et al., Analytical modeling of alpha-particle emission rate at wafer-level, *IEEE Trans. Nucl. Sci.* 58, 2798–2803. © (2011) IEEE. With permission.)

where ϕ_{α_U} and X_{ppb_U} are the emissivity and contamination level, respectively, for uranium and ϕ_{α_Th} and X_{ppb_Th} are the emissivity and contamination level for thorium. The complete characteristics of the different alpha emitters for both thorium and uranium decay chains are detailed in Adamiec and Aitken (1998) and Gedion et al. (2011).

The results of these calculations are shown in Figure 3.11, which represents an emissivity nomogram for silicon layers of different thicknesses. This nomogram can be directly used for a fast estimation of the contamination level from emissivity measurements at wafer level using a low-background alpha-particle detector.

In the same way, Figure 3.12 shows the evolution of emissivity for different materials commonly used at BEOL level (copper, lead, aluminum, and silicon). As expected, for the same contamination level of uranium or thorium and at fixed thickness, the resulting emissivity strongly depends on the considered material. This is due to the fraction of energy lost by emitted alpha particles in the considered material, which, of course, strongly depends on the atomic nature of the layer. For example, the emissivity is larger for aluminum than for copper because the alpha-particle range for a given energy in aluminum is greater than that in copper. Table 3.10 indicates the maximum range for alpha particles of 8 MeV in the considered BEOL materials. Such values thus clearly explain the orders of magnitude of emissivity reported in Figure 3.11.

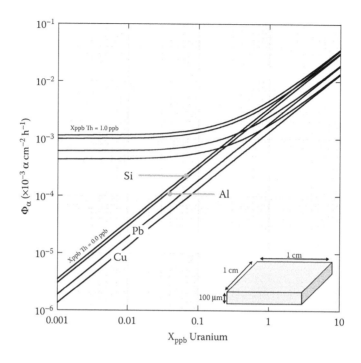

FIGURE 3.12

Alpha-particle emissivity (copper, lead, aluminum, and silicon) versus uranium contamination for different levels of thorium contamination. The threshold energy for the alpha-particle detection system is fixed at 2 MeV. (Reprinted from Martinie, S. et al., Analytical modeling of alpha-particle emission rate at wafer-level, *IEEE Trans. Nucl. Sci.* 58, 2798–2803. © (2011) IEEE. With permission.)

TABLE 3.10

Maximum Range of Alpha Particles for Different Materials (SRIM)

	Maximum Range for Alpha Particles of 8 MeV (μm)
Silicon (Si)	49.3
Aluminum (Al)	43.0
Lead (Pb)	26.4
Copper (Cu)	19.3

Source: Data from Martinie, S. et al., *IEEE Trans. Nucl. Sci.* 58, 2798–2803, 2011.

References

Adamiec, G. and Aitken, M. 1998. Dose-rate conversion factors: Update. *Ancient TL* 16:37–50.

Cabral, C., Rodbell, K.P., and Gordon, M.S. 2007. Alpha particle mitigation strategies to reduce chip soft error upsets. *Journal of Applied Physics* 101:014902.

Gedion, M. 2012. Contamination des composants électroniques par des éléments radioactifs [Contamination of electronic components by radioactive elements]. PhD Thesis, Université de Montpellier II [in French].

Gedion, M., Wrobel, F., and Saigné, F. 2010. Natural radioactivity consideration for high-κ dielectrics and metal gates choice in nanoelectronic devices. *Journal of Physics D: Applied Physics* 43:275501.

Gedion, M., Wrobel, F., Saigné, F., and Schrimpf, R. 2011. Uranium and thorium contributions to soft error rate in advanced technologies. *IEEE Transactions on Nuclear Science* 58:1098–1103.

Gordon, M.S., Heidel, D.F., Rodbell, K.P., Dwyer-McNally, B., and Warburton, W.K. 2009. An evaluation of an ultralow background alpha-particle detector. *Transactions on Nuclear Science* 56:3381–3386.

Gordon, M.S., Rodbell, K.P., Heidel, D.F., et al. 2010. Alpha-particle emission energy spectra from materials used for solder bumps. *IEEE Transactions on Nuclear Science* 57:3251–3256.

Heidel, D.F., Rodbell, K.P., Cannon, E.H., et al. 2008. Alpha-particle-induced upsets in advanced CMOS circuits and technology. *IBM Journal of Research and Development* 52:225–232.

Kumar, S., Agarwal, S., and Jung, J.P. 2013. Soft error issue and importance of low alpha solders for microelectronics packaging. *Reviews on Advanced Materials Science* 34:185–202.

Magill, J. and Galy, J. 2005. *Radioactivity-Radionuclides-Radiation: Including the Universal Nuclide Chart on CD-Rom*, vol. 1. Berlin: Springer.

Martinie, S., Autran, J.L., Munteanu, D., Wrobel, F., Gedion, M., and Saigné, F. 2011. Analytical modeling of alpha-particle emission rate at wafer-level. *IEEE Transactions on Nuclear Science* 58:2798–2803.

Martinie, S., Saad-Saoud, T., Moindje, S., Munteanu, D., and Autran, J.L. 2014. Behavioral modeling of SRIM tables for numerical simulation. *Nuclear Instruments and Methods in Physics Research Section B* 322:2–6.

Narula, A.K., Chauhan, R.P., and Chakarvart, S.K. 2010. Testing permeability of building materials for radon diffusion. *Indian Journal of Pure and Applied Physics* 48:505–507.

Valkovic, V. 2000. *Radioactivity in the Environment*. Amsterdam: Elsevier.

Wilkinson, J. and Hareland, S. 2005. A cautionary tale of soft errors induced by SRAM packaging materials. *IEEE Transactions on Device and Materials Reliability* 5:428–433.

Wong, R., Su, P., Wen, S.-J., McNally, B., and Coleman, S. 2010. The effect of radon on soft error rates for wire bonded memories. In *Proceedings of the IEEE International Integrated Reliability Workshop Final Report (IRW)*, pp. 133–134. 17–21 October, Stanford Sierra, CA, IEEE.

Wrobel, F. 2008. Fiabilité des composants électroniques en environnement atmosphérique: Contraintes radiatives et prédiction [Reliability of electronic components in atmospheric environment: Radiative constraints and prediction], HDR Thesis, Université de Montpellier II [in French].

Wrobel, F., Gasiot, J., and Saigné, F. 2008a. Hafnium and uranium contributions to soft error rate at ground level. *IEEE Transactions in Nuclear Science* 55:3141–3145.

Wrobel, F., Gasiot, J., Saigné, F., and Touboul, A.D. 2008b. Effects of atmospheric neutrons and natural contamination on advanced microelectronic memories. *Applied Physics Letters* 93:064105.

Wrobel, F. and Saigné, F. 2011. MC-Oracle: A tool for predicting soft error rate. *Computer Physics Communications* 182:317–321.

4

Alpha-Radiation Metrology in Electronic Materials

4.1 Introduction

As introduced in Chapter 3 and explained in detail in Chapter 5, soft errors can be caused in part by energetic alpha particles (<10 MeV), emitted from ultratraces of radioactive contaminants in circuit materials, that can penetrate transistor junctions of semiconductor devices. As device dimensions have decreased and circuit complexity has grown, the probability of alpha-induced soft error has mechanically increased, justifying the efforts of the microelectronics industry to select and use lower-alpha-emitting materials at both wafer and packaging levels. Ultralow-background alpha-particle counting has become an important issue for the semiconductor industry, a domain for which alpha-emissivity levels (i.e., the rates of emission of alpha radiation measured in counts per unit area and per unit time) are typically several orders of magnitude lower than for other disciplinary fields, such as geology.

This chapter presents a few dedicated methods and instruments for measuring alpha emissivity typically less than 10 α kh^{-1} cm^{-2} in materials used in the manufacturing of semiconductor circuits. The text introduces, in the first part, a few basic notions concerning detection of alpha radiation. In the following, it focuses on the design, operation, and simulation of a state-of-the-art ultralow-background alpha-particle (gas-filled) counter recently introduced as a new standard of measurement at wafer level. In the last section, other detection techniques are briefly reviewed and discussed, always in the particular framework of semiconductor circuit fabrication.

4.2 Alpha-Particle Detection Techniques: Terms and Definitions

Alpha-particle detection includes indication of the presence of alpha particles as well as their counting and measurement of their energy (spectroscopy). As none of the human senses is sensitive to the radiation emitted by radioactive substances, one can detect this radiation only by using interactions between alpha particles and matter. Alpha radiation can be detected by various means: the goal may be to determine the global alpha radioactivity of a sample or to know the nature, and precisely quantify the activity, of alpha-emitting radionuclides: counting or spectrometry of alpha radiation, respectively. A detection system consists of a sensor wherein the interaction of the radiation with matter by ionization or atomic excitation results in the emergence of an electrical signal or light; a measuring system then processes this signal (amplification, counting, etc.). The different types of detectors can be divided into several categories depending on the nature of the radiation's interaction with the detector:

- In the case of ionization or proportional chambers, or Geiger–Müller and semiconductor detectors, the detection is based on the ionization of matter by the incident radiation.
- In the case of scintillation detectors, atomic excitation plays an important role in the production of light.

Table 4.1 summarizes the main techniques that may be used to analyze the alpha emitters; the choice depends on the required information in terms of identification, quantification, limit of detection, and so on (Ansoborlo et al. 2012). Two types of techniques exist: radiometric techniques (exploiting alpha radiation) and nonradiometric methods.

Radiometric techniques include (1) gas-filled counters and chambers, (2) scintillator-based detectors, and (3) semiconductor detectors. In view of the importance of large-area gas counters in microelectronics, they will be the subject of the next section; the two other solutions will be introduced in Section 4.5. Whatever the mode of operation of a radiometric detector and, therefore, the principle on which radiation detection is based, a detector is always composed of the following elements: (i) a sensor in which the radiation level interacts with matter; (ii) an amplification system that formats and amplifies the signal generated by the sensor; (iii) optionally, a signal processing system; and (iv) a display system that indicates (a) the particle flow (counter); (b) the particle energy (spectrometer); or (c) the absorbed dose or absorbed dose rate (dosimeter or flowmeter). A number of characteristic parameters are defined for each type of radiation detector (JEDEC 2011): (i) the detection efficiency, which is the ratio between the number of alpha particles detected and the actual number of events occurring, that is, the number of particles received by the detector (the

TABLE 4.1

Main Techniques Used for Analysis of Alpha-Particle Emitters

Type of Measurement	Technique	Constraint	Advantages	Disadvantages
Global counting	Gas-filled counter	Gas	Sensitivity Robustness	Reproducibility Nonisotopic technique
	Solid-state scintillation (ZnS)	Contamination	High efficiency	Detector fragility Nonisotopic technique
	Liquid scintillation	Liquid scintillator	Easy sample processing Sensitivity	Organic liquid wastes Nonisotopic technique
Spectrometry	Gas-filled counting chamber (proportional, ionization)	Gas (pure Ar, Ar/CH$_4$, Ar/CO$_2$, etc.)	Solid angle 2π Isotopic technique	Deconvolution of spectra Resolution Risk of contamination
	Silicon detector	Vacuum measurements	Resolution Detector size Isotopic technique	Small solid angle Deconvolution of spectra
	PERALS® system	Liquid scintillator	Sensitivity Solid angle 4π Isotopic technique	Organic liquid wastes Resolution Deconvolution of spectra

Source: Adapted from Ansoborlo, E. et al., *Mesure du rayonnement alpha* [*Alpha Radiation Measurement*], Lavoisier, Paris, 2012.

efficiency depends on the nature and energy of the radiation); (ii) the dead time, that is, the smallest time interval between two consecutive events that are taken into account by the system; (iii) the detector background, that is, the counting rate recorded in the absence of any source of radiation; and, finally, (iv) the geometric characteristics that define the detector shape, the importance of its sensitive surface and its directivity.

Nonradiometric techniques are divided into two main categories: analysis techniques with laser sources and techniques for which the detector is a mass spectrometer. The principle of mass spectrometry is based on measuring the ratio of mass to charge of ions generated by the ionization source. Several types of mass spectroscopy exist: secondary-ion mass spectroscopy (SIMS), resonant ionization mass spectroscopy (RIMS), inductively coupled plasma mass spectroscopy (ICP-MS), and so on. This last technique will be briefly introduced in Section 4.5, since it has the best detection efficiency for the most pertinent radionuclides to characterize (U, Th).

4.3 Gas-Filled Counters

In the semiconductor industry, alpha-particle emission levels are most commonly measured using sensitive, large-area, gas-filled detectors operating as counters (detection of rare events). The detector active area has a significant and direct effect on the accuracy and precision of the calculated alpha emissivity (JEDEC 2011). Through careful sample preparation and control of detector parameters, it is believed to be possible to measure alpha-emission rates of materials down to 1 α kh^{-1} cm^{-2} (one emitted particle per square centimeter per 1000 h). For many alpha-emitting isotopes, this corresponds to concentrations in parts per trillion (ppt) or lower (Wilkinson et al. 2011).

4.3.1 Principle of Operation

A gas-filled counter uses the ionization phenomenon for the detection of particles; more precisely, it measures the amount of ionization produced by charged particles passing through the gas volume. Neutral particles can also be detected by this device via the production of secondary charged particles resulting from the interaction of the primary particles with electrons or nuclei (Grupen and Shwartz 2008). Table 4.2 shows the ionization potential and the mean energy required to create an ion pair for different common gases

TABLE 4.2

Ionization Potential and Mean Energy Required to Create an Ion Pair for Different Common Gases

Gas	First Ionization Potential (eV)	W-Value (eV per ion pair)	
		Fast Electrons	Alpha Particles
Ar	15.7	26.4	26.3
He	24.5	41.3	42.7
H$_2$	15.6	36.5	36.4
N$_2$	15.5	34.8	36.4
O$_2$	12.5	30.8	32.2
CH$_4$	14.5	27.3	29.1

with fast electrons and alpha particles. This last value is termed the *W-value*. Due to the competition mechanism of the energy loss, that is, excitation, the W-value is always greater than the ionization energy. Note that the energy required to create an ion pair in a gaseous medium is approximately 10 times higher than the energy of creation of an electron–hole pair in a semiconductor (see Section 4.5.1).

Figure 4.1 illustrates the architecture of a typical planar ionization gas-filled detector: it includes a chamber containing a gas or gas mixture and two electrodes (plates, wires, or grids, as a function of counter type; see below) between which a potential difference U_0 is applied. This voltage provides a uniform electric field $E| = U_0 d^{-1}$ between the two electrodes separated by the distance d. When an ionizing particle is partially or totally absorbed in the chamber, the ionization products (ion pairs) created along the particle track are guided toward the anode and the cathode as a function of their electrical charge. If the gas absorbs all the energy of the incident particle, this type of detector can measure the particle energy (Grupen and Shwartz 2008). In the absence of voltage between the two electrodes ($U_0=0$ V), the ions formed are subjected to thermal agitation and recombine after a certain time to form neutral atoms. For a nonzero biasing, the more massive ions drift slowly toward the negative cathode, while the lighter electrons drift toward the positive anode about 1000 times more quickly (Knoll 1989). When the detector is operated as a counter in pulse mode, the current induced in the anode by the drifting electrons is amplified and integrated so that each ionization track produces a single output pulse and is counted individually.

As a function of the electrode design, the type and pressure of the fill gas, and the strength of the electric field between the electrodes, the detector's response to ionizing radiation varies. To illustrate these differences in counter operation, we consider a simple cylindrical gas-filled detector with the metal chamber forming the cathode and an axial wire electrode, as illustrated in Figure 4.2. We denote by n the average number of ion pairs created by the passage of a given ionizing particle through the active region of the counter. We examine in the following the variation of the number N of collected charges (electrons) on the anode as a function of the applied potential difference U_0 between the counter electrodes. The corresponding curve is given in Figure 4.2. Five distinct regions can be distinguished:

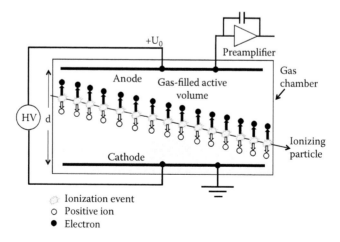

FIGURE 4.1
Schematic representation of a typical planar ionization chamber.

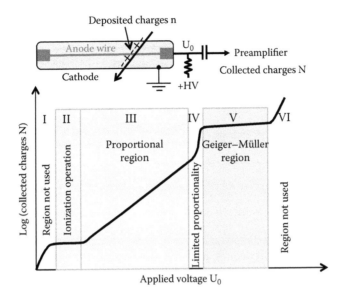

FIGURE 4.2
Variation of the collected charge as a function of the applied voltage for a wire cylinder gaseous radiation detector.

- In Region I, the applied field is too low and the recombination of ion pairs drastically reduces the charge collection on the electrodes.
- In Region II, the electric field is low but sufficient to separate all formed electron–ion pairs and to cause electrons to drift toward the anode and ions toward the cathode. All charges produced by the primary ionization are then collected; that is, the number of collected charges is equal to the number of charges produced by the primary ionization: there is no amplification and the ratio N/n is close to 1. The detector operates in the *ionization mode*: for a given particle type and energy, the pulse height is constant in this region. In this regime, it is possible to discriminate ionizing particles according to their ionizing power.
- In Region III, N becomes greater than n ($N/n \gg 1$) due to an internal amplification phenomenon that occurs in the immediate vicinity of the anode wire. At this distance of only a fraction of millimeter from the anode wire, the local electric field strength becomes large enough to cause electron multiplication via a phenomenon termed *electron avalanche*. This effect is schematically illustrated in Figure 4.3. The avalanche is a cascade reaction initiated by a free electron that gains sufficient energy from the electric field to liberate an additional electron when it collides with a gas atom or molecule. The two electrons then continue to travel toward the anode, gaining sufficient energy to cause impact ionization when the next collisions occur, and so on. The total number of electrons reaching the anode is equal to the number of electrons generated during this chain reaction plus the single initiating free electron. As a consequence, the collected charge is a multiple of the charge due to the primary ionization: Region III is thus termed the *proportional counting region*. The amplification factor, equal to the ratio N/n, varies with U_0 but is independent of n.
- In Region IV, this proportionality progressively disappears as U_0 continues to increase and the pulse height tends to become independent of the ionization.

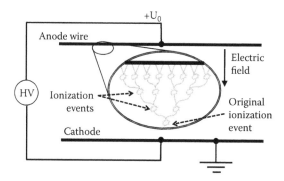

FIGURE 4.3
Schematic illustration of an electron avalanche occurring in the vicinity of the anode wire in a gas-filled counter.

- In Region V, the collected charge is totally independent of the initial ionization. In fact, a kind of saturation occurs, and the detector outputs the maximum number of electrons that it is capable of providing, whatever the initial ionization. The detector can then be used to count the number of radioactive particles that interact without being able to discriminate between them according to their energy. This is the *Geiger–Müller regime*.

- In Region VI, the counter becomes unstable and a permanent discharge is established. The potential difference between the two electrodes is too high. Any ionization phenomenon inside the counter may produce a discharge.

From the different regimes defined in Figure 4.2 and by adjusting the instrument characteristics (electrodes, gas mixture, electric field strength), it is possible to select the operation mode and to optimize the detection response of three different types of gas-filled detectors: (1) ionization chambers; (2) proportional counters; and (3) Geiger–Müller tubes. In the following we briefly describe the first two, specifically employed for the detection of alpha particles in semiconductor-circuit fabrication.

4.3.2 Ionization Counters

Large-area ionization-mode counters used to detect alpha particles are simply gas-filled volumes fitted with electrodes so that an electric field can be applied to the volume and any charges generated therein collected. These counters are filled with purified gas (generally argon) and operate at atmospheric pressure (vented chamber) under low electric-field strength, well below the threshold of avalanche. Charge carriers (Ar^+ and e^-) created by the passage of an ionizing particle drift toward the high-voltage electrodes in the chamber. As the electrons drift under the action of the electric field in the bulk volume, they induce a time-varying charge on the anode that is seen as a current by a preamplifier, which integrates it to produce an output pulse that can be digitized and then analyzed. This mode is termed *pulse mode* operation. It requires a sophisticated low-noise preamplifier and pulse-processing electronics, since the signal amplitude is extremely small. Particular precautions can be taken for the strict control of moisture inside the measurement chamber. Indeed, moisture can disturb the operation of the instrument and affect its accuracy, modifying the gas-transport properties and potentially inducing leakage currents. Section 4.4 will detail a particular type of ionization chamber, the UltraLo-1800, a state-of-the-art instrument for low-background alpha-particle counters that employs electronic background suppression to drive achievable background rates to 0.0001 α cm^{-2} h^{-1} and below. This is better by a factor of 50 than can be achieved by the conventional proportional counter systems described in Section 4.3.3.

4.3.3 Proportional Counters

Proportional counters are very similar to ionization counters, but they use gas avalanche gain to increase the number of electrons produced (strictly proportional to the initial deposited charges) and thus to increase the signal-to-noise ratio. Very large electric fields are required for avalanche multiplication to occur, of the order of 1–10×10^6 V m^{-1}, which are usually produced by applying a voltage of the order of 1–2 kV to a wire whose dimension is typically 0.02–0.08 mm in radius (Warburton et al. 2004). Since the electric field falls off in inverse proportion to the distance from the wire center, avalanching can occur only within about 100 μm of the wire surface, which, in turn, provides the limitation required to assure gain proportionality (Knoll 1989). Further, because essentially all the avalanche charge is produced close to the wire, there are no drifting electron-induced charge effects in proportional counters, so that output pulse amplitude and charge are proportional to the initial charge in the ionization track, independently of its original location within the counter (Warburton et al. 2004). Proportional counters are commonly operated in single-pulse counting mode. Because the avalanche process is very fast, it lasts only as long as it takes for the ionization track to arrive at the anode wire. In a well-designed counter, this time is short compared with the time it takes the ions formed in the avalanche to drift away from the anode wire, typically a few microseconds. As it is this latter process that induces the detector output signal current in the anode, all output pulses in such well-designed detectors have approximately the same shape (Warburton et al. 2004).

Three main types of proportional counters were originally developed several decades ago for use by the nuclear industry: (1) sealed detectors, (2) Frisch grid detectors, and (3) gas-flow proportional counters (Alpha Science 2009). The sealed detector has a window that is relatively thick, and consequently cannot be used for alpha detection, since alpha particles cannot pass through the window. The Frisch grid detector is generally used for spectroscopy applications, in which the energy of the particle is required for radioisotope identification. It requires a large volume of counting gas to fully absorb the energy of the alpha particles. This large volume of gas may degrade the detector background, which limits its use for ultralow-background analysis. The third type of proportional counter, the gas-flow system, is the most commonly developed instrument for large-area samples (Alpha Science 2009). Current state-of-the-art low-background gas-flow instruments include multiwire proportional counters (see Figure 4.4) equipped with a single chamber (e.g., Ordela model 8600A) or with two chambers (windowed counter) separated by a very thin Mylar film (e.g., Alpha Sciences Inc. model 3950). Both models operate with P-10 gas mixture (90% argon + 10% methane as quench gas to ensure that each pulse discharge terminates) in flow mode at atmospheric pressure. Typical backgrounds may vary from 2 to 10 counts per hour, depending on ambient radon concentrations, interfering electric fields, altitude, and purity of counting gas. In practice, these counters can achieve sensitivities of about 0.005 α cm^{-2} h^{-1}. Their performances are therefore limited, because they lose any information about where pulses originated inside the system (see Figure 4.4). Pulses originating from the sample (α_1) look the same as those from the sidewalls (α_5), the ceiling (α_2), the window (α_4), or the electrode wires (α_3). This is because the vast majority of the signal comes from the time between the first electrons of a track entering the multiplication region and the final multiplication ending. The duration of that signal is dependent

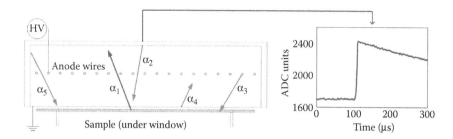

FIGURE 4.4

Schematic representation of a multiwire proportional counter with two chambers (sample and detection) separated by a very thin Mylar film (window). Pulses originating from the sample (α_1) look the same as those from the sidewalls (α_5), the ceiling (α_2), the window (α_4), or the electrode wires (α_3). (Courtesy of B.D. McNally.)

only on the orientation of the track relative to the anode wires, and not on where the track originated in the detector (McNally 2013). This background counting rate can therefore be significantly reduced by constructing all the counter components from materials having very low alpha emissivity. This approach not only adds significantly to the difficulty and expense of constructing such counters, but becomes exponentially more difficult as ever lower backgrounds are sought. After 20 years of development, the approach appears to have reached its natural limits (Warburton et al. 2004).

One additional loss mechanism for windowed counters, recently investigated in Wilkinson et al. (2014), is due to the gap between the emitting surface of the sample and the counter's entrance window. The alpha counter can only register events that enter the counting gas volume above the entrance window with enough residual energy to create a signal that can be distinguished from noise. Alpha particles emitted at shallow angles may either fail to impact the entrance window or be absorbed within it. It is also possible for low-angle alpha particles to lose all their energy in the gas volume below the window after traversing approximately 4 or 5 cm laterally. For all these cases, an increase in the gap dimension leads to increased losses of particles, particularly those emitted at shallow angles (Wilkinson et al. 2014).

4.4 Ultralow-Background Alpha Counter

This section describes the design, operation, modeling, and numerical simulation of an ultralow-background alpha counter operated by the authors since 2010 in the framework of research works dedicated to the alpha-emissivity metrology of semiconductor and packaging materials. We successively evaluated a prototype ionization counter and used the commercial version of the same counter named UltraLo-1800, an instrument patented (Warburton et al. 2004), designed, and manufactured by XIA LLC (formerly X-Ray Instrumentation Associates). The prototype instrument was used in 2010–2011 to evaluate the impact of terrestrial cosmic rays on the counter operation, in the framework of a qualification campaign conducted by XIA LLC (McNally 2011) at Marseille, on the ASTEP platform, and at the Underground Laboratory of Modane (see Section 7.4.1 for the description of these two test platforms). The commercial instrument has been in continuous operation since November 2011 at Aix-Marseille University (near sea level).

4.4.1 Design and Operation of the UltraLo-1800

The UltraLo-1800 is a windowless instrument, in which samples are inserted directly into a gas-filled, large-area, active counting chamber. In this system, samples are arranged on a tray that is then moved into measurement position via an electromechanical stage, thereby ensuring repeatable positioning from measurement to measurement. Figure 4.5 shows a global view of the instrument with the same tray fully ejected and loaded with a 300 mm bare silicon wafer.

The UltraLo-1800 operates in a fundamentally different manner from the gas proportional counters previously described. In essence, it is an ionization chamber without internal gain whose geometry intentionally exaggerates differences between signals from alpha particles generating ionization tracks that originate from its different surfaces and those from the gas bulk (McNally 2013). The counter is filled with pure argon at atmospheric pressure coming from the boiloff of liquid argon contained in a large dewar. This provides a high-purity gas source with a flow rate of about 4 L min^{-1}. Argon is used since it has the smallest energy requirement to create an ion pair (W-value) among the most common laboratory gases (see Table 4.2).

Figure 4.6 shows a schematic cross section of the UltraLo-1800. The internal counter elements comprise an active volume filled with argon; a lower grounded electrode, which is a conductive tray with a square geometry and a surface of 1800 cm^2 (called the sample tray)

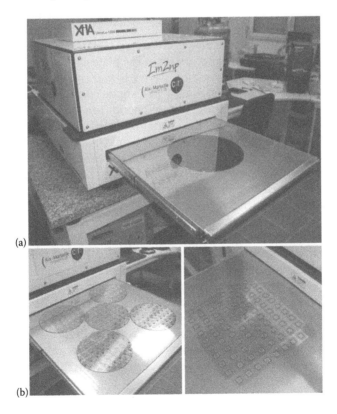

FIGURE 4.5
(a) Global view of the XIA UltraLo-1800 alpha counter installed at Aix-Marseille University (sample tray opened with a 300 mm bare silicon wafer). (b) Two examples of samples positioned on the counter sample tray (multi-wafer measurements—200 mm wafers, and metallization strips for circuit packaging).

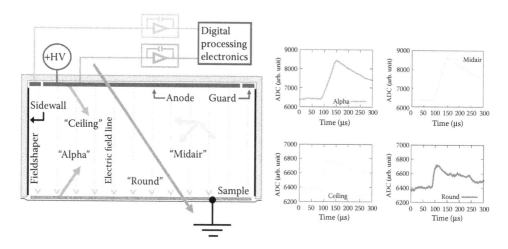

FIGURE 4.6
Schematic cross-section view of the UltraLo-1800 with different classes of events and their resulting measurement signals.

holding the sample; and an upper pair of positively biased electrodes: the anode and the guard. The first sits directly above the sample, while the second surrounds and encloses the anode. They are held at positive voltage, typically $U_0 = 1000$ V. This arrangement produces an electric field between the sample tray and the top electrodes that has the value $E = U_0 D^{-1} \approx 67$ V cm^{-1}, where D is the distance between the plate electrodes (15 cm). Note that, in the UltraLo-1800, the anode is composed of two parts that can be selected by the user as a function of the sample to be characterized: the *full* anode corresponds to a square 1800 cm^2 electrode configuration and the *wafer* anode corresponds to a 707 cm^2 circular electrode configuration that fits the surface of a 300 mm wafer centered on the sample tray. Finally, the sides of the chamber, called the sidewalls, hold the fieldshapers, which are printed circuit boards (PCBs) containing strips of copper separated by resistors that keep the electric field lines parallel throughout the volume. Both electrodes are connected to charge-integrating preamplifiers whose output signals are digitized and then processed by a digital-pulse-shape analyzer. When an ionization track is detected in the chamber, the resultant signal waveforms are captured by onboard digital electronics and their pulse shapes are analyzed to determine the location from which the track originated (McNally 2011). Figure 4.6 also shows typical signal waveforms of different tracks originating from the sample under test, from the internal surfaces of the counter, and from external origin (terrestrial cosmic rays, including atmospheric neutrons reacting in the bulk gas and charged particles traversing the chamber). As may be seen, the waveforms differ significantly in both amplitude and risetime between these different cases. The following section will examine this point on the basis of a simple analytical model. The UltraLo-1800 software recognizes these differences and actively rejects nonsample tracks, thereby driving background rates down by an order of magnitude or more compared with what can be achieved in the best commercially available gas proportional counters.

4.4.2 Signal Generation and Rejection

A simple analytical model has been developed by XIA (McNally 2011, 2013; Warburton et al. 2004) for describing the counter operation and performing pulse shape analysis. We survey in this paragraph the most important results derived from this analytical formulation. The following text is largely inspired by McNally (2013). Figure 4.7 shows an internal cross-section view of the counter with two different alpha-emission events and the corresponding pulse profiles delivered by the output electronic chain. Before describing these two events, it is important to understand how the ionizing products, that is, the ions and electrons created along the considered alpha particle track, move inside the gas volume and how they can induce a current in the attached electronics. Being charged, these ions and electrons drift in the applied electric field: the electrons toward positive voltage on the electrode and the ions toward the sample tray. Considering only the electrons in the following analysis, each drifts with a velocity (v_e) equal to its mobility (μ_e) in the gas multiplied by the electric field ($E = U_0 D^{-1}$), which is uniform in the counter chamber: $v = \mu_e \times E = \mu_e \times U_0 D^{-1}$. Considering an electron freed at distance d away from the electrode (see Figure 4.7), its drift time to reach the anode is $t_e = d\, v_e^{-1} = d \times D\, (\mu_e \times U_0)^{-1}$. When d is

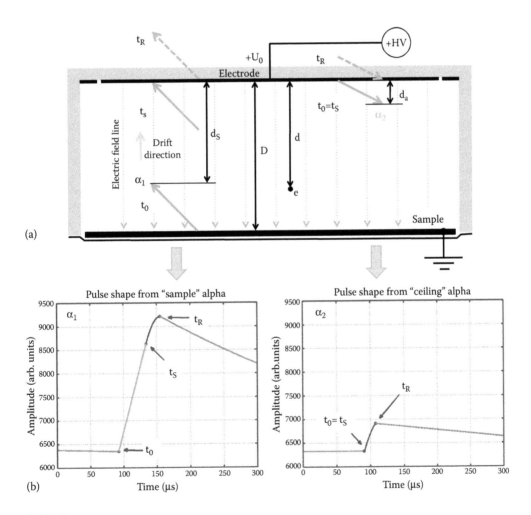

FIGURE 4.7
(a) Definition of different times and distances for two alpha emissions from the sample under test and from the top counter electrode (*ceiling*). (b) Corresponding pulse shapes of the integrated signal delivered by the output electronic chain. (Courtesy of B.D. McNally.)

equal to the height of the entire chamber, $D = 15$ cm, and $U_0 = 1000$ V, then t_e is found to be approximately 70 µs in pure argon gas at atmospheric pressure and at room temperature.

Because all the electrons originating from the same alpha-particle track drift at the same velocity, the initial geometry of the particle track is roughly preserved from the moment it is created (t_0) until the first electron reaches the electrode (t_S), as illustrated in Figure 4.7 for the track labeled α_1. It must be noted (McNally 2013) that there is some diffusion of electrons out from the track, but this effect can be neglected at this modeling level. As the track continues to drift, the electrode absorbs the electrons as they arrive, until they are all absorbed (t_R). The time between t_0 and t_R is called the risetime.

Once the transport of charges in the chamber has been modeled, the next step is to evaluate the response of the counter electronics and thus to understand how the signal is generated. At this level, it is important to understand that the current is not due to charge collection by the electrode but is induced by the movement of electrons in the chamber toward the electrode. The Shockley–Ramo theorem states that the instantaneous current (i) induced on a given electrode due to a charge's motion is given by $i = q \times v_e \times E_v$, where q is the charge of the particle, v_e is its instantaneous velocity, and E_v is the component of the electric field in the direction of v at the charge's instantaneous position, under the following conditions: charge removed, given electrode raised to unit potential (1 V), and all other conductors grounded. It is important to note that this Ramo induction only happens while the electron is traveling; once it reaches the electrode, its electrostatic influence disappears, and therefore the induced current goes to zero.

In the counter chamber, tracks consist of hundreds of thousands of electrons distributed along straight segments of length corresponding to the particle range in argon. The range and number of ion–electron pairs (N) created in the gas increase with the energy of the alpha particle. In that case, the total observed current is the sum of the individually induced currents. If the track is not parallel with the electrode (general case), some electrons in the track will induce current for a longer time than others, and, once the track begins to hit the electrode (at t_S), the rate of induction slows. The rate at which induction slows will be proportional to the angle of the track relative to the electrode. This slowing shows up as a rounding of the pulse from t_S to t_R, as can be seen in Figure 4.7b.

In the UltraLo-1800, the electrode is connected to a charge-sensitive preamplifier that integrates the current. The preamplifier outputs a signal (in volts) that is proportional to the total current induced. As discussed above, for a track originating on the sample (α_1 in Figure 4.7), there are two distinct regions in time: one from t_0 until t_S, where the electrons drift unchanged, and another from t_S until t_R, where the electrons are disappearing linearly in time. The resultant integrated signal $S_S(t)$ is thus linear until t_S and then parabolic until t_R, the chamber maximum transit time:

$$S_S(t) = \frac{Nq\mu_e V}{C_f D^2}(t - t_0) \quad \text{from } t_0 \text{ to } t_S$$

$$S_S(t) = \frac{Nq\mu_e V}{C_f D^2}\left(t - \frac{(t - t_S)^2}{2(t_R - t_S)}\right) \quad \text{from } t_S \text{ to } t_R \tag{4.1}$$

where:

C_f = detector capacitance

q = charge of an electron

and

$$t_S(t) = \frac{d_S D}{\mu_e V}, \quad t_R(t) = \frac{D^2}{\mu_e V}, \quad \text{and} \quad S_{SMAX} = \frac{Nq}{2C_f}\left(1 + \frac{d_S}{D}\right) \tag{4.2}$$

where d_S is the distance from the topmost electron of the track to the electrode (see Figure 4.7a). The resultant pulse is shown in Figure 4.7b. Note that S_{SMAX} scales with N, so that the final pulse amplitude is proportional to the energy of the alpha particle.

However, when a uniform charge track originates from the electrode, charge immediately starts disappearing linearly in time, so that the resultant signal $S_a(t)$ is a parabola given by

$$S_a(t) = \frac{Nq\mu_e V}{C_f D^2}\left(t - \frac{(t - t_0)^2}{2t_a}\right) \tag{4.3}$$

The risetime t_a and maximum amplitude S_{aMAX} are found to be

$$t_a(t) = \frac{Dd_a}{\mu_e V} \quad \text{and} \quad S_{aMAX} = \frac{Nq}{2C_f}\frac{d_a}{D} \tag{4.4}$$

where d_a is the track length normal to the electrode.

The important lesson to be drawn from this comparison of the two track types is that electrode pulses look different from sample pulses. Both the risetime and the maximum amplitude are much longer for sample pulses. Because the ratio of electrode to sample risetimes (t_a/t_R) is $d_a D^{-1}$, if the sample chamber is several times longer than the maximum range of an alpha particle, the risetimes of the two cases will always be separated. The UltraLo-1800 is designed so that this ratio will be about 1/3 for a 5 MeV alpha particle emitted perpendicular to the anode. Similarly, the ratio of maximum amplitudes (S_{aMAX}/S_{SMAX}) for two identical alpha decays is $d_a (D + d_s)^{-1}$. This ratio is dependent on the angle of emission relative to the anode, but it will always be greater than or equal to the ratio between risetimes. Since D has already been chosen to exaggerate the difference in risetimes, the difference in amplitudes is exaggerated as well.

The second key design feature that allows the UltraLo-1800 to be an ultralow-background instrument is its veto, or guard, electrode. The electrodes are arranged as shown in Figure 4.7a and are read out simultaneously, with the interior, active portion called the anode and the exterior portion called the guard. Tracks originating on the counter sidewalls or on the tray outside the sample region will induce signals on the guard and can be rejected. When combined with the risetime discrimination discussed earlier, the system is capable of rejecting events that originate on any surface other than the sample (McNally 2011). This is a key feature of the UltraLo-1800 as compared with gas proportional counters (see Section 4.3.3), which lose information about where pulses originated in the detector.

4.4.3 Pulse and Event Classification

As shown in Section 4.4.2, a typical signal resulting from charge collection in the counter has four distinct regions (Figure 4.7): the baseline of the signal, which occurs before the ionization track is formed; a linear rise, which occurs while all charges in the ion track are drifting toward the anode; a parabolic curve, which starts when the first charges in the ion track are incident upon the anode and ends when the last charges are collected from the track; and a decay region, which occurs when charge collection is complete (Gordon et al. 2013). The duration of the parabolic portion of the signal is directly related to the vertical projection of the charge track on the anode. Ionization tracks normal to the anode plane will be maximally parabolic, while ionization tracks parallel to the anode plane will have zero parabolic time (Gordon et al. 2013).

The UltraLo-1800's control software includes a specific *alpha analysis engine* for analyzing and classifying all the measured signals. The goal of this program is to identify the signature of events corresponding to an effective emission of an alpha particle from the sample surface. The program uses a multilevel algorithm to progressively eliminate nonalpha events based on their characteristics. It uses the analytical description of the pulses to fit signals with cuts based on amplitude, risetime, and other timing results. At the end of the analysis, the remaining signals can be classified into four main classes of events (McNally 2013):

- *Alpha*: If an event is classified as an alpha, this means that, with respect to the best ability of the discrimination algorithm, it was caused by emission of an alpha particle from the sample area under the anode.
- *Midair*: Midairs are events whose risetime is too short to be an alpha but too long to be a ceiling. They are caused by interactions in the bulk of the counting gas, mostly from radon but also from cosmic rays (see Section 4.4.4).
- *Round*: Rounds are events whose peaks are too rounded to be caused by an alpha particle. These are the result of particles that can cause longer ionization tracks, most likely cosmic rays.
- *Ceiling*: An alpha decay emanating from the electrode.

Figure 4.6 illustrates these different kinds of events in terms of typical particle track and corresponding signals measured by the counter electronics. The origin of *round* and *midair* events is specifically discussed in Section 4.4.4.

4.4.4 Cosmogenics and Radon Issues

Terrestrial cosmic rays and radon have been identified as two sources of rare events that produce traces that can be only partially discriminated from alpha particles originating from the sample in the UltraLo-1800. Consequently, these sources contribute to a background rate that is presently estimated to be about 0.0005 counts h^{-1} cm^{-2} for a counter at sea level with no significant overhead shielding.

The contribution of both cosmic rays and radon to the counter response has been extensively investigated in recent years, from measurements performed in different environments and from modeling and simulation approaches. All these important results have been published in a series of key papers and presentations, notably in Gordon et al. (2009, 2010, 2012, 2013) and McNally (2011).

In 2010–2011, XIA (McNally 2011) performed a series of emissivity measurements of a silicon substrate using a beta prototype of the UltraLo-1800, transported and operated for the same 48 h measurement time, in an identical configuration at several locations. These locations include the XIA offices in Hayward (CA), a cave laboratory in Stanford (CA), the Soudan underground laboratory (MI), and the ASTEP platform and the Modane

TABLE 4.3

Number of Events Detected by the UltraLo-1800 (Prototype Version) for the Same Measurement Duration (48 h) and Identical Configuration at Several Locations

Location	Depth/Altitude	Alphas	Ceilings	Midairs	Rounds
XIA	0 m	93	71	92	286
Stanford (CA)	−17 mwe[a]	62	49	17	19
Soudan (MN)	−2060 mwe	58	37	35	2
LSM	−4800 mwe	50	65	22	7
ASTEP	+2552 m	126	139	347	882

Source: Data from McNally, B.D., Recent advancements in next-generation alpha counting technology. *Third Annual IEEE-SCV Soft Error Rate (SER) Workshop,* Santa Clara, CA, 2011.

[a] mwe, meter water equivalent.

underground laboratory (LSM), both in the French Alps (see Section 7.4.1). Results of these measurements are reported in Table 4.3. In addition, Figure 4.8 shows the complete raw measurements for the ASTEP and LSM locations in the form of scatter plots of risetime as a function of pulse height, illustrating the different categories of events identified by the software. The data shown in Table 4.3 present a clear altitude dependence for all classes of events, including alphas, which are more numerous in altitude than at sea level or underground. Once underground, the alpha-particle yield apparently saturates with increasing depth. As analyzed by Gordon et al. (2012), alpha-particle emission scales with the neutron flux. To explain this result, the same authors modeled the interactions of cosmogenic neutrons, protons, and pions in the materials within the XIA UltraLo-1800 alpha-particle counter. They demonstrated by Monte Carlo numerical simulation that the interactions of terrestrial neutrons with silicon atoms in bulk silicon wafers, and in the argon gas within the UltraLo-1800, produce appreciable alpha-particle emissivities. Their results show, in particular, that the neutrons are responsible for 90% of the alpha particles generated (Gordon et al. 2012). The total alpha-particle emissivity from the reactions on the silicon substrate and in the argon gas is 0.34 α kh^{-1} cm^{-2}. This result leads to an important conclusion: at sea

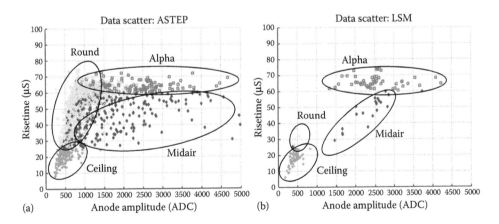

FIGURE 4.8

Scatter plot of risetime as a function of pulse height for the events detected and their classification. Measurements have been performed at mountain altitude on the ASTEP platform (a) and underground at the LSM laboratory (b). (Courtesy of B.D. McNally.)

level (New York City), in a lightly shielded laboratory, one would not expect to observe an emissivity less than 0.34 α kh⁻¹ cm⁻² on a blank silicon wafer, that is, with no contamination. For proportional gas counters, the active counter volume being smaller than that of the ionization counter, the effect of the interactions on the argon gas would necessarily be smaller than the calculations shown above for the ionization detector (Gordon et al. 2012).

Concerning the alphas observed during measurements at the underground sites, they are understood to be due to radon production from the internal components of the prototype instrument used during the experiment. Gordon et al. (2012) examined the effects of radon contaminants, which we found to be significant, even at very low concentration levels of ~0.1 pCi L⁻¹. They presented some measurements obtained after a 45 min purge cycle for samples stored under different conditions where they would be exposed to different amounts of laboratory air. The data clearly show 7.7 MeV alpha particles from the decay of ^{214}Po, which dies off in less than 4 h. Alpha particles from the decay of ^{212}Po and ^{212}Bi, however, at 8.8 and 6.1 MeV, respectively, can take several days to die off due to the 10.6 h half-life of the parent, ^{212}Pb. They showed that this effect could be greatly reduced by storing the sample in a pillbox, which limited its exposure to laboratory air. Exposure to air should be kept to a minimum, including the time between manufacturing and arrival at the measurement laboratory. If these times are large, samples should be kept in closed containers, preferably hermetically sealed in an inert gas. A Monte Carlo model of radon plateout on a 300 mm diameter surface shows that only a dozen radon atoms are required to produce an alpha-particle emissivity of a few tenths of α kh⁻¹ cm⁻² (Gordon et al. 2012).

Recently, the same group of authors presented the initial results of their model showing that the terrestrial neutrons can produce a measurable proton flux within the structure of the UltraLo-1800 (Gordon et al. 2013). The majority of these protons come from interactions of the neutrons with the plastic material, which makes up the detection volume. The peak proton flux from these neutron-induced reactions is on a par with the terrestrial proton flux, showing that this reaction might contribute to the alpha-particle counter's background. As expected, modeling of the linear energy transfer (LET) of terrestrial particles in argon counter gas showed that protons had about 10 times lower LET than alpha particles and about five times greater LET than either pions or muons. The LET for electrons (or beta particles) was several orders of magnitude lower than for alpha particles. Additionally, the authors showed that the protons could only lose a few MeV, at most, in the counter gas. This implies that, above the MeV level, there is probably very little contribution from cosmic ray particles. Early results from a beam of high-energy protons show that, while the detector registers their presence, due to the active signal rejection, nearly all (95%) of the protons are rejected and only a few are misidentified as alpha particles.

In complement to the previous works, Moindjie et al. (2014) recently investigated the specific response of the UltraLo-1800 to atmospheric charged particles. They compiled all the events detected by the counter during several weeks (~1000 h at sea level in a lightly shielded laboratory), calculated the corresponding hourly rate for rounds and midair events, and compared these signals with the responses of two other instruments installed in the same room: a charge-coupled device (CCD) camera and a neutron monitor (NM), the clone instrument of the PdBNM presented in Chapter 2. The CCD camera continuously takes video frames in complete darkness; dedicated control and analysis software isolates pixels or group(s) of pixels with an electrical charge clearly above the background and meeting certain criteria. These detected events correspond to the interaction of charged particles (muons and protons) with the CCD, as demonstrated by extensive characterization, modeling, and simulation work, the contribution of atmospheric neutrons being totally negligible at sea level (Saad Saoud et al. 2014). On the contrary, the NM is only

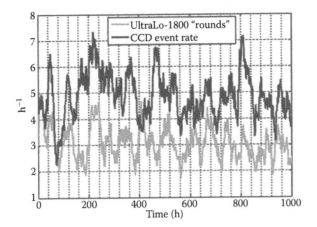

FIGURE 4.9
Correlation between the experimental response of a CCD camera (detection of events in complete darkness) and the hourly rate of *round* events detected by the UltraLo-1800.

sensitive to atmospheric neutrons; we showed in Section 2.2.4.4 that the contribution of charged particles at sea level in the NM response is limited to a few percent. Figure 4.9 shows the good correlation between the CCD pixel event rate and the roundsignal of the UltraLo-1800. This result emphasizes the role of atmospheric protons and muons in the production of round events. It is in very good agreement with the experiment conducted by Gordon et al. (2013) using a proton beam, showing that 95% of the protons are rejected by the counter and mostly classified as round events. On the contrary, the comparison between the midair event rate and the NM signal does not show a clear correlation, suggesting that the interaction of atmospheric neutrons within the gas molecules in the counter chamber is not the dominant mechanism in the production of midair events.

4.4.5 Example of Measurements

Figure 4.10 shows an example of measurements obtained with the UltraLo-1800 on a fully processed 300 mm wafer. The UtraLo-1800 control and analysis software facilitates reporting by having several different files available for export, most notably a full measurement report (PDF format) that corresponds to this figure (the header of the report, which contains sample name, description, date, and run characteristics, has been removed for confidentiality reasons). This export can be performed either immediately, at the end of a measurement run, or later, after reanalyzing a file. Figure 4.10a shows the evolution of the emissivity, and the error in the emissivity, versus measurement time. The last value (the rightmost value on the plot) corresponds to the final emissivity value that is the result of the measurement. This plot is useful diagnostically as it allows the user to look for systematic changes over time. For instance, a slow rise in emissivity could indicate that there were problems with moisture early in the run, and a sharp drop could indicate radon exposure. Figure 4.10b is a histogram of the energies of all alpha particles observed in the measurement. The energy scale ranges from 0 to 10 MeV. This graph is also useful for the possible identification of alpha emitters from the energy peaks that can appear on such an energy distribution. Figure 4.10c is a histogram of alpha counts in time. This plot depicts the number of alphas observed in a given interval of time and is similar to the emissivity-versus-time plot.

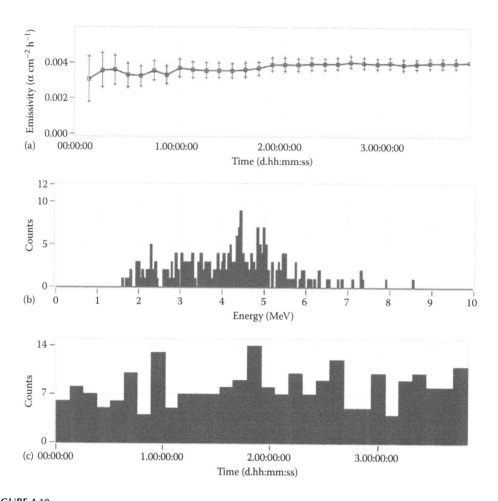

FIGURE 4.10
Typical measurement report exported by the UltraLo-1800 control and analysis software in the form of a PDF file (header not shown). The measurements shown have been obtained on a fully processed 300 mm wafer (without solder bumps).

Other detailed examples of measurements using the UltraLo-1800 can be found in literature. For example, we can cite Wong et al. (2010), who investigated emissivity measurements of fully processed wafers from different vendors. This study showed that fully-processed-wafer alpha emission typically varies from 0.001 to 0.004 α cm^{-2} h^{-1} and does not show a technology or a wafer-supplier trend. Since many wafers exhibit nonnegligible alpha-emission rates, the authors concluded that it is imperative to consider the wafer contribution to the alpha flux used to evaluate the soft error rate of the fabricated circuits (Wong et al. 2010). This point will be illustrated later in this book, notably in Chapters 7 (static random-access memories [SRAM]) and 11 (nonvolatile memories).

4.4.6 Multicenter Comparison of Alpha-Particle Measurements

To conclude this part of the chapter dedicated to the UltraLo-1800, we summarize in this section the results of a round-robin field assessment conducted between different laboratories, each participant being equipped with an UltraLo instrument. This recent study

(McNally et al. 2014) explored how much variation is observed between participants in the measurement of a set of standard samples and determined the degree to which any observed variations are random or systematic by participant. This study follows initial work conducted in 2010 by the Alpha Consortium (Wilkinson et al. 2011), a multicenter comparison study conducted by nine international centers to quantify variability in alpha-emission measurements using a shared set of samples. Four samples representing low-alpha (LA, between 2 and 50 α cm^{-2} kh^{-1}) and ultralow-alpha (ULA, <2 α cm^{-2} kh^{-1}) materials were counted by each participating laboratory in a blinded trial. The study results, however, unexpectedly showed that emissivity values measured at the LA level varied by factors of two or more that correlated with the site, while measurement uncertainties for the ULA samples were so large that it was not possible to determine whether their values showed similar site-to-site variations. The authors hypothesized that the observed variability at the LA level could be explained by differences between the counters' low-energy discriminator settings, while variability at the ULA level was probably affected both by these settings and by the counters' different backgrounds. A follow-up study (Wilkinson et al. 2014) using both an LA sample and a calibrated point source was conducted to explicitly determine the role of low-energy discriminator settings in site-to-site measurement variability. The results indicated that it plays a minor role, if any, and continued to show site-to-site measurement variability by a factor of two. The authors then suggested a different source of site-to-site variation, that is, differences in the separation between samples and counter entrance windows, which cause variations in counter efficiency because alpha particles emitted at low angles are less likely to enter the counter as the distance to the entrance window increases. The UltraLo-1800 counter, whose design eliminates the entrance window and has demonstrated the ability to measure samples at the sub-ULA (SULA, <1 α cm^{-2} kh^{-1}) level, is now available in enough laboratories around the world to allow the Alpha Consortium's round-robin experiment to be repeated with enough new instruments to test this hypothesis. The details of the complete experimental protocol followed by the participants to perform measurements are not given here but can be found in MacNally et al. (2014).

In contrast to the initial Alpha Consortium measurements, the data presented in this new study show minimal site-to-site variability for LA-level emissivity measurements using the UltraLo-1800. An example of measurement results is shown in Figure 4.11. Good agreement among the participants is evident; six of the seven values are within 1 σ of the consensus mean value of 38.0 α cm^{-2} kh^{-1}, while Location 3 records a value just outside this limit, approximately 1.09 σ from the mean. The complete set of values is fully consistent with a normal distribution of measurement errors. The relative uncertainty (defined here as $\sigma \mu^{-1}$, where μ is the observed emissivity) for all measurements is approximately 4%. This suggests that it may be possible to also achieve site-to-site uniformity using windowed gas proportional counters, provided that the community can agree on a standard sample-to-window separation and develop a practical methodology to repeatedly achieve it, particularly in the presence of sample-to-sample height variations.

At the ULA level, the large uncertainties reported for nearly all results from the initial Alpha Consortium study make it difficult to deduce specific sources of the observed variation. When making emissivity measurements of a sample that is at or below an instrument's background rate, an accurate determination of counter background is critical in order to produce a statistically supportable result. In the absence of a true zero-emissivity sample that can be placed at a standard sample-to-window separation, it appears difficult to design a standardized background measurement for windowed gas proportional

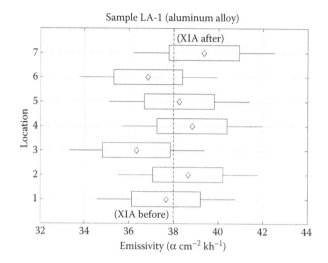

FIGURE 4.11
Emissivity measurements for Sample LA-1 obtained by the six participants of the multicenter study conducted by McNally et al. (2014). The diamonds represent the measured emissivities, the boxes bound ±1 sigma, and the whiskers bound ±2 sigma. The consensus mean value of 38.00 α cm^{-2} kh^{-1} is depicted by the vertical dashed line. (Reprinted from *Nuclear Instruments and Methods in Physics Research A*, A750, McNally, B.D., Coleman, S., Warburton, W.K., et al., Sources of variability in alpha emissivity measurements at LA and ULA levels, a multi-center study, 96–102, Copyright (2014), with permission from Elsevier.)

counters that is conceptually capable of producing a result at the accuracy needed for ULA measurements that are reproducible from site to site.

While measurement data at ULA and SULA levels using the new UltraLo-1800 instrument still exhibit site-to-site variations, as shown in Figure 4.12 for two samples, these have been reduced to about 0.8 α cm^{-2} kh^{-1}, lower than the earlier work by a factor of about three (Wilkinson et al. 2011). These remaining variations may be primarily due to a class of background events in the counter, of cosmogenic origin, that cannot yet be fully identified and rejected by the XIA pulse shape analysis software. In particular, it was observed during the study that Location 2 consistently measures higher activity levels than the other locations (Figure 4.12). Investigations and additional measurements have suggested that the Location 2 environment is the more likely source of the observed increased counting rates (McNally et al. 2014). The most likely candidate for this increase is the cosmogenic background, presented and discussed in Section 4.4.4. To estimate this effect, the study participants were asked to provide a description of the facility where their instrument was located, including elevation and facility construction. From this information, a simple *over-burden* model was constructed to estimate the mass thickness above the instrument at each location. The results of this survey and calculation have immediately shown evidence that Location 2 has significantly less overburden than the other sites. This model could also explain why the measurements at Locations 1 and 7 are also consistently high for sample SULA-Si-1. It does strongly suggest that a residual cosmogenic background term is the sole remaining source of site-to-site measurement differences made at the ULA level using the XIA counter. The same cosmogenic background term may also explain some, or perhaps all, of the residual activity found for sample SULA-Si-1, whose activity was expected to be zero. Further research, perhaps including underground measurements, will be needed to distinguish between cosmogenic background and other possible residual terms, such as radon emitted from the counter materials (McNally et al. 2014).

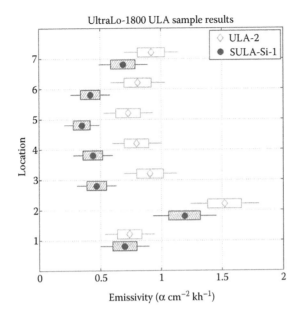

FIGURE 4.12
Comparison of emissivity results from ULA and SULA samples used in the multicenter study conducted by McNally et al. (2014). The gray diamonds represent the measured emissivity of sample ULA-2, and the black circles represent the measured emissivity of sample SULA-Si-1. In both cases, the boxes bound ±1 sigma, and the whiskers bound ±2 sigma. (Reprinted from *Nuclear Instruments and Methods in Physics Research A*, A750, McNally, B.D., Coleman, S., Warburton, W.K., et al., Sources of variability in alpha emissivity measurements at LA and ULA levels, a multicenter study, 96–102, Copyright (2014), with permission from Elsevier.)

4.5 Other Techniques

We describe in this section three analysis techniques alternative or complementary to gas-filled counters, used in the semiconductor industry to quantify ultratraces of alpha-particle-emitter contaminants in circuit and packaging materials: silicon alpha spectrometers, liquid or solid-state scintillators, and a particular type of mass spectrometry coupled with a wafer decomposition technique. The content of this part is limited to a brief description of these different approaches; the reader is invited to refer to specialized literature and cited references for further reading.

4.5.1 Silicon Alpha Detectors

Silicon alpha detectors are a particular type of ionization chamber with silicon as the counting medium. Charged particles or photons produce electron–hole pairs in silicon; an electric field applied across the domain where electron–hole pairs are generated allows these latter to be separated and collected. Because of their high density compared with gaseous detectors, solid-state semiconductor detectors can absorb particles of correspondingly higher energies (Grupen and Shwartz 2008). The use of a solid-state medium is also attractive since the energy required to produce electron–hole pairs in silicon (3.6 eV) is about 10 times lower than the formation energy of a ion–electron pair in the counting gas (see Table 4.2), so that the statistics of charge generation are much better (Warburton et al. 2004).

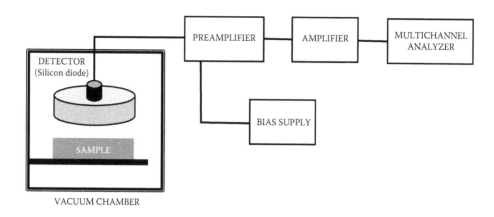

FIGURE 4.13
Block diagram of a typical spectrometer setup for alpha particle counting and spectrometry.

Silicon detectors are generally large-area (up to a few square centimeters) p-i-n diodes that are reverse biased and connected to a charge-sensitive preamplifier and amplifier (see Figure 4.13). No internal amplification is involved: the charges generated within the silicon material by alpha particles are simply collected by the internal electric field (developed by the p-i-n junction) and amplified by the electronics. The major advantages of silicon alpha spectrometers are their good energy resolution and relative robustness (energy resolutions of 1%–2% can readily be obtained from such detectors). However, they have several limitations. First, it is not practical to make them with large areas, because their capacitance becomes too large and spoils their energy resolution (silicon detectors are reverse-biased diodes with parallel, planar electrodes and therefore have the capacitance of the corresponding parallel-plate capacitor). Another size limitation comes from the required high quality of silicon, which is not available in large areas, typically above a few square centimeters. Progress in the field now allows the fabrication of silicon diodes up to 20 cm^2 (Gordon et al. 2008; HAMAMATSU 2014). The second limitation is the need to perform measurements in a vacuum chamber (alpha particles lose their energy in air) with the sample well positioned at a minimum distance in front of the detector, depending on the sample geometry. The development of large-area vacuum chambers (e.g., compatible with 300 mm wafers) using a matrix of assembled ultralow-background silicon detectors may be a promising future solution for alpha-radiation metrology at wafer level.

Two major types of silicon diode detectors are produced and commercialized by the main manufacturers in the domain (Canberra, Ortec, Hamamatsu, etc.): ion-implanted and surface-barrier technologies. The two processes are complementary in that each technique is better for manufacturing certain types of detectors (ORTEC). Figure 4.14 shows simplified representations of the two manufacturing processes. There are several advantages to using ion-implanted devices, which are the detectors of choice for alpha spectroscopy: (i) thinner and more rugged front contact, with better energy resolution for some alpha spectroscopy applications; (ii) lower electronic noise; (iii) higher geometric efficiency for some alpha spectroscopy applications; and (iv) operation possible up to 60°C and bakeout at 200°C.

An issue of particular importance in alpha spectroscopy is the need to perform low-background measurements. The irreducible background for these detectors is clearly set by cosmic radiation. Measurements performed on state-of-the-art ion-implanted detectors have confirmed that the ultimate limit to the low-background performance of silicon

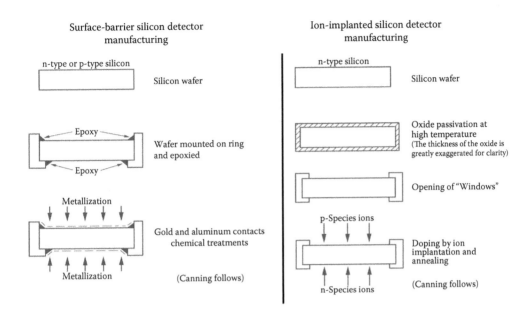

FIGURE 4.14
Simplified representations of the two manufacturing processes for silicon detectors: surface barrier and ion-implanted detectors. (Adapted from ORTEC, Introduction to charged particle detectors. Available at: http://www.ortec-online.com.)

detectors, when manufactured and packaged with special materials and following strict cleaning procedures, is associated with the omnipresent cosmic radiation (ORTEC). This limit in the energy range from 3 to 8 MeV is 0.05 counts h^{-1} 10^{-2} cm^{-3} of active volume. This means that, for 100 μm thick, low-background-grade detectors with an active area of 450 mm^2, a background counting rate of about six counts per day is expected (ORTEC). To achieve such a low level, extreme precautions must be taken both concerning vacuum chamber contamination and in detector handling procedures.

4.5.2 Liquid and Solid-State Scintillators

Scintillation methods are among the oldest methods of particle detection, very sensitive and generally used for special environmental surveys and as laboratory instruments (Grupen and Shwartz 2008). They can detect a wide variety of radiation, including gamma, neutron, alpha, and beta radiation, with good quantum efficiency and can measure both the intensity and the energy of incident radiation. The basic principle of all scintillation techniques is the use of a special material (crystal, plastic, or organic liquid) that glows or *scintillates* when radiation interacts with it. A sensitive photon detector, generally a photomultiplier tube (or a photodiode), converts the light to an electrical signal, which is processed and analyzed using dedicated electronics (see Figure 4.15).

Different scintillator materials are used depending on the radiation to be detected. Solid scintillators are mostly inorganic crystals, for example cesium iodide (CsI) for the detection of protons and alpha particles, sodium iodide (NaI) containing a small amount of thallium for the detection of gammas, silver-activated zinc sulfide (ZnS(Ag)) for the detection of alpha particles, lithium iodide (LiI) for the detection of neutron particles, and so on (Grupen and Shwartz 2008).

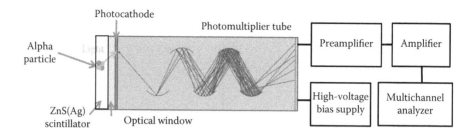

FIGURE 4.15
Schematic illustration of a ZnS(Ag) scintillator setup for alpha particle counting and spectrometry.

ZnS(Ag) has a very high scintillation efficiency, comparable to that of NaI(Tl). It is only available as a polycrystalline powder and has a maximum in the scintillation emission spectrum at 450 nm. Thicknesses greater than about 25 mg cm^{-2} become unusable because of the opacity of the multicrystalline layer to its own luminescence. Its use is limited to thin screens used primarily for detection of alpha particles or other heavy ions. A wide variety of detectors and commercial systems based on ZnS scintillators are frequently used for the detection and spectrometry of alpha particles, including for measuring concentrations of radon in air (Philipsborn and Just 2005). The main advantages of this type of detector are the absence of a window on the entrance and the low cost of the detector itself. An active area up to a few hundred square centimeters can be implemented (see, e.g., the range of detectors and instruments proposed by Ludlum Measurements). This offers the possibility to measure thick or irregularly shaped samples, as reported in Gordon et al. (2008) for the characterization of plating, ceramic substrates, or bulk materials. In this work, several ZnS scintillators 125 mm in diameter and approximately 100 μm thick were assembled in a light-tight environment filled with pure N$_2$ gas to displace radon. Under these controlled conditions, typical backgrounds for these counters are in the order of 3 counts h^{-1}, well below the background of standard ZnS detectors, which exhibit a relatively poor background of a few tens of counts h^{-1}.

Liquid scintillation offers another way to detect alpha particles with no sample self-absorption or geometry problems and with 100% counting efficiency. Sample preparation may include extraction of the alpha emitter of interest by a specific organic-phase-soluble compound directly into the liquid-scintillation counting medium. Detection electronics use energy and pulse-shape discrimination to yield alpha spectra without beta and gamma background interference. Different liquid-scintillation analyzers are commercially available. We can mention, for example, the Ordela system, based on photon-electron-rejecting alpha liquid scintillation (PERALS®), which is an instrument for efficient, rapid, and accurate counting and spectrometry of alpha particles from alpha-emitting nuclides in appropriate extractive scintillators.

4.5.3 ICP-MS and VPD ICP-MS

ICP-MS is undoubtedly the fastest-growing trace-element technique available today in industry and research (Thomas 2008). This particular mass spectrometry technique is able to detect ultratraces of metals and several nonmetals at concentrations as low as one part in 10^{12} ppt. Schematically, an ICP-MS (Figure 4.16) combines a high-temperature inductively coupled plasma (ICP) source with a mass spectrometer (MS). The ICP source converts the atoms of the elements in the sample to ions. These ions are then separated and detected by the MS. The sample is typically introduced into the ICP plasma as an aerosol, by aspirating

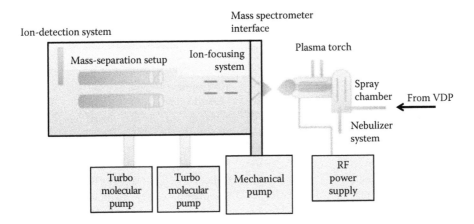

FIGURE 4.16
Schematic of an ICP-MS.

a liquid or dissolved solid sample into a nebulizer. Once the sample aerosol is introduced into the ICP torch, it is completely desolvated, and the elements in the aerosol are first converted into gaseous atoms and then ionized toward the end of the plasma. For coupling to mass spectrometry, the ions from the plasma are extracted through a series of cones into a MS, usually a quadrupole mass filter. The ions are separated on the basis of their mass-to-charge ratio and a detector receives an ion signal proportional to the concentration.

In the domain of alpha-radiation metrology, ICP-MS is an ideal technique to measure concentrations of ^{238}U and ^{232}Th in a wide variety of electronic materials. In Kobayashi et al. (2009), for example, this technique has been used to characterize the traces of these radioactive contaminants in packaging materials. From the measured concentrations, the authors used a model to estimate the alpha emissivity of different package resins, assuming that both the ^{238}U and ^{232}Th decay chains were in secular equilibrium. They correlated the level of alpha emissivity with the soft error rate of SRAM circuits evaluated during underground tests. The link between ultratraces of alpha-particle emitters and soft errors will be presented and discussed in Chapter 7.

When coupled with vapor-phase decomposition (VPD), VPD ICP-MS is the most widely used analytical technique to monitor ultratraces of metals of bare wafers and wafers with silicon oxide or nitride films. Other additional methods of analysis have been proposed for the monitoring of a broader variety of process films, including metal films, low-κ and high-κ dielectric films, and SiGe and polysilicon layers (ChemTrace 2012). Taking the example of the analysis of a bare silicon wafer, this is placed in the VPD chamber and exposed to HF vapor to dissolve the SiO_2 surface layer. The wafer surface is then scanned with a typical 100 μL extraction droplet. This can be done manually or with an automated wafer scanner. The extraction droplet picks up the contents of the dissolved silicon layer as it is moved across the wafer surface. The extraction droplet is then pipetted from the wafer surface and transferred to the ICP-MS nebulization chamber for analysis. This technique, VPD ICP-MS, provides accurate measurement of up to 60 elements and detection limits in the range of 10^6–10^{10} atoms cm^{-2} on the silicon wafer. Table 4.4 summarizes typical values of detection limit (expressed in atoms cm^{-2}) for a wafer surface scan performed on 300 mm silicon wafers (ChemTrace 2012). Note the excellent detection limits for uranium (3×10^8) and thorium (3×10^7), both in the sub-parts per trillion range, and also the excellent recovery for these two radioactive nuclides (84% and 100% respectively).

TABLE 4.4

Method Detection Limit (atoms cm^{-2}) for Wafer Surface
Scan VPD ICP-MS on 300 mm Silicon Wafers

Aluminum	Al	1×10^9	Magnesium	Mg	5×10^8
Antimony	Sb	1×10^7	Manganese	Mn	1×10^8
Arsenic	As	2×10^9	Molybdenum	Mo	1×10^7
Barium	Ba	8×10^6	Niobium	Nb	5×10^7
Beryllium	Be	1×10^9	Nickel	Ni	3×10^8
Bismuth	Bi	5×10^6	Potassium	K	1×10^9
Boron	B	2×10^{10}	Rubidium	Rb	5×10^8
Cadmium	Cd	2×10^7	Sodium	Na	1×10^9
Calcium	Ca	1×10^9	Strontium	Sr	5×10^7
Cesium	Cs	3×10^7	Tantalum	Ta	2×10^7
Chromium	Cr	3×10^8	Thallium	Tl	3×10^7
Cobalt	Co	3×10^8	Thorium	Th	3×10^7
Copper	Cu	1×10^8	Tin	Sn	1×10^8
Gallium	Ga	2×10^7	Titanium	Ti	3×10^8
Germanium	Ge	5×10^7	Tungsten	W	5×10^6
Hafnium	Hf	3×10^7	Uranium	U	3×10^8
Iron	Fe	5×10^8	Vanadium	V	5×10^7
Lanthanum	La	1×10^7	Yttrium	Y	1×10^8
Lead	Pb	2×10^7	Zinc	Zn	3×10^8
Lithium	Li	3×10^8	Zirconium	Zr	5×10^7

Source: Data from ChemTrace, VPD ICP-MS method detection limits and recoveries for trace metals contamination analysis of silicon wafers, Fremont, CA, 2012.

References

Alpha Sciences. 2009. How are alpha particles counted? Alpha Sciences. http://www.alphacounting.com.

Ansoborlo, E., Aupiais, J., and Baglan, N. 2012. *Mesure du rayonnement alpha* [*Alpha Radiation Measurement*]. Paris: Lavoisier.

ChemTrace. 2012. VPD ICP-MS method detection limits and recoveries for trace metals contamination analysis of silicon wafers. Fremont, CA: ChemTrace. Available at: http://www.chemtrace.com/files/VPD%20ICP-MS%20Method%20Detection%20Limits%20Rev0%20May%202012.pdf.

Gordon, M.S., Heidel, D.F., Rodbell, K.P., Dwyer-McNally, B., and Warburton, W.K. 2009. An evaluation of an ultralow background alpha-particle detector. *IEEE Transactions on Nuclear Science* 56:3381–3386.

Gordon, M.S., Rodbell, K.P., Heidel, D.F., Cabral, C., Cannon, E.H., and Reinhardt, D.D. 2008. Single-event-upset and alpha-particle emission rate measurement techniques. *IBM Journal of Research and Development* 52:265–273.

Gordon, M.S., Rodbell, K.P., Heidel, D.F., et al. 2010. Alpha-particle emission energy spectrum from material used for solder bumps. *IEEE Transactions on Nuclear Science* 57:3251–3256.

Gordon, M.S., Rodbell, K.P., Tang, H.H.K., Yashchin, E., Cascio, E.W., and McNally, B.D. 2013. Selected topics in ultra-low emissivity alpha-particle detection. *IEEE Transactions on Nuclear Science* 60:4265–4274.

Gordon, M.S., Rodbell, K.P., Tang, H.H.K., et al. 2012. Ultra-low emissivity alpha-particle detection. *IEEE Transactions on Nuclear Science* 59:3101–3109.

Grupen, C. and Shwartz, B.A. 2008. *Particle Detectors*, 2nd edn. Cambridge: Cambridge University Press.

HAMAMATSU. 2014. Si detectors for high energy particles. Available at: https://www.hamamatsu.com/resources/pdf/ssd/e10_handbook_for_high_energy.pdf.

JEDEC. 2011. Alpha radiation measurement in electronic materials. JEDEC Standard JESD211, JEDEC Solid State Technology Association, 2011.

Knoll, G. F. 1989. *Radiation Detection and Measurement*, 2nd edn. Hoboken, NJ: Wiley.

Kobayashi, H., Kawamoto, N., Kase, J., and Shiraish, K. 2009. Alpha particle and neutron-induced soft error rates and scaling trends in SRAM. In *Proceedings of the IEEE International Reliability Physics Symposium*, pp. 204–211. 26–30 April, Montreal, QC, IEEE.

McNally, B.D. 2011. Recent advancements in next-generation alpha counting technology. *Third Annual IEEE-SCV Soft Error Rate (SER) Workshop*, Santa Clara, CA. Available at: http://ewh.ieee.org/soc/cpmt/presentations/cpmt1110w-9.pdf.

McNally, B.D. 2013. UltraLo-1800 alpha particle counter user's manual, Version 0.3, XIA LLC.

McNally, B.D., Coleman, S., Warburton, W.K., et al. 2014. Sources of variability in alpha emissivity measurements at LA and ULA levels, a multicenter study. *Nuclear Instruments and Methods in Physics Research A* A750:96–102.

Moindjie, S., Saad Saoud, T., Autran, J.L., and Munteanu, D. 2014. Detection and monitoring of atmospheric protons and muons with an ultra-low background alpha-particle counter. Unpublished.

ORTEC. Introduction to charged particle detectors. ORTEC. Available at: http://www.ortec-online.com/download/introduction-charged-particle-detectors.pdf.

Philipsborn, H. and Just, G. 2005. Fast retrospective determination of radon exposure with a sensitive alpha scintillation probe. *Journal of Radiological Protection* 25:299.

Saad Saoud, T., Moindjie, S., Autran, J.L., et al. 2014. Use of CCD to detect terrestrial cosmic rays at ground level: Altitude vs. underground experiments, modeling and numerical Monte Carlo simulation. In *IEEE Nuclear and Space Radiation Effects Conference*, Paris, France, July 14–18.

Thomas, T. 2008. *Practical Guide to ICP-MS*, 2nd edn. Boca Raton, FL: CRC Press.

Warburton, W.K., Wahl, J., and Momayezi, M. 2004. Ultra-low background gas-filled alpha counter. US Patent 6732059 B2.

Wilkinson, J.D., Clark, B.M., Wong, R., et al. 2011. Multicenter comparison of alpha particle measurements and methods typical of semiconductor processing. In *Proceedings of the IEEE International Reliability Physics Symposium*, pp. 5B.3.1–5B.3.10. 10–14 April, Monterey, CA, IEEE.

Wilkinson, J.D., Clark, B.M., Wong, R., et al. 2014. Follow-up multicenter alpha counting comparison. *IEEE Transactions on Nuclear Science* 61(4):1516–1521.

Wong, R., Wen, S.J., Su, P., and McNally, B.D. 2010. Alpha emission of fully processed silicon wafers. In *Proceedings of the IEEE International Integrated Reliability Workshop*, pp. 34–36. 17–21 October, Stanford Sierra, CA, IEEE.

Section II

Soft Errors

Mechanisms and Characterization

5

Particle Interactions with Matter and Mechanisms of Soft Errors in Semiconductor Circuits

5.1 Interactions of Neutrons with Matter

The neutron was discovered in 1932 by Sir James Chadwick (1891–1974), a British physicist (Cambridge), who was awarded the 1935 Nobel Prize in Physics for this fundamental discovery in the domain of particle physics. The neutron is an uncharged elementary particle, with a mass (1.67493×10^{-27} kg) slightly larger than that of the proton (1.67262×10^{-27} kg). It is also unstable as a free particle; it disintegrates into a proton, an electron, and an antineutrino with a lifetime of 886.7 s.

Since neutrons are uncharged, they do not experience Coulomb's interaction, unlike charged particles: they do not interact with orbital electrons, and can pass through electronic clouds without being affected. Consequently, they cannot directly ionize matter. In contrast to electrons, photons, and charged particles, neutrons undergo extremely weak electromagnetic interactions. Therefore, neutrons are highly penetrating particles, and they can travel a significant distance through matter largely without obstruction.

As we will see later in this section, neutrons can interact with atomic nuclei following several mechanisms depending on their energy, mainly elastic scattering, inelastic scattering, neutron capture, and nuclear fission. Even though these interactions have a low probability associated with them, they absolutely cannot be neglected, since these interactions are precisely at the origin of single-event effects (SEEs) via the production of secondary charged particles. This mechanism is called indirect ionization.

Neutrons are classified according to their energy E, generally as follows:

- Cold: E < 1 meV
- Thermal: E < 0.5 eV
- Epithermal: 0.5 eV < E < 50 keV
- Fast: 50 keV < E < 1 MeV
- Medium energy: 1 MeV < E < 10 MeV
- High energy: E > 10 MeV

In other classifications the energy limits of each category may differ, or they may be more detailed concerning a specific category (e.g., the cold-neutron category may be divided into three subcategories: ultracold neutrons, very cold neutrons, and cold neutrons).

5.1.1 Cross Section

The concept of *cross section* has been introduced to express the probability of a particular event occurring between a neutron and a nucleus. The cross section has the dimension of a surface and is usually denoted by the symbol σ. It represents the effective area of the nucleus for interaction with a neutron. The cross section is usually expressed in barns (1 barn = 10^{-24} cm²).

When a large number of neutrons of the same energy strike a material (thin layer), several events may occur: (i) some neutrons can pass through the material without interaction, (ii) other neutrons may have interactions that change their directions and energies, and (iii) other neutrons may not leave the material. Each of these events is characterized by a certain probability. For example, the probability that a neutron does not get out of the material (in other words, the probability of being captured or absorbed) is the ratio between the number of neutrons that do not leave the material and the number of initial incident neutrons (Rinard 1991). The absorption cross section is given by the probability of the neutrons being absorbed divided by the density of atoms per unit area (i.e., the number of target atoms per unit area of the layer). In silicon, the total neutron cross section decreases from 1.95 barn at E = 40 MeV to 0.6 barn at E = 200 MeV.

As explained in Gaillard (2011), it is also interesting to calculate the mean free range of neutrons in the silicon without interaction. For this, it is necessary to calculate the number, a, of nuclei per volume unit, as follows:

$$a = N/(AM/d) \tag{5.1}$$

where:
  N   = 6.02×10^{23} = the Avogadro number
  AM = 28 = silicon atomic mass
  d   = 2.33 g cm⁻³ = silicon density

The number of nuclei per unit volume in silicon is then 5×10^{22} at cm⁻³. Considering a cross section of interaction of 1 barn, the neutron mean free range is given by

$$\frac{1}{10^{-24} \times 5 \times 10^{22}} = 20 \text{ cm} \tag{5.2}$$

This calculation confirms the fact that neutrons can cross a considerable distance into the material without interaction, as explained previously.

5.1.2 Types of Neutron–Matter Interactions

Interactions of neutrons with atomic nuclei can be divided into two major mechanisms (Rinard 1991).

  1. *Scattering*: When a neutron is scattered by a nucleus, its speed and direction change, but the number of protons and neutrons in the nucleus remains the same as before the interaction. The nucleus will have some recoil velocity, and it may be left in an excited state that will lead to the eventual release of gamma radiation.
  2. *Capture or nonelastic scattering*: When a neutron is captured or absorbed by a nucleus, a wide range of radiations can be emitted, or fission may be induced.

Scattering interactions themselves can be divided into two subcategories: elastic and inelastic scattering. These interactions are schematically summarized in Figure 5.1. The type of interaction mainly depends on the incident neutron's energy. During interaction, neutrons give a part of their energy as kinetic energy to the recoils. In some materials, such as ^{10}B, low-energy neutron capture will give an exoenergetic reaction, but generally either the energy is conserved (elastic collisions) or a fraction of it is changed into mass, so that the total kinetic energy of the recoils is lower than the kinetic energy of the incident neutron (Gaillard 2011). These issues will be addressed in the following sections.

To describe an interaction and its products, a simple notation can be used. Considering the interaction of a neutron n and a nucleus B, resulting in a recoil nucleus C and which releases an outgoing particle g:

$$n + B \rightarrow C + g \tag{5.3}$$

this interaction is denoted by B(n, g)C. In this notation, on the left is the symbol of the target nucleus, the first symbol in parentheses indicates the projectile (bombarding particle), the second is the emitted particle, and the symbol of the product is indicated on the right (Valkovic 2000). This means that the system of reactants is shown on the left side of the comma and the system of products on the right. The lighter elements are written in parentheses to describe the reaction, while heavy nuclei are shown outside the parentheses. Before and after the nuclear reaction, both the sum of mass number and the sum of atomic number remain unchanged (Valkovic 2000). The heavier element resulting from the interaction is obtained by the equilibrium of the number of neutrons and protons before and after the reaction. To describe a type of interaction, without considering the nuclei involved, only the part in parentheses is indicated. For example, if ^{28}Si is considered as the target nucleus, the (n, p) reaction results in a proton and an Al recoil, or the reaction (n, α) results in Mg and He recoils.

In the following, we succinctly describe the main neutron–matter interactions.

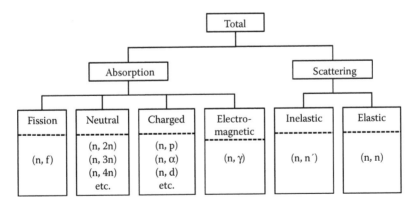

FIGURE 5.1
Classification of possible interactions of neutrons with matter. The reactions are indicated by the lighter elements of the reaction without considering the heavier nuclei involved. The particle to the left of the comma is the incoming particle (neutron); the particle at the right of the comma is the outgoing particle. (Adapted from Rinard, P. Neutron interactions with matter. Los Alamos Technical Report. Available at: http://www.fas.org/sgp/othergov/doe/lanl/lib-www/la-pubs/00326407.pdf, 1991.)

5.1.2.1 Elastic Scattering

In elastic scattering, the nature of the interacting particles is not modified; the recoil nucleus is thus the same as the target nucleus. In this type of scattering, the total kinetic energy of the neutron and nucleus is conserved after the interaction. In the case of ^{28}Si, the reaction is the following:

$$n + {}^{28}\text{Si} \rightarrow n + {}^{28}\text{Si} \tag{5.4}$$

which is also denoted by ^{28}Si(n, n)^{28}Si or (n, n).

During the interaction, a fraction of the neutron's kinetic energy is transferred to the nucleus. The maximum energy that can be given to the recoil is (Gaillard 2011)

$$E_{max} = 4 \times E \times \frac{A}{(A+1)^2} \tag{5.5}$$

where A is the atomic number of the target nucleus. The probability to transfer a given energy between 0 and E_{max} is obtained by the differential elastic cross section (Gaillard 2011).

Concerning the silicon recoils produced by elastic scattering, most of them have a low energy. For example, with an initial neutron energy of 125 MeV, only 5% of the interactions lead to a recoil with energy larger than 0.27 MeV, and 0.1% give a recoil energy larger than 1.5 MeV (Gaillard 2011).

5.1.2.2 Inelastic Scattering

Inelastic scattering is similar to elastic scattering except that the target nucleus undergoes an internal rearrangement into an excited state, from which it eventually releases radiation (Rinard 1991). The incident neutron is absorbed in the target nucleus, and a short time later a neutron is ejected with a lower energy; the emission is considered to be isotropic (Gaillard 2011). The total kinetic energy of the outgoing neutron and nucleus is less than the kinetic energy of the incoming neutron. Part of the kinetic energy of the incident neutron is used to place the recoil nucleus into the excited state. In the case of ^{28}Si, this type of reaction is denoted by ^{28}Si(n, n)^{28}Si* or (n, n').

5.1.2.3 Capture (Nonelastic Scattering)

Instead of being scattered by a nucleus, the neutron may be absorbed or captured, as shown in Figure 5.1. Many reactions are possible, and a large variety of particles can be emitted. This type of interaction is also called nonelastic interaction. In such reactions, the kinetic energy is not conserved because part of the incident kinetic energy is used to change the nature of the target nucleus or its excitation energy. As explained in Rinard (1991), after interaction, the nucleus may change its internal structure and release one or more gamma rays. Charged particles may also be emitted, such as protons, deuterons, and alpha particles. The nucleus may also get rid of excess neutrons. If only a single neutron is emitted, the reaction is indistinguishable from a scattering event. Where more than one neutron is emitted, the number of neutrons traveling through the material is larger than the initial number of neutrons before interaction. In this case, the number of neutrons is said to have been multiplied. Finally, as shown in Figure 5.1, there may be a fission event, which results in two or more fission fragments (nuclei with intermediate atomic mass) and more neutrons (Rinard 1991).

In this type of interaction, the sum of masses of generated particles is greater than the sum of masses of particles that initially interact. A portion of the incident kinetic energy is converted into a mass variation. Producing a reaction requires the incident neutron to have a certain energy. The minimum kinetic energy to produce a given reaction is called the threshold energy. The number of possible responses is larger when the energy of the incident neutron increases. In the case of silicon, for an incident energy of 10 MeV, there are five possible reactions; at 20 MeV, there are 18; and at 50 MeV, there are nearly 150 (Wrobel 2002). All these reactions give a wide range of particles, as will be shown in the next section. For a given neutron energy, different reactions are possible, but their relative probability varies with the neutron energy.

5.1.3 Recoil Products

As a general rule, the different recoils produced in the interaction of neutrons with matter extend from the proton or neutron to the nucleus of the target atom (Gaillard 2011). For example, in the case of a silicon target nucleus, the possible reactions and emitted particles are summarized in Table 5.1 for an incident neutron with energy of 20 MeV (Wrobel 2002). The threshold energy for each reaction is also indicated (elastic scattering has a zero threshold and is therefore always possible).

The study of SEEs and their impact on soft-error rate (SER) in integrated circuits requires a thorough knowledge of the interaction mechanisms of particles with materials that compose the circuit. Thus, recoils that are produced by neutron interactions with circuit materials must be clearly identified by their nature, energy, and direction. For the accurate calculation of the circuit SER, it is necessary to consider all the recoils that can pass through a sensitive node or those that can reach its proximity (these recoils must be produced at a distance from the sensitive node lower than their range). One computational method frequently used to perform such complex calculations is to build a complete nuclear database that contains all the information concerning the recoil products resulting from the interaction of a great number of incident neutrons and a given target material. Two detailed examples of nuclear databases will be given and analyzed in the following:

- In Section 5.1.5, for the interaction of atmospheric thermal, epithermal, and low-energy neutrons with undoped and boron-doped silicon.
- In Section 6.5, to investigate the neutron–silicon (n–Si) nuclear events resulting from the interactions with silicon of neutrons produced by different sources (monoenergetic and broad spectrum, natural and artificial). We will also illustrate in Section 6.5 how the databases built by extensive Monte Carlo simulation can be used to compute the SER of test-vehicle circuits and to compare, at circuit level, the capability of several artificial neutron sources to mimic the natural radiation background.

5.1.4 Interaction of Thermal Neutrons with ^{10}B

A notable interaction for the radiation response of electronic devices is the interaction between thermal neutrons and the ^{10}B isotope of boron. In electronic devices, in addition to silicon, various materials are involved. The cross section for thermal neutrons and for several materials used in semiconductor devices is shown in Figure 5.2. Except for boron,

TABLE 5.1

Products of Reaction and Threshold Energy for
Different Nuclear Reactions between Incident Neutrons
(<20 MeV) and a Silicon Target

Products of Reaction	Threshold Energy (MeV)
$^{28}Si + n$	0
$^{28}Si^* + n$	1.78
$^{25}Mg + \alpha$	2.75
$^{28}Al + p$	4.00
$^{27}Al + d$	9.70
$^{24}Mg + n + \alpha$	10.34
$^{27}Al + n + p$	12.00
$^{26}Mg + ^3He$	12.58
$^{21}Ne + 2\alpha$	12.99
$^{27}Mg + 2p$	13.90
$^{24}Na + p + \alpha$	15.25
$^{26}Al + t$	16.74
$^{15}N + ^{14}N$	16.97
$^{12}C + ^{16}O + n$	17.35
$^{27}Si + 2n$	17.80
$^{26}Mg + p + d$	18.27
$^{12}C + \alpha + ^{13}C$	19.65
$^{20}Ne + n + 2\alpha$	20.00

Source: Data from Wrobel, F., Elaboration d'une base de données
de particules responsables des dysfonctionnements dans
les composants électroniques exposés à des flux de protons ou neutrons. Application au calcul de taux d'erreurs
dans les mémoires SRAM en environnement radiatif
[Database of particles induced failure in electronic components in proton and neutron environment. Application
to the calculation of soft-error rate in SRAM memories].
PhD Thesis, Université de Montpellier II, 2002.

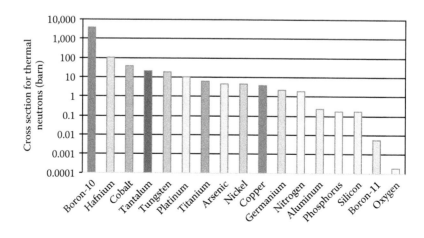

FIGURE 5.2
Cross section for thermal neutrons (2200 m s⁻¹) and several materials used in the fabrication of microelectronic
devices. (Data from *Neutron News*, 3(3), 29–37, 1992.)

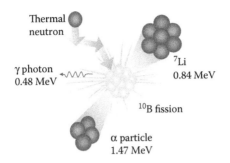

FIGURE 5.3
Schematic of ^{10}B–thermal neutron interaction. The products of the dominant reaction are an energetic alpha particle (1.47 MeV) and a lithium nucleus (0.84 MeV).

the value for each element has been averaged from the cross-section value of its isotopes weighted by their relative abundance. From this figure, it is clear that the cross section of ^{10}B may reach several thousands of barns and is $30–10^6$ times more able to capture a thermal neutron than other materials.

Boron is widely used in the fabrication of integrated circuits, particularly for substrate doping and in borophosphosilicate glasses (BPSG) used in the back end of line (BEOL). Boron has two isotopes: ^{11}B (80% in natural abundance) and ^{10}B (20% in natural abundance). ^{11}B is not a problem for the thermal neutron sensitivity of circuits because its cross section is extremely small and when ^{11}B captures a neutron it emits a photon, which cannot cause a single-event upset. In contrast, not only does ^{10}B exhibit a very high neutron-capture cross section but, upon capturing a neutron, the ^{10}B nucleus has a high probability of fissioning into two highly ionizing particles, each of which can cause a soft error (Fleischer 1983; JESD89A 2006; Oldham et al. 1986). As already introduced in Section 2.1.2, the dominant (93%) nuclear reaction is

$$n_{th} + {}^{10}B \rightarrow {}^{7}Li\left(0.84\text{ MeV}\right) + {}^{4}He\left(1.47\text{ MeV}\right) + \gamma\left(0.48\text{ MeV}\right) \tag{5.6}$$

This reaction creates energetic Li and He particles as well as gamma rays (Figure 5.3), which can lead to soft errors. A key study published in Baumann et al. (1995) showed the importance of thermal neutrons in the occurrence of failures in integrated circuits at ground level. In a typical BPSG layer of 600 nm, the emission rate of alpha particles and ^{7}Li was estimated to be 0.022 particles $cm^{-2}\,h^{-1}$ (Baumann et al. 1995). This rate is two orders of magnitude higher than the typical rate of cosmic rays and three orders of magnitude higher than the typical rate of alpha particles issued from radioactive impurities incorporated during the fabrication processes.

Most modern semiconductor processes have completely eliminated the presence of ^{10}B in BPSG or the use of BPSG, considered as the principal reservoir of ^{10}B and the dominant source of boron fission in circuits (Baumann and Smith 2001). However, ^{10}B remains present at silicon level, since bulk-substrate doping and source/drain implantation are not selective for isotope and continue to use natural boron. ^{10}B can also subsist within the BEOL structure, for example as a coating over tungsten plugs. The issue of thermal-neutron sensitivity of current technologies is still relevant and remains open, in particular for ultrascaled technologies in the natural terrestrial environment at ground level and also in the atmosphere. Recent work (Olmos et al. 2006; Wen et al. 2010) demonstrated substantial

SER sensitivity with low neutron energies for many static random-access memory (SRAM) circuits in the 0.25 μm–45 nm technology range. In Section 9.4, we will illustrate the capability of the Monte Carlo simulation code TIARA-G4 to simulate the impact of thermal and low-energy neutrons on the SER of 40 nm SRAMs. Comparison with accelerated tests using a thermal neutron reactor and also with real-time tests conducted at altitude will also be presented and discussed.

5.1.5 Atmospheric Neutron–Silicon Interaction Databases

In order to investigate the interaction of atmospheric thermal, epithermal, and low-energy neutrons with silicon, we performed in Autran et al. (2012) Monte Carlo simulations with Geant4, a radiation-transport simulation toolkit for the simulation of the passage of particles through matter (Agostinelli et al. 2003). In this study, we considered a 1 cm^2×20 μm target of undoped silicon and silicon doped with natural boron (i.e., composed of the two isotopes ^{10}B and ^{11}B with natural abundance: 19.9% and 80.1%, respectively) with atomic concentrations ranging from 10^{16} to 3×10^{20} cm^{-3}. Details of the simulation procedure can be found in Autran et al. (2012). Incident neutrons were generated perpendicularly to the target layer; their energy distribution follows the full atmospheric spectrum at ground level, shown in Figure 1.11. As defined in Section 1.3.1.2, this spectrum can be conveniently partitioned into three energy intervals: Part 1 (<1 eV), Part 2 (1 eV–1 MeV), and Part 3 (>1 MeV). These three regions correspond, respectively, to 17.4% (7.6 neutrons cm^{-2} h^{-1}), 36.7% (16 cm^{-2} h^{-1}), and 45.9% (20 cm^{-2} h^{-1}) of the total neutron flux at sea level (43.6 cm^{-2} h^{-1}). The upper limit chosen for delimiting *thermal* or, more exactly, *low-energy* neutrons is quite arbitrary and is fixed in this work at 1 eV.

Table 5.2 summarizes the main characteristics of the different nuclear-reaction databases computed in this work. The ensemble constitutes a global event database for 1.09×10^9 atmospheric neutrons, corresponding to 25×10^6 h of natural irradiation at sea level. Data in Table 5.2 show that the three regions of the atmospheric spectrum contribute very differently in terms of number of induced nuclear reactions and generated secondary products.

For natural (i.e., undoped) silicon, a total of 2,212 + 10,018 + 66,364 = 78,594 nuclear reactions (including all elastic and inelastic events for a total of 105,946 generated products, including silicon recoil nuclei and secondary ions) is generated in the bulk target. The contributions of the three domains of the atmospheric neutron spectrum to these results correspond, respectively, to 2.8% for Part 1, 12.7% for Part 2, and 84.4% for Part 3 in terms of number of nuclear reactions and 2.1% for Part 1, 9.5% for Part 2, and 88.5% for Part 3 in terms of generated products. This justifies why one can generally consider that, at first order, only high-energy neutrons (>1 MeV) induce soft errors in silicon-based circuits. Low-energy neutrons (<1 eV) are found to quasi-exclusively induce ^{28}Si(n,γ)^{29}Si radiative captures, and intermediate-energy neutrons (Part 2) induce elastic scattering reactions (Si recoil nuclei). For neutrons corresponding to the higher-energy domain (Part 3), we obtain 41% of inelastic events and 59% of elastic events. Detailed analysis of the database in terms of secondary-ion production, energy histogram of produced ions, nuclear-reaction-induced shower multiplicity, and ratio of elastic/inelastic processes will be conducted in Section 9.4, in combination with the analysis of additional databases corresponding to different monoenergetic and broad-spectrum artificial sources of neutrons.

Considering p-type boron-doped silicon target with natural boron does not significantly change the results except for the contribution of the thermal part (Part 1): a factor of 10 is observed in the number of nuclear reactions when the boron doping level is increased from 10^{16} (2,212 reactions) to 10^{20} atoms cm^{-3} (23,619 reactions). Such a range of values for

TABLE 5.2

Summary of Main Characteristics of the Computed Nuclear Reaction Databases with Geant4

Neutron Source	Target (1 cm² × 20 μm)	Number of Incoming Neutrons	Total Number of Nuclear Reactions in the Database[a]	Total Number of Generated Products[b]	Comments
Part 1 (<1 eV)	Natural Si	1.9×10^8 (17.4%)	2,212	2,212	93% of
	p-type Si [B] = 10^{16} cm⁻³		2,276	2,276	^{28}Si(n,γ)^{29}Si
	p-type Si [B] = 10^{20} cm⁻³		23,619	44,771	~90% of ^{10}B(n,α)^{7}Li
Part 2 (1 eV–1 MeV)	Natural Si	4×10^8 (36.7%)	10,018	10,018	98.5% of elastic reaction (Si recoil nuclei)
	p-type Si [B] = 10^{16} cm⁻³		10,072	10,072	
	p-type Si [B] = 10^{20} cm⁻³		11,401	12,292	~8% of ^{10}B(n,α)^{7}Li; ~91% of elastic events
Part 3 (>1 MeV)	Natural Si	5×10^8 (45.9%)	66,364	93,716	41% of inelastic events; 59% of elastic events
	p-type Si [B] = 10^{16} cm⁻³		66,770	95,068	
	p-type Si [B] = 10^{20} cm⁻³		66,216	94,470	
		Total: 1.09×10^9 (100%)			

Source: Data from Autran, J.L., et al., *IEEE Trans. Nucl. Sci.* 59, 2658–2665, 2012.

Note: These databases correspond to the irradiation of a target of silicon material (undoped or doped with natural boron) with neutron distributions defined in Figure 1.11.

[a] Elastic + inelastic reactions.

[b] All silicon recoil nuclei + secondary ions.

the boron concentration practically covers the entire range of real p-type doping level in most complementary metal-oxide semiconductor (CMOS) standard processes. Figure 5.4 details the analysis of the database for this thermal-energy domain. In contrast to the two other energy domains (Parts 2 and 3), only three types of events can occur at this thermal energy level: neutron captures by silicon or boron nuclei and ^{10}B fissions. These latter reactions produce ionizing secondary products that represent a real threat for the circuit in terms of SEEs. From the data of Figure 5.4, we calculated the normalized graph of Figure 5.5, which directly gives the number of ^{10}B fissions in 10^9 h induced by thermal neutrons of the atmospheric spectrum in a normalized layer of natural silicon of 1 mm² × 1 μm, as a function of the natural boron concentration in the layer. We obtained a linear graph in log–log scale which leads to $N = A \times [^{10}B]$, where N is the number of ^{10}B fissions per mm² × μm × 10^9 h, [^{10}B] is the concentration in ^{10}B in cm⁻³, and A is the linear coefficient evaluated to 4.2510×10^{-18} (for these units) from Geant4 simulation data of Figure 5.4.

5.2 Interactions of Charged Particles with Matter

When a charged particle enters matter, it will interact with the electrons and nuclei of the target material and begins to lose energy as it penetrates into the matter. The main forces

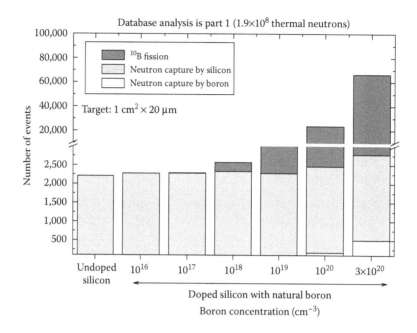

FIGURE 5.4
Analysis of the different nuclear-reaction databases defined in Table 5.3 (Part 1 of the atmospheric spectrum; see Figure 1.11). (Reprinted from Autran, J.L., et al. Soft-error rate induced by thermal and low energy neutrons in 40 nm SRAMs, *IEEE Trans. Nucl. Sci.* 59, 2658–2665. © (2012) IEEE. With permission.)

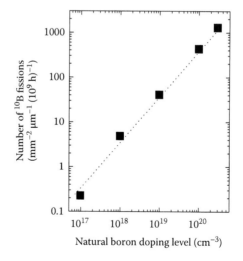

FIGURE 5.5
Normalized graph giving the number of ^{10}B fissions in 10^9 h induced by thermal neutrons of the atmospheric spectrum (Figure 1.11, Part 1) in a layer of natural silicon of 1 mm$^2\times$1 µm, as a function of the natural boron concentration in the layer (19.9% ^{10}B, 80.1% ^{11}B). (Reprinted from Autran, J.L., et al. Soft-error rate induced by thermal and low energy neutrons in 40 nm SRAMs, *IEEE Trans. Nucl. Sci.* 59, 2658–2665. © (2012) IEEE. With permission.)

involved are Coulombian, because nuclear forces are lower due to a relatively weak probability of interaction (the capture cross section is very small, around 1 barn). The interaction can be generally considered as a collision between the charged particle and the atomic electron or nucleus (considered separately). The released energy primarily results in an ionization phenomenon (production of electron–ion pairs in the matter). It can also emerge as electromagnetic radiation, a process known as bremsstrahlung (braking radiation), as explained later. Different mechanisms reflect the energy loss of the ionizing particle passing through the matter: inelastic electron collisions, elastic nuclear collisions, excitations, and nuclear interactions. We briefly describe below the most significant mechanisms for the study of SEE in integrated circuits.

1. Inelastic collisions between the incident particle and atomic electrons. This is the main process of energy transfer, since the probability of a direct collision with the atomic nucleus is very low in the initial phase of the charged particle's interaction with matter. The charged particle passes through a medium having a large number of electrons, so it undergoes a large number of interactions with electrons; during these interactions the particle gradually loses its energy. Collisions between the particle and atomic electrons lead to excitation of the atomic electrons (still bound to the nucleus) and to ionization (electron stripped off the nucleus). The ionization phenomenon is schematically described in Figure 5.6. The electron is ejected from the deep electronic layers with an energy equal to that transferred by the incident particle minus the binding energy.

2. Nuclear elastic collisions between the incident particle and a nucleus of the material. Such a collision results in a transfer of energy between the incident particle and the atomic nucleus and may cause the displacement of the latter and the diffusion of the incident particle. This phenomenon is found at the end of the particle's travel through the matter, when the particle has already lost a large amount of its energy.

It is important to note that, if the charged particle is a proton or any other charged hadron, it can also undergo a nuclear interaction (Tavernier 2010). A hadron is a compound of subatomic particles (quarks) held together by the strong force. There are two categories of hadrons: baryons (such as protons and neutrons, made of three quarks) and mesons (such as pions, made of one quark and one antiquark). In brief, at high energy almost all

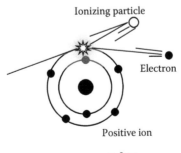

Ionizing particle

Electron

Positive ion

Ionization: $Z \rightarrow Z^+$

FIGURE 5.6
Illustration of the ionization mechanism occurring when an energetic charged particle passes through matter.

hadrons undergo a nuclear interaction after a distance approximately equal to a mean free path (called the hadronic interaction length), which is in the range 10–100 cm in solids (Tavernier 2010). For example, when a very high-energy proton enters a material, it will first lose its energy by ionization and its range will be longer than the hadronic interaction length. As explained in Tavernier (2010), the proton may undergo a nuclear interaction in which the target nucleus will be broken up. In addition, a number of additional hadrons are produced, 90% of the secondary particles being pions, with approximately equal numbers of π^+, π^-, and π^0 (Tavernier 2010). If the energy of the primary proton is large enough, these pions and other hadrons will also have sufficient energy to produce further nuclear interactions, and an avalanche of hadrons is produced. This type of interaction of charged hadrons is ignored in the present section, but the reader may find a detailed description in Tavernier (2010).

5.2.1 Ionization

In the initial phase of the passage of the charged particle in matter, collisions with atomic electrons are the principal mode of energy loss in a very wide range of energies of the incident particle (Figure 5.6). These interactions gradually slow down the particle. In the final phase, the particle's slowing and stopping are due to collisions with nuclei. The main mechanism that leads to energy loss and slowing down of the charged particle is then the ionization phenomenon. But the energy transferred during each collision is small compared with the initial energy of the incident particle, so its trajectory deviates only very slightly. The ionization induces the generation of a large number of excited electrons, which generally have sufficient energy to ionize other atoms. An electronic cascade is activated in which the number of free electrons continues to increase while their average energy decreases. During the passage of the ionizing particle, a highly ionized channel of very small diameter (typically 0.5 µm) develops around the track of the particle. Very rapidly, the excited electrons in the plasma will lose their kinetic energy in excess by a series of elastic collisions with electrons of the lattice to finally reach an energy close to the binding energy of the material. Simultaneously, the positively charged ionized atoms rearrange their electrons, resulting in creation of holes in the valence band. A high-density column of electron–hole pairs is then formed in a narrow region around the particle track.

It is important to note that the ionization phenomenon and the generation of electron–hole pairs previously described occur when a charged particle travels in matter. In an electronic device, it is necessary to take into account in addition other physical phenomena, such as the presence of an electric field in any type of structure containing p-n junction(s). The column of electron–hole pairs generated by the passage of the charged particle will therefore evolve in the electronic component (under the influence of the electric field or thermal agitation) following three mechanisms that contribute to reduction of the density of excess carriers: carrier recombination, ambipolar diffusion, and separation of carriers under the effect of the local electric field. Released charges are transported and collected at device electrodes, inducing a parasitic current transient, which may disturb the circuit. These mechanisms are at the origin of SEEs in microelectronic devices and circuits, addressed in detail in Section 5.6.

In order to characterize the interaction of a charged particle with matter, two key quantities are defined: the stopping power and the range, that is, the distance traveled by the particle in matter from the point of entry to the point where it is definitely stopped; these quantities are introduced in the following.

5.2.2 Stopping Power

The amount of energy lost by a particle in the matter per unit length is called stopping power. The stopping power is a quantity that characterizes the way a particle loses energy. The total stopping power is decomposed into two components:

1. The electronic stopping power, corresponding to the loss of energy of the particle due to collisions with atomic electrons of the target material
2. The nuclear stopping power, corresponding to the loss of energy of the particle due to collisions with the nuclei of atoms of the target material

The electronic stopping power thus characterizes the creation of electron–hole pairs by ionization of the target material, while nuclear stopping power describes the atomic displacement of the target material. The stopping power is usually expressed in megaelectronvolts per micrometer. It can be estimated from numerical simulation, for example using the reference and extremely popular Stopping and Range of Ions in Matter (SRIM) software (http://www.srim.org) (Ziegler et al. 2008; Ziegler et al. 2010). Figure 5.7 shows the example of the electronic and nuclear stopping powers of the ion [13]Al into silicon. This figure shows that the ionization phenomenon (creation of electron–hole pairs) is dominant for ion energies typically beyond the megaelectronvolt range. The nuclear contribution is important only at low energies.

The variation of electronic stopping power can be explained using three theories (Wrobel 2002). At low speed, the stopping power is calculated by the method of Lindhard and Scharff (1961). In their formalism, the expression of stopping power depends linearly on the speed, but in practice the experimental results are fitted by a power law of energy. The dielectric theory is used to describe the region of intermediate energy (Wrobel 2002), wherein the maximum of the electronic stopping power, called the Bragg peak, is located. This theory assumes that the passage of the particle locally modifies the dielectric constant of the material. This results in a change of the electric field (polarization), which is

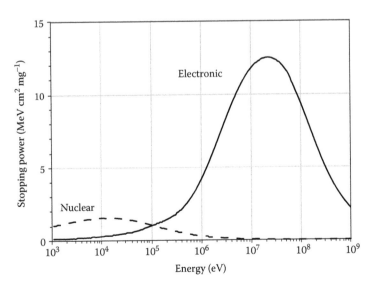

FIGURE 5.7
Electronic and nuclear stopping powers of the ion [13]Al into silicon calculated with SRIM software (http://www.srim.org).

opposed to the particle's movement and therefore causes its slowdown. The basic equation of this theory is the Poisson equation. Finally, at higher energies (beyond the Bragg peak), Bethe's quantum theory is used to calculate the electronic slowdown. In the nonrelativistic case, the Bethe–Bloch formula gives

$$\left|\frac{dE}{dx}\right| \propto \frac{z^2 \times Z}{v^2} \tag{5.7}$$

where:
   z = atomic number of the incident charged particle
   Z = atomic number of the target material
   v = speed of the particle

Concerning the issue of SEEs in electronic circuits, the creation of electron–hole pairs is the essential phenomenon that leads to electrical instabilities in these circuits. So, the contribution of nuclear stopping power, which is considered to be low, is often neglected, and only the electronic stopping power is taken into account.

The electronic stopping power is also called linear energy transfer (LET). The LET corresponds to the energy deposited by an ionizing particle along its track when it passes through the device. This deposition of ionizing energy per unit length corresponds to the creation of electron–hole pairs in the device. The LET is different from the total stopping power, but it is often mistakenly considered that the two stopping powers, that is, the LET and the total stopping power, are identical.

The LET is given by the following equation:

$$LET = -\frac{\Delta E}{\Delta x} \tag{5.8}$$

It represents the energy loss (ΔE) per unit length (Δx). In Equation 5.8, the LET is expressed in megaelectronvolts per micrometer. In studies of radiation effects on electronic components, the usual unit is picocoulombs per micrometer; this allows the charge generated by the ionization mechanism when the particle passes through the device to be directly defined. A weighted LET is also used, defined as the ratio between the LET and the density ρ of the target material:

$$LET = -\frac{1}{\rho}\frac{\Delta E}{\Delta x} \tag{5.9}$$

The unit of this weighted LET is megaelectronvolts square centimeter per milligram. To obtain equivalences between these three units, the target material must be known. We consider here the case of silicon, the most commonly used material in integrated circuits. A first equivalence between the energy deposited per unit length in megaelectronvolts per micrometer and LET expressed in megaelectronvolts square centimeter per milligram can be obtained using the silicon density (2.32 g cm^{-3}), as follows:

$$1\,\text{MeV}\,\mu\text{m}^{-1} \leftrightarrow \frac{1\,\text{MeV}}{10^{-4}\text{cm} \times 2.32 \times 10^{3}\text{mg cm}^{-3}} = 4.31\,\text{MeV cm}^2\,\text{mg}^{-1} \tag{5.10}$$

The amount of charge generated per unit length is also an interesting quantity. It is calculated as follows, taking into account that an average energy of 3.6 eV is needed to create an electron–hole pair in silicon:

$$1\,\text{MeV}\,\mu\text{m}^{-1} \leftrightarrow \frac{1.6 \times 10^{-7}\,\text{pC} \times 10^6\,\text{eV}}{3.6\,\text{eV} \times 1\,\mu\text{m}} = 0.0446\,\text{pC}\,\mu\text{m}^{-1} \tag{5.11}$$

$$1\,\text{MeV}\,\text{cm}^2\,\text{mg}^{-1} \approx 10\,\text{fC}\,\mu\text{m}^{-1} \tag{5.12}$$

The LET depends on the nature of the ion, its energy, and the target material. The SRIM software (Ziegler et al. 2008, 2010) provides the value of LET for each charged particle, depending on its energy and the target material. An example is shown in Figure 5.8 for several ions that may be generated during the interaction between a neutron and silicon. It should be noted that *hydrogen ion* corresponds to a proton and *helium ion* corresponds to an alpha particle. Phosphorus ion is the potential product of the n–Si interaction that has the highest LET (this reaction is still very rare; it corresponds to ^{30}Si capturing the incident neutron and decaying by beta emission). This figure shows that the higher the charge on the ion, the higher are the stopping power and the Bragg peak. The LET variation is the same for all particles. Thus, the LET increases with the energy of the particle until it reaches its maximum at the Bragg

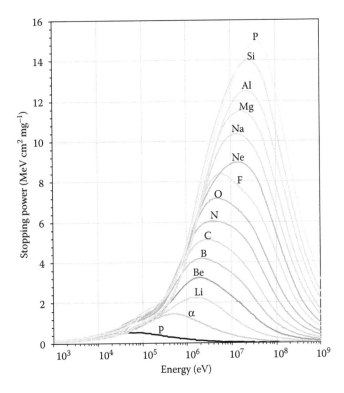

FIGURE 5.8
Variation of LET as a function of energy for several ions that may result from neutron interactions with silicon.

peak. The LET then decreases as the energy of the particle increases beyond the Bragg peak. For particles considered in this example, the LET at the Bragg peak may vary from about 0.6 MeV cm² mg⁻¹ for the hydrogen ion (proton) to about 15 MeV cm² mg⁻¹ for the phosphorus ion.

5.2.3 Range

Another important quantity that characterizes the interaction of a charged particle with matter is the range of the particle. The range is the distance traveled by the ionizing particle in matter before its final stopping; it is calculated from the stopping power as

$$\text{Range} = \int_E \frac{1}{\frac{\Delta E}{\Delta x}} \, dE \tag{5.13}$$

The SRIM software (Ziegler et al. 2008; Ziegler et al. 2010) allows the variation of the range to be obtained as a function of the initial energy. Figure 5.9 shows these variations for the particles considered above (particles potentially issued from n–Si interactions). This figure shows that, for a given energy, the lighter the particle, the longer its range. For example, for an energy of 1 GeV, the range varies from about 700 µm for the phosphorus

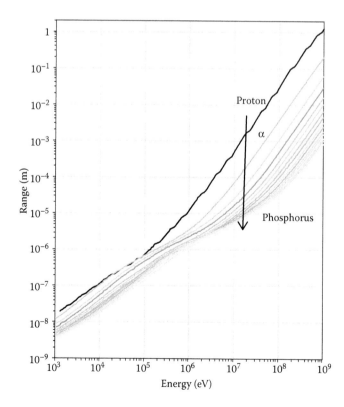

FIGURE 5.9
Range as a function of initial energy for ions considered in Figure 5.8.

ion to almost 2 m for the hydrogen ion (proton). To summarize, light particles have a low LET and a long range, while heavy particles have a higher LET and a shorter range.

In the domain of SEEs on electronic devices, it is often interesting to plot the LET variation as a function of particle penetration (depth) in matter. This representation allows better characterization of the ionizing behavior of a given particle and how dangerous it is with respect to the circuit of interest. An example is shown in Figure 5.10: the LET variation is plotted as a function of particle depth in silicon for both He (alpha) and silicon ions. In the case of the silicon ion, which is a heavy particle, the LET is high

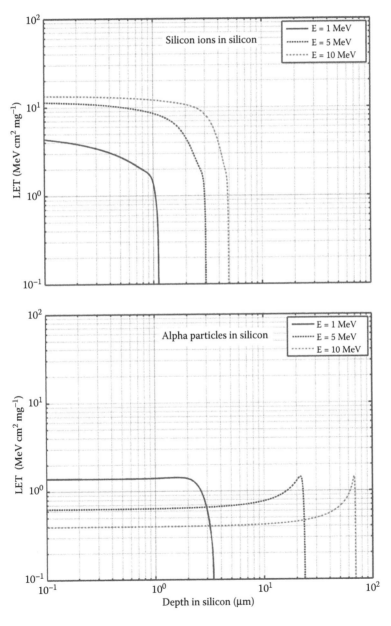

FIGURE 5.10
Energy loss of silicon ions and alpha particles of different incident energies in silicon.

but its range in matter may not exceed a few microns. In contrast, for the lighter alpha particle, the energy deposition per unit length will be lower, but its range in silicon may be of the order of a few tens of microns. This figure shows that the maximum LET (Bragg peak) of an alpha particle may appear at tens or hundreds of microns from its generation point. A light particle, such as an alpha particle, may therefore induce SEE in a given circuit even if it has been generated far from the circuit's sensitive node(s). A heavy ionizing particle will travel a distance of only a few microns into the matter (so it will be active close to its generation point), but it has a strong ability to locally ionize the target material.

5.2.4 Alpha Particles

As already mentioned, an alpha particle corresponds to the nucleus of an atom of helium (He). With its two protons and two neutrons, it is a very stable configuration of particles. In microelectronic devices, alpha particles can be emitted through alpha decay by radioactive isotopes (from the uranium and thorium series) present in the form of (ultra)traces in common materials used in device fabrication or by radioactive isotopes of chemical elements recently introduced into ultraminiaturized CMOS technologies (e.g., hafnium, platinum). These isotopes, called alpha emitters, have been described in Chapter 3.

As also explained in Chapter 3, alpha particles emitted by radioactive isotopes of the uranium and thorium series have initial energies between 4 and 9 MeV. For this energy range, the nuclear stopping power of alpha particles in silicon is far below the electronic stopping power, as shown in Figure 5.11. Alpha-particle energy loss occurs to a large extent by interaction with atomic electrons, except at the end of the particle's travel in matter.

Figure 5.12 shows the variation of the alpha particle's LET as a function of its energy; curves for electrons and protons are also plotted for comparison. For the energy range between 1 and 10 MeV (corresponding to energies of alpha particles emitted by different disintegration series), the alpha particle's LET is higher than that of electrons and protons. The alpha-particle range in silicon as a function of its initial energy is plotted

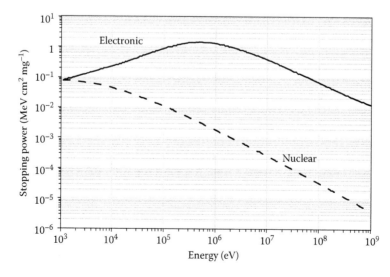

FIGURE 5.11
Electronic and nuclear stopping powers of an alpha particle in silicon as a function of its energy.

in Figure 5.13. This figure shows that an alpha particle with an energy of 10 MeV travels a distance of about 70 μm in silicon. The trajectory of alpha particles is approximately straight (Tavernier 2010). Figure 5.14 shows that the alpha particle's energy loss is not uniform during its travel in matter. The maximum LET at the Bragg peak is 1.5 MeV $(\text{mg cm}^{-2})^{-1}$.

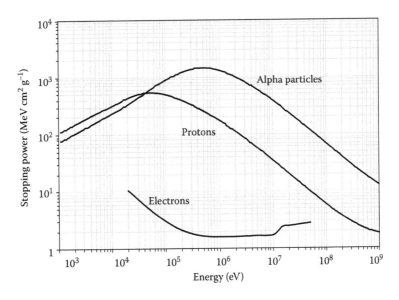

FIGURE 5.12
Stopping powers for electrons, protons, and alpha particles in silicon as a function of energy.

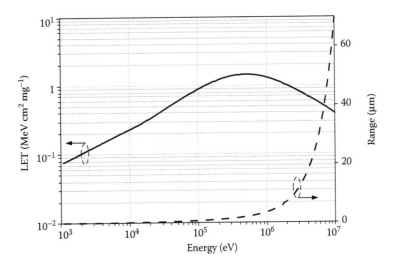

FIGURE 5.13
LET and range of an alpha particle in silicon.

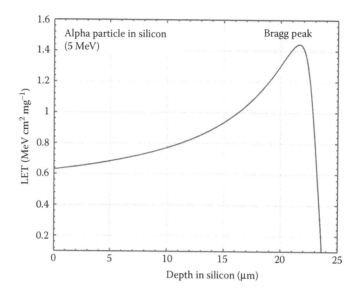

FIGURE 5.14
Energy loss of an alpha particle of 5 MeV in silicon.

5.2.5 Heavy Ions

Heavy ions correspond to a category of particles characterized by one or more units of electric charge and a mass exceeding that of the helium-4 nucleus (alpha particle). Heavy ions (or nuclear fragments) produced by nuclear reactions are energetic charged particles, which interact with matter by three processes, depending on their energy:

1. Electron capture at low energies
2. Collisions with atomic electrons at medium and high energies
3. Nuclear collisions and nuclear reactions at very high energies

Heavy ions produced by nuclear reactions interact with matter as charged particles, but they behave somewhat differently from alpha particles. As explained in Tavernier (2010), heavy ions tend to pick up electrons as they travel in the matter. As they slow down, the fragments pick up more and more electrons, and the energy loss decreases rather than increases. For alpha particles, this electron pickup only occurs at the very end of the range.

Heavy ions show extremely high ionization, and therefore their range is typically only a few microns long (Tavernier 2010). The LET and range for the different ions which may be produced by n–Si interactions are given in Figures 5.8 and 5.9, respectively.

5.2.6 Electrons

The passage of energetic electrons through matter is similar to that of heavy charged particles in that the Coulomb interaction plays a dominant role. Electrons, especially at low energies, lose energy by exciting and ionizing atoms along their trajectory. Losses of energy through these mechanisms are called collisional losses. In contrast to heavy particles, electrons lose a large fraction of their energy in a single collision with an electron. At high energies, electrons predominantly lose a significant fraction of their energy by bremsstrahlung (Tavernier 2010). Bremsstrahlung is a general process in which electromagnetic radiation is emitted whenever a charged particle undergoes a substantial acceleration. The acceleration of a heavy charged particle, by the electromagnetic force of a nucleus in an atomic collision, is low. This is different for the electron, which, having a small mass, can be accelerated strongly by the same electromagnetic force within the atom, and hence it can emit bremsstrahlung radiation. While ionization loss rates rise logarithmically with energy, bremsstrahlung losses rise nearly linearly (Nakamura et al. 2010).

The stopping power of an electron in matter has two components:

1. A collision stopping power, $(dE/dx)_c$, relative to the mechanism of energy loss by ionization
2. A radiative stopping power, $(dE/dx)_r$, relative to the energy loss by bremsstrahlung radiation

The total stopping power is then the sum of the collision stopping power and the radiative stopping power. Figure 5.12 illustrates the variations of this stopping power for electrons as a function of their energy; curves for alpha particles and protons are also plotted for comparison.

As explained in Tavernier (2010), electrons typically travel several centimeters in matter before losing all their energy, but the distance traveled according to a straight line is usually much shorter than the actual length of the trajectory. This is due to the trajectories of electrons, which are erratically twisted due to multiple scattering. Electrons do not have a well-defined range. A quantity called the *continuous slowing down approximation* (CSDA) range is usually defined for electrons:

$$R_{CSDA} = \int_0^{E_0} \left(\frac{dE}{\rho dx} \right)^{-1} dE \tag{5.14}$$

where E_0 is the initial energy of the electron. The unit of R_{CSDA} is grams per square centimeter.

5.3 Interaction of Protons with Matter

The proton is a subatomic particle with a positive elementary electric charge. Protons are present in the nucleus of each atom (the nucleus of hydrogen, H^+, is a single proton), but they are also stable outside atomic nuclei (the spontaneous decay of free protons has never been observed). The proton is not a fundamental particle; like the neutron, it is composed of three quarks. Protons and neutrons are both nucleons, which may be bound together by the nuclear force to form atomic nuclei. Free protons of high energy and velocity make up 90% of cosmic rays, which propagate in vacuum for interstellar distances. Free protons are emitted directly from atomic nuclei in some rare types of radioactive decay. Protons also result (along with electrons and antineutrinos) from the radioactive decay of free neutrons, which are unstable, as explained in Section 5.1.

The interaction of protons with matter depends on proton energy. Low-energy protons (<1–10 MeV) behave as charged particles and interact with matter through Coulomb interaction, with either atomic nuclei or atomic electrons orbiting nuclei. Interactions with electrons and nuclei present in matter will give rise to very different effects. Interactions with nuclei are elastic scatterings: the proton will transfer some of its energy to the nucleus and its direction will be changed. Since the proton is much lighter than most nuclei, the proton will lose little energy, but its direction can be changed completely (Tavernier 2010). Proton interactions with atomic electrons are generally inelastic: the proton loses a large amount of energy by ionizing atoms in the matter (creation of electron–hole pairs), but the direction of the proton can only be slightly changed. As a result, most of the energy loss of the proton is due to collisions with electrons, and most of the change of direction is due to collisions with nuclei (Tavernier 2010). In the ionization mechanism, similarly to other charged particles, the maximum energy loss occurs at the end of the proton range in matter (at the Bragg peak); this property has applications in proton therapy. Compared with other charged particles, protons ionize matter much more than electrons but less than alpha particles, as confirmed by the LET curves in silicon plotted in Figure 5.12.

High-energy protons lose their energy by nuclear interactions, like neutrons. The characteristics of high-energy proton–nucleus and neutron–nucleus reactions are very similar, typically above 50–100 MeV. At low energies, because of the Coulomb interaction, the proton and neutron cannot be treated as the same particle. But, at high energies, this interaction becomes almost negligible, and the nuclear reactions are practically the same for neutrons and protons (Tang 1996). In the particular case of light elements (such as Si and O) commonly used in semiconductor materials, the nuclear interactions of protons and neutrons are very similar for incident-particle energies above ~20 MeV (JESD89A 2006). As mentioned in Tang (1996), the proton–neutron similarity is frequently exploited in SER-related experimental works: high-energy-proton (accelerator) experiments are performed to investigate the high-energy-neutron response of device and circuits.

Finally, as explained in Tavernier (2010), protons travel a long distance through matter before they undergo a significant interaction through either nuclear interaction with nuclei, for high-energy protons, or Coulombian forces, for low-energy protons. The proton range in solids is of the order of 1 mm, and its trajectory is approximately straight (Tavernier 2010).

5.4 Interaction of Pions with Matter

The pion (denoted π) is a subatomic particle of the meson particle family, which consists of a quark and an antiquark. Three categories of pions exist: π^0, π^+, and π^-. Pions are the lightest mesons, and they play an important role in explaining the low-energy properties of the strong nuclear force; in particular, the pion can be thought of as one of the particles that mediate the interaction between a pair of nucleons (protons and neutrons). Charged pions, π^+ and π^-, are unstable particles with a mass of 139.6 MeV c^{-2} and a mean lifetime of 26 ns. They decay due to the weak interaction. The neutral pion π^0 has a slightly smaller mass of 135 MeV c^{-2} and a much shorter mean lifetime of 8.4×10^{-17} s. This pion decays in an electromagnetic force process.

The charged pions may lose their energy in the same way as ions by electromagnetic interaction, but, since pions are hadrons, they may have strong interaction, like neutrons.

As shown in Figure 1.9, the flux of pions at sea level is much lower than that of neutrons, and they are generally neglected. However, as explained in Ziegler and Puchner (2004), pion capture is a particular interaction that has the capability to disturb microelectronic circuits. This phenomenon has been described in Ziegler and Puchner (2004) and was confirmed experimentally by Dicello et al. (1983, 1985). In brief, pion capture occurs at the end of its travel in matter, when it has lost almost all its energy. The negatively charged pion attaches itself to a nucleus and begins to orbit like an electron, with the difference that the pion is not a fermion and does not obey the Pauli exclusion principle. Thus, the pion is not excluded from the electron shells. It quickly cascades to tighter orbits, emitting x-rays, until a 1-s orbit, where it couples with a nucleon and all its almost 140 MeV c^{-2} of mass is transformed into nucleonic energy (Ziegler and Puchner 2004). Nuclear fission follows, and a considerable charge (estimated at 22 nC in Ziegler and Puchner [2004]) may be released, which can induce large SEEs in electronics. The pion-capture phenomenon occurs at a rate of about 8.5 cm^{-3} $year^{-1}$ in terrestrial materials, as estimated from geology studies (Ziegler and Puchner 2004). This rate multiplied by the active volume of an integrated circuit gives a rough projection of the number of events per year at chip level.

5.5 Interaction of Muons with Matter

The muon is an elementary particle similar to the electron, with a unitary negative electric charge and a mass about 200 times the mass of an electron. The muon, denoted by μ^- and also called the *negative muon*, has a corresponding antiparticle of opposite charge and equal mass: the antimuon, often called the *positive muon* (μ^+). Both negative and positive muons are unstable particles with a mean lifetime of 2.2 μs. Independently of any interaction with matter, they spontaneously decay into three particles:

$$\mu^- \to e^- + \bar{\upsilon}_e + \upsilon_\mu$$

$$\mu^+ \to e^+ + \upsilon_e + \bar{\upsilon}_\mu$$

(5.15)

These reaction products deposit very little energy through electronic stopping (electrons, positrons) and rarely interact with matter (neutrinos); their effect on electronics can thus be neglected.

As shown in Section 1.3.1.1, atmospheric muons represent an important part of the natural radiation constraint at ground level. Muons belong to the meson or *hard* component of atmospheric cosmic-ray cascades and are the products of the decay of charged pions (charged mesons π^+ and π^-) via the weak interaction. In spite of their short lifetime, since they are relativistic, these particles are easily able to penetrate the atmosphere; they constitute the most preponderant charged particles at sea level. Muons do not interact with matter via the strong force but only through the weak and electromagnetic forces; they can travel large distances in matter, thus deeply penetrating into material circuits.

Ziegler and Lanford (1979) have precisely described how muons can interact with matter at relatively low incident primary energies. They decompose the interaction into three primary processes:

1. Direct ionization wake. A charged muon loses its kinetic energy passing through semiconductor material by excitation of bound electrons and frees electron–hole pairs along its path as a result.

2. Electromagnetic scattering, which induces energetic coulomb silicon nucleus recoil.

3. Capture of the negative muons by atomic nuclei when they are quasi-stopped in matter. This complex capture mechanism releases recoiling heavy nuclei with a simultaneous emission of light particles (neutrons, protons, deuterons, α-particles, etc.).

Similarly to protons, direct ionization is significant for low-energy muons. Calculations for energy loss through electronic stopping indicate that the effect is primarily dominated by the charge and velocity of the particle. Although muons and protons have different masses, a muon with the same velocity as a proton will therefore cause the same ionization (Sierawski et al. 2010). Figure 5.15 shows the LET for a muon and a proton in silicon. Whereas the Bragg peak for a proton in silicon is roughly 55 keV, it is approximately 8 keV for a muon (Sierawski et al. 2010).

Also, at low energies, the mechanism of negative-muon capture can be at the origin of the production of large particle showers in matter. In general, since a negative muon rapidly slows down in stopping target material, it can be captured in an outer atomic orbit. Fast electromagnetic cascades ensue, bringing the muon down to the innermost 1s Bohr-level orbit. Afterwards, there is a delay until the muon disappears, either by decay or by nuclear capture as the muon comes closer to the nucleus (Mukhopadhyay 1977; Singer 1974). More precisely:

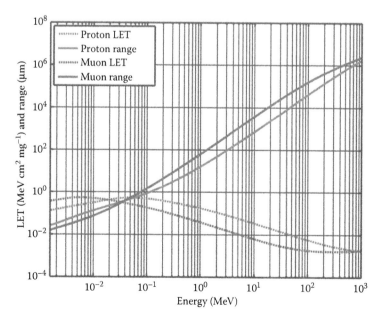

FIGURE 5.15
LET and range of protons and muons in silicon.

- Since the energy of the incoming muon is higher than 2 keV, the muon's velocity is greater than the velocity of the valence electrons (Bohr velocity), and the muon crossing the target material knocks out electrons.

- In the low-energy phase (lower than 2 keV) to rest, the muon's velocity is less than that of the valence electrons: the muon can exchange energy with the degenerate electron gas in arbitrarily small steps, and rapidly (in about $\sim 10^{-13}$ s) come to a stop.

- Once the muon reaches a state of no kinetic energy, it is captured by the host atom into high-orbital-momentum states, forming a *muonic atom*. Since all low-lying muonic states are unoccupied, the muon electromagnetically cascades down rapidly to the lowest quantum state (1s) available. Fermi and Teller showed that the time taken by a muon trapped in an atom to cascade down to the lowest Bohr orbit (1s) is negligible compared with its natural lifetime (Fermi and Teller 1947).

- After the muon has reached the 1s orbit, it either decays or is captured by the nucleus via the weak interaction. Around $Z = 11$, the capture probability is approximately equal to the decay probability. In heavy nuclei ($Z \sim 50$), the ratio of capture to decay probabilities is about 25. In the highly interesting case of silicon for microelectronics ($Z = 14$), the ratio of capture to decay probabilities has been evaluated at between 1.72 and 1.93 (Suzuki et al. 1987).

Historically, the study of nuclear muon capture can be said to have truly begun with the paper by Tiomno and Wheeler (1949). As a result of the weak interaction, the following nuclear reaction occurs:

$$\mu^- + A_Z^N \rightarrow \nu_\mu + X \tag{5.16}$$

where the detectable product X consists of a residual heavy nucleus and light particles. The mean excitation energy in nuclear muon capture is around 15 to 20 MeV. This is well above the nucleon-emission threshold in all complex nuclei. Thus, the daughter nucleus (A^*, $Z-1$) can de-excite by emitting one or more neutrons, or charged light particles, as well as via the electromagnetic mode (Singer 1974). In intermediate and heavy nuclei, the light particles are neutrons γ rays, or both in most cases. The few percent of charged light particles observed are mainly protons, but deuterons and α-particles have also been observed in still smaller quantities. Charged-particle emission is more probable for light nuclei than for heavy nuclei.

Among all the possible channels of nuclear muon-capture reactions, neutron emission is preferred. This emission of neutrons can be approximately classified as either direct or from an intermediate *compound nucleus* formed after the muon-capture process. Direct emission refers to the neutron created in the elementary process:

$$\mu^- + p \rightarrow n + \nu_\mu \tag{5.17}$$

which succeeds in leaking out of the nucleus. These neutrons have fairly high energies, from a few megaelectronvolts to as high as 40–50 MeV (Sundelin and Edelstein 1973). Most of the neutrons emitted after capture seem, however, to be *evaporation neutrons*. In intermediate and heavy nuclei, the excitation energy acquired by the neutron formed in the capture process is shared with the other nucleons of the nucleus, and a *compound nucleus* is formed. The intermediate excited nuclear state then loses energy by boiling off mainly low-energy neutrons (with γ-rays) until a ground state is reached (Singer 1974).

The following physical picture involving a two-step process can be used: the muon is captured by a quasi-free nucleon, whose acquired energy is distributed among the nucleons of the nucleus, and a compound nucleus is thus formed:

$$\mu^- + A_Z^N \rightarrow \nu_\mu + (A_{Z-1}^{N+1})^* \tag{5.18}$$

The excited nuclear state then loses energy by evaporating nuclear particles (mainly neutrons) and γ-rays till a ground state is reached.

The particular case of negative-muon capture in silicon is important for its implications in regard to the production of SEEs in circuits. When negative muons stop in silicon, about 35%, on average, decay into an electron and two neutrinos. The remaining 65% are captured (Suzuki et al. 1987). If an intermediate state is assumed, the reaction is

$$\mu^- + {}^{28}\text{Si} \rightarrow {}^{28}\text{Al}^* + \nu_\mu + 100.5\,\text{MeV} \tag{5.19}$$

Sobottka and Wills (1968) have measured the energy spectrum for charged-particle emission resulting after muon capture in ^{28}Si following some modes of de-excitation of the ^{28}Al recoiling nucleus:

$$^{28}\text{Al}* \rightarrow {}^{27}\text{Al} + \text{n} \ (12.4 \text{ MeV})$$

$$\rightarrow {}^{27}\text{Mg} + \text{p} \ (14.2 \text{ MeV})$$

$$\rightarrow {}^{24}\text{Na} + \alpha \ (15.5 \text{ MeV})$$

$$\rightarrow {}^{26}\text{Mg} + \text{d} \ (18.4 \text{ MeV}) \tag{5.20}$$

where the energy listed with each final state is the ground-state energy with respect to the ^{28}Si ground state.

According to the compilation of several works (Budyashov et al. 1971; Macdonald et al. 1965; Sobottka and Wills 1968; Sundelin and Edelstein 1973; Sundelin et al. 1968; Wyttenbach et al. 1978), among all the muons that are captured, 28% result in no particle emission; 15% result in charged-particle emission (~10% protons, 5% deuterons, and < 1% tritons or α-particles—in some cases, there may be several percent of α-particles); and 67% result in neutron emission, with 10% emission of both charged particles and neutrons.

Finally, Measday (2001) has suggested a global pattern of muon-capture reactions for ^{28}Si, expressed as a percentage of all captures. These results are reported in Table 5.3 and give an exhaustive view of negative-muon capture in silicon. Such capture is thus able to produce secondary heavy nuclei that can deposit important charge in silicon; in Chapter 10,

TABLE 5.3

Global Patterns for Negative-Muon
Capture Reactions in ^{28}Si Target Material

Reaction	%	Reaction	%
$(\mu^-,\nu)^{28}$Al	26	$(\mu^-,\nu p)^{27}$Mg	2.0
$(\mu^-,\nu n)^{27}$Al	45	$(\mu^-,\nu pn)^{26}$Mg	4.9
$(\mu^-,\nu 2n)^{26}$Al	12	$(\mu^-,\nu p2n)^{25}$Mg	1.4
$(\mu^-,\nu 3n)^{25}$Al	1	$(\mu^-,\nu p3n)^{24}$Mg	0.6
		$(\mu^-,\nu p4n)^{23}$Mg	0.1
$\Sigma(n)$	**84**	$\Sigma(p)$	**9.0**
$(\mu^-,\nu d)^{26}$Mg	3.1	$(\mu^-,\nu 2pn)^{25}$Na	0.1
$(\mu^-,\nu dn)^{25}$Mg	1.0	$(\mu^-,\nu \alpha)^{24}$Na	1.0
$(\mu^-,\nu d2n)^{24}$Mg	0.3	$(\mu^-,\nu \alpha n)^{23}$Na	0.8
		$(\mu^-,\nu \alpha 2n)^{22}$Na	0.5
		$(\mu^-,\nu \alpha 3n)^{21}$Na	0.2
$\Sigma(n)$	**4.4**	$\Sigma(p)$	**2.6**

Source: After Measday, D.F., *Physics Reports*
354, 243–409, 2001.

we will illustrate by simulation the importance of this capture mechanism on the soft-error occurrence in a static memory circuit.

5.6 Basic Mechanisms of Single-Event Effects on Microelectronic Devices

The physical mechanisms related to the production of SEEs in microelectronic devices schematically consist of three main successive steps:

1. Charge deposition by the energetic particle striking the sensitive region in the active region of the circuit
2. Transport of the released charge into the device
3. Charge collection in the sensitive region of the device

Figure 5.16 illustrates these successive steps in the case of the passage of a high-energy ion through a reverse-biased n$^+$/p junction. In the following, we succinctly describe these different mechanisms; for a detailed presentation we invite the reader to consult Baumann (2005a), Dodd (1996, 2005), and Dodd and Massengill (2003).

5.6.1 Charge Deposition (or Generation)

As largely explained in previous sections, when an energetic particle strikes a semiconductor device, an electrical charge along the particle track can be deposited by one of the following mechanisms: direct ionization by the interaction of the incoming particle with the material or indirect ionization by secondary ionizing particles issued from nuclear reactions with the atoms of the struck material.

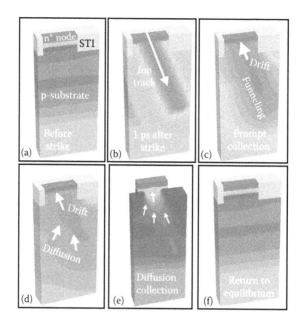

FIGURE 5.16
(a–f) Charge generation, transport, and collection phases in a reverse-biased junction caused by the passage of an energetic ionizing particle. (TCAD screenshots courtesy of P. Roche.)

Direct ionization by alpha particles, atmospheric charged particles (muons, protons), and heavy ions ($Z \geq 2$) (notably in the space environment) is particularly important. They interact with the target material mainly by inelastic interactions and transmit a large amount of energy to the electrons of the struck atoms. These electrons produce a cascade of secondary electrons, which thermalize and create electron–hole pairs along the particle path (Figure 5.16b). In a semiconductor or insulator, a large amount of the deposited energy is thus converted into electron–hole pairs, the remaining energy being converted into heat and a very small quantity into atomic displacements.

Other particles, such as the neutrons of the terrestrial environment, do not interact directly with the atomic electronics of the target material and so do not ionize the matter on their passage, as explained in Section 5.1. However, due to their probability of nuclear reaction with the atoms of materials that compose the microelectronic devices, they can produce SEEs by indirect ionization; the charged products resulting from a nuclear reaction can deposit energy along their tracks, in the same manner as direct ionization (the creation of the column of electron–hole pairs of these secondary particles is similar to that of ions).

5.6.2 Charge Transport

As explained in Section 5.2.1, the high-density column of electron–hole pairs generated by either direct or indirect ionization is subjected in the considered device to the influence of the electric field, in addition to other phenomena that reduce excess carrier density. Three basic mechanisms govern the evolution of the generated column of electron–hole pairs:

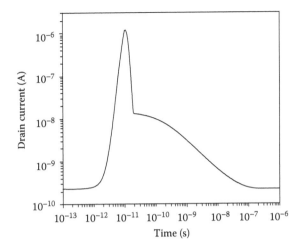

FIGURE 5.17
Transient current pulse extracted from a reverse-biased junction and caused by the passage of an energetic ionizing particle.

1. Carrier recombination (Shockley–Read–Hall and Auger recombination)
2. Ambipolar diffusion (electrons and holes diffuse together, thus preserving electrical neutrality)
3. Separation of carriers under the effect of local electric field

The released carriers are quickly transported in the device by two main mechanisms (Figure 5.16c–e): charge drift in regions where an electric field exists and charge diffusion. They are next collected by elementary biased structures (e.g., p-n junctions), as explained in Section 5.6.3.

5.6.3 Charge Collection

The charges transported in the device induce a parasitic current transient (Figure 5.17), which may induce disturbances in the device and associated circuits. The devices most sensitive to ionizing-particle strikes are generally devices containing reverse-biased p-n junctions, because the strong electric field existing in the depletion region of the p-n junction allows very efficient collection of the deposited charge. The effects of ionizing radiation are different according to the intensity of the current transient, as well as the number of circuit nodes impacted. If the current is sufficiently important, it can induce permanent damage to gate insulators (single-event gate rupture [SEGR]) or single-event latchup (SEL) of the device. In the usual low-power circuits, the transient current generally induces only an eventual change of the logical state (cell upset).

5.7 SEU Mechanisms in Memories (Single-Bit Upset and Multiple-Cell Upset)

Dynamic random-access memories (DRAMs), but also SRAM cells and SRAM-based programmable logic devices, are subject to single-event-upset (SEU) mechanisms. Unlike capacitor-based DRAMs, SRAMs are constructed of cross-coupled devices, in which the capacity of each cell is significantly less. The probability of occurrence of an upset is greater when the capacity of a cell is lower. Given that supply voltage and cell size are reduced with each technological generation, the capacity of SRAM cells continues to decrease, making the cell more vulnerable to more types of particles (i.e., particles with lower LET). An ionizing particle that strikes the sensitive region of a memory cell deposits a dense track of electron–hole pairs. If the collected charge at a particular sensitive circuit node exceeds the minimum charge that is needed to flip the value stored in the cell, a soft error occurs. An error due to a hit of a single particle is called an SEU.

Figure 5.18 illustrates the occurrence of an SEU in a standard single-port SRAM cell composed of two CMOS inverters and two access transistors connecting the storage nodes to the bitlines. When the wordline (WL) is low (access transistors in the off state), the cell is holding its stored data using the back-to-back inverter configuration. If the particle strike causes a transient on one of the nodes, the disturbance can propagate forward through the CMOS inverter and induce a transient in the second node. The second node, in its turn, leads the first node toward a wrong value, and consequently the two nodes will flip. Then, the memory cell will reverse its state and will store a false value (Karnik et al. 2004); there is no mechanism to restore its state other than explicitly rewriting the state via the two complementary bitlines. In this sense, the SEU is a reversible phenomenon that does not lead to the destruction of the cell.

Note that SEUs can also occur when the particle strikes the bitline (Karnik et al. 2004; Rajeevakumar et al. 1988). During the read operation, a bitline is discharged through a small current from a memory cell. The bit of information is read as "0" or "1" based on the voltage differential developed on the bitline during the access period of the memory cell. This voltage differential can be easily disturbed if a particle strikes close to a diode of an access transistor of any cell on this bitline.

The minimum amount of collected charge that results in a soft error is called the critical charge (Q_{crit}) of the SRAM cell (see also Section 10.3). The collected charge in the junction depends on many factors, primarily the node capacity, the supply voltage, and the characteristics of the incident particle (such as energy, path, and charge). In SRAM cells and flip-flops, the critical charge depends also on the strength of the feedback transistors. The emergence of a soft error in the circuit following the impact of a particle depends on the charge-deposition pattern of the incident particle, the geometry of the impact, and the design of the logic circuit. For simple isolated junctions (such as DRAM cells in storage mode), the impact of a particle induces a soft error if the collected charge is higher than the critical charge. In SRAM and logic circuits with active feedback, a soft error occurs only when the collected charge is greater than the critical charge by a certain factor that depends on the compensation current from the feedback. Generally, a higher critical charge means fewer soft errors, but, at the same time, a higher critical charge also means higher power dissipation and a slower logic gate.

The rate at which soft errors occur is called the SER and is typically expressed in terms of failures in time (FIT) (the number of failures per 10^9 h of operation).

A practical equation used during accelerated or real-time testing, for example, is

$$ SER = \frac{N}{AF \times \Sigma_r} \times 10^9 \quad (FIT/Mbit) \tag{5.21} $$

where:

N_r = number of bit flips observed at time T_r

Σr = number of Mbit×h cumulated at time T_r

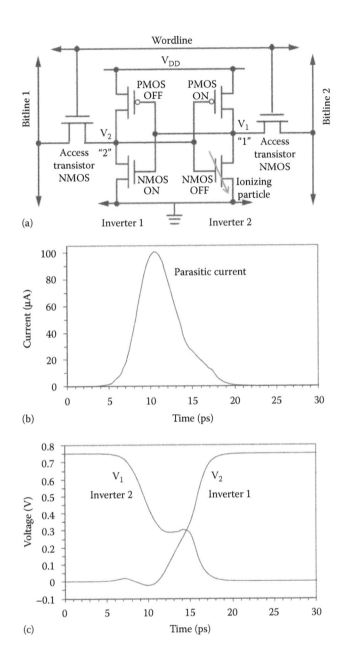

FIGURE 5.18
(a) Schematic circuit for an SRAM cell; (b) transient current induced by an ionizing particle striking the NMOS transistor (OFF-state) of the second inverter; (c) variation of voltages on first and second inverters.

AF = acceleration factor (i.e., the amplification factor of the particle flux defined in Section 1.3.1.2) with respect to New York City

Another important SEU mechanism in memories is related to multiple-cell upsets (MCU) and multiple-bit upsets (MBU). Their increasing importance for state-of-the-art memories comes from the fact that the reduction of circuit feature sizes increases the probability that a single particle strike will simultaneously affect several adjacent cells, depending on the particle-track structure and characteristic dimensions. The topological shape of MCUs detected in a given memory plan results from a complex combination of the memory layout (alternating structure of vertical P-wells and N-wells) with the test pattern considered to fill the memory plan and with the particle-induced perturbation. Most modern memories now interleave logical bits from different words so that bits from the same logical word are never physically adjacent in the memory plan, strongly reducing the occurrence of MBUs.

5.8 SEE Mechanisms in Digital Circuits

With the continuous decreasing of the CMOS feature size, it is well established that single-event transients (SETs) have become a significant error mechanism and are of great concern for digital circuit designers. CMOS scaling is accompanied by higher operating frequencies, lower supply voltages, and lower noise margins, which render the sensitivity of circuits to SET increasingly high (Baumann 2005b; Benedetto et al 2004, 2005, 2006; Buchner and Baze 2001; Dodd 2005; Dodd et al. 2004; Eaton et al. 2004; Gadlage et al. 2004; Massengill 1993; Narasimham et al. 2006).

Digital single-event transients (DSETs) constitute a temporary voltage or current transient generated by the collection of charge deposited by an energetic particle (Ferlet-Cavrois et al. 2013). Even if this transient does not induce an SEU in the struck circuit, it can propagate through the subsequent circuits and may be stored as incorrect data when it reaches a latch or a memory element (Roche 2006). Unlike an SRAM cell (where an SEU occurs as a *persistent* error when a SET with sufficient charge impacts a critical node), in a combinational logic node an SET with sufficient charge may become manifested as a persistent error only if it propagates through the circuit and is latched into a static cell (Benedetto et al. 2006). DSETs must fill a certain number of conditions in order to induce an error within a memory element (Dodd 2005; Diehl et al. 1983):

1. The ion strike must produce a transient able to propagate in the circuit.
2. There must be an open logic path by which the DSET can propagate to reach a latch or a memory element.
3. The DSET must have sufficient amplitude and duration to change the latch/memory state.
4. In synchronous logic, the DSET must reach the latch during a clock pulse enabling the latch. Then the probability of capturing an SET increases with increasing clock frequency.

Digital circuits are constituted from sequential elements (e.g., latches, flip-flops, register cells) and combinational logic (e.g., NAND and NOR gates). The effects of single-event-induced transients in these two types of circuits are succinctly described in the following.

5.8.1 Sequential Logic

Typical sequential elements in the core logic are a latch (Figure 5.19a), a domino cell (Figure 5.19b), or a register file cell (Figure 5.19c). State changes can occur in core logic

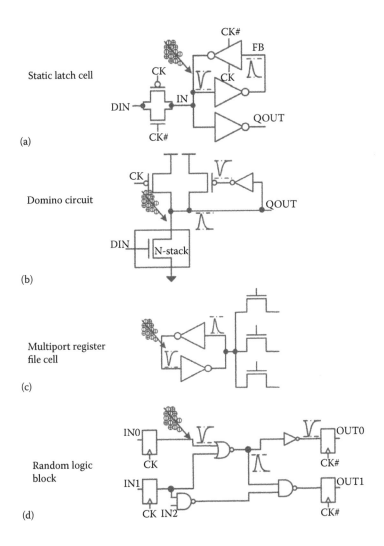

Static latch cell

(a)

Domino circuit

(b)

Multiport register
file cell

(c)

Random logic
block

(d)

FIGURE 5.19
(a–d) Illustration of typical sequential logic (latch, domino, register) and combinational circuits (random logic block). (Reprinted from Karnik, T., et al., Characterization of soft errors caused by single-event upsets in CMOS processes, *IEEE Trans. Dependable Secure Comput.* 1, 128–143. © (2004) IEEE. With permission.)

similarly to memory elements. In sequential logic (as in SRAM), the soft-error rate has been found to be independent of the clock frequency of the circuit (Buchner et al. 1997). For example, the latch state can be flipped by the charge deposited by a particle strike on a circuit node regardless of the state of the clock signal.

Flip-flop circuits (Figure 5.20) are other typical sequential logic circuits. With technology scaling, flip-flops have become more susceptible to soft errors, mainly due to the decrease in supply voltage and in their node capacitances. The simplified schematics of Figure 5.20 shows that flip-flop circuits are similar to SRAM cells, as both apply feedback loops of cross-coupled inverter-pairs. As previously mentioned, the soft-error sensitivity of this class of circuits is determined by the critical charge Q_{crit} (and the so-called collection efficiency; see Section 10.3). In an SRAM cell, Q_{crit} is mainly the same for the two storage nodes because the cell is symmetrical. In flip-flops, the inverters are sized differently and have

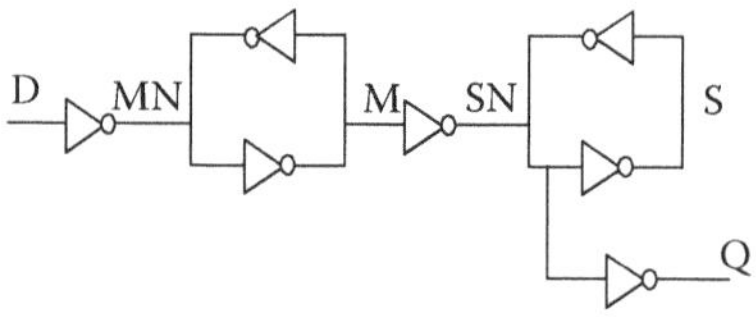

FIGURE 5.20
Simplified schematic of the flip-flop circuit. The sensitive nodes are labeled MN, M, SN, and S.

different fan-outs, which makes the flip-flop circuit asymmetric compared with the SRAM cell. Thus, the individual storage nodes in a flip-flop have a different critical charge than in a SRAM, and their SER sensitivity can vary by several orders of magnitude (Heijmen et al. 2004).

5.8.2 Combinational Logic

Any node in a combinational circuit can be impacted by an SEU and cause a voltage transient that can propagate through the combinational stages (Figure 5.19d) and cause an error if latched by a sequential element, such as a memory cell. In combinational logic, a certain number of transients will not be latched and, even when latched, some of these data will not be perceived as errors for the software operation. A transient error in a logic circuit might not be captured in a memory circuit because it could be masked by one of the following three phenomena (Karnik et al. 2004; Shivakumar et al. 2002):

1. Logical masking (Shivakumar et al. 2002; Cha and Patel 1993) occurs when a particle strikes a portion of the combinational logic that cannot affect the output due to a subsequent gate whose result is completely determined by its other input values. For example, if the strike happens on an input to a NAND (NOR) gate (as illustrated in Figure 5.21a) but one of the other inputs is in the controlling state (e.g., 0[1] for a NAND [NOR] gate), the strike will be completely masked and the output will be unchanged (i.e., the particle strike will not cause a soft error).

2. Temporal masking (or latching-window masking) occurs when the pulse resulting from a particle strike reaches a latch, but not at the clock transition where the latch captures its input value (Shivakumar et al. 2002). This is explained in Figure 5.21b: when the transient propagates toward a sequential element (a latch in the present case), the disturbance on node DIN may be outside the latching window (Liden et al. 1994). Hence, the error will not be latched, and there will be no soft error.

3. Electrical masking occurs for transients with bandwidths higher than the cutoff frequency of the CMOS circuit. These transients will then be attenuated (Dahlgren and Liden 1995). The pulse amplitude may reduce, the rise and fall times may increase, and, eventually, the pulse may disappear (as shown in Figure 5.21c). On the other hand, since most logic gates are nonlinear circuits with substantial voltage gain, low-frequency pulses with sufficient initial amplitude will be amplified (Karnik et al. 2004).

Due to these masking effects, the SER in combinational logic has been found to be significantly lower than expected (Karnik et al. 2004; Shivakumar et al. 2002; Liden et al. 1994). In addition to these masking mechanisms, two key factors impact the SER in combinational logic: the clock frequency and the SET pulse width (Baumann 2005b). With increasing clock

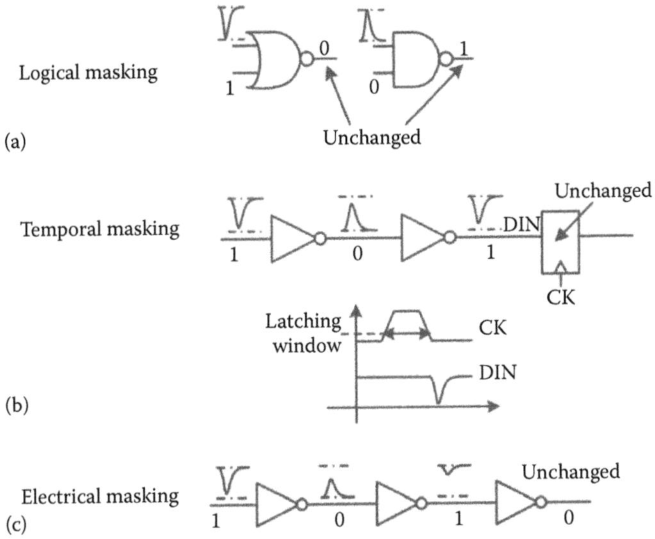

FIGURE 5.21
(a–c) Illustration of the masking phenomena in combinational logic. (Reprinted from Karnik, T., et al., Characterization of soft errors caused by single-event upsets in CMOS processes, *IEEE Trans. Dependable Secure Comput.* 1, 128–143. © (2004) IEEE. With permission.)

frequency, there are more latching clock edges to capture a pulse, and then the error rate increases. The pulse width is a key parameter that determines both the distance the SET will travel through the combinational chain and the probability that the SET will be latched in a memory element as wrong data (Buchner and Baze 2001). The wider the SET pulse width, the greater probability it has of arriving on the latching edge of the clock. If the transient becomes longer than the time period of the clock, every induced transient will be latched (Gadlage et al. 2004). The SET pulse width and amplitude depend on both process and circuit parameters (substrate or epitaxial layer doping, circuit capacitance, etc.) (Benedetto et al. 2006).

References

Agostinelli, S., Allison, J., Amako, K., et al. 2003. Geant4—A simulation toolkit. *Nuclear Instruments and Methods in Physics Research Section A: Accelerators, Spectrometers, Detectors and Associated Equipment*, 506:250–303. See also http://geant4.cern.ch.

Autran, J.L., Serre, S., Semikh, S., Munteanu, D., Gasiot, G., and Roche, P. 2012. Soft-error rate induced by thermal and low energy neutrons in 40 nm SRAMs. *IEEE Transactions on Nuclear Science* 59:2658–2665.

Baumann, R., Hossain, T., Murata, S., and Kitagawa, H. 1995. Boron compounds as a dominant source of alpha particles in semiconductor devices. In *Proceedings of the IEEE International Reliability Physics Symposium*, pp. 297–302, Piscataway, NJ, IEEE.

Baumann, R.C. 2005a. Radiation-induced soft errors in advanced semiconductor technologies. *IEEE Transactions on Device and Materials Reliability* 5:305–316.

Baumann, R.C. 2005b. Single event effects in advanced CMOS technology. In *IEEE Nuclear and Space Radiation Effects Conference*. Short Course Text, 11–15 July, 2005, Seattle, WA.

Baumann, R.C. and Smith, E. 2001. Neutron-induced ^{10}B fission as a major source of soft errors in high density SRAMs. *Microelectronics Reliability* 41:211–218.

Benedetto, J., Eaton, P., Avery, K., Mavis, D., Gadlage, M., and Turflinger, T. 2004. Heavy ion induced digital single-event transients in deep submicron processes. *IEEE Transactions on Nuclear Science* 51:3480–3485.

Benedetto, J.M., Eaton, P.H., Mavis, D.G., Gadlage, M., and Turflinger, T. 2006. Digital single event transient trends with technology node scaling. *IEEE Transactions on Nuclear Science* 53:3462–3465.

Benedetto, J.M., Eaton, P.H., Mavis, D.G., Gadlage, M., and Turflinger, T. 2005. Variation of digital SET pulse widths and the implications for single event hardening of advanced CMOS processes. *IEEE Transactions on Nuclear Science* 52:2114–2119.

Budyashov, Y.G., Zinov, V.G., Konin, A.D., Mukhin, A.I., and Chatrchyan, A.M. 1971. Charged particles from the capture of negative muons by the nuclei ^{28}Si, ^{32}S, ^{40}Ca, and ^{64}Cu. *Soviet Journal of Experimental and Theoretical Physics* 33:11.

Buchner, S. and Baze, M. 2001. Single-event transients in fast electronic circuits. In *IEEE Nuclear and Space Radiation Effects Conference*. Short Course Text, 16–20 July, Vancouver, Canada, IEEE.

Buchner, S., Baze, M., Brown, D., McMorrow, D., and Melinger, J. 1997. Comparison of error rates in combinational and sequential logic. *IEEE Transactions on Nuclear Science* 44:2209–2216.

Cha, H. and Patel, J.H. 1993. A logic-level model for α-particle hits in CMOS circuits. In *Proceedings of the IEEE International Conference on Computer Design*, pp. 538–542, 3–6 October, Cambridge, MA, IEEE.

Dahlgren, P. and Liden, P. 1995. A switch-level algorithm for simulation of transients in combinational logic. *Proceedings of the IEEE International Symposium on Fault-Tolerant Computing*, 207–216, 27–30 June, Pasadena, CA, IEEE.

Dicello, J.F., McCabe, C.W., Doss, J.D., and Paciotti, M. 1983. The relative efficiency of soft-error induction in 4K static RAMs by muons and pions. *IEEE Transactions on Nuclear Science* NS-30:4613–4615.

Dicello, J.F., Schillaci, M.E., McCabe, C.W., et al. 1985. Meson interactions in NMOS and CMOS static RAMs. *IEEE Transactions on Nuclear Science* NS-32:4201–4205.

Diehl, S.E., Vinson, J.E., Shafer, B.D., and Mnich, T.M. 1983. Considerations for single event immune VLSI logic. *IEEE Transactions on Nuclear Science* NS-30:4501–4507.

Dodd, P.E. 1996. Device simulation of charge collection and single-event upset. *IEEE Transactions on Nuclear Science* 43:561–575.

Dodd, P.E. 2005. Physics-based simulation of single-event effects. *IEEE Transactions on Device and Materials Reliability* 5:343–357.

Dodd, P.E. and Massengill, L.W. 2003. Basic mechanisms and modeling of single-event upset in digital microelectronics. *IEEE Transactions on Nuclear Science* 50:583–602.

Dodd, P.E., Shaneyfelt, M.R., Felix, J.A., and Schwank, J.R. 2004. Production and propagation of single-event transients in high-speed digital logic ICs. *IEEE Transactions on Nuclear Science* 51:3278–3284.

Eaton, P., Benedetto, J., Mavis, D., et al. 2004. Single event transient pulse width measurements using a variable temporal latch technique. *IEEE Transactions on Nuclear Science* 51:3365–3368.

Ferlet-Cavrois, V., Massengill, L.W., and Gouker, P. 2013. Single event transients in digital CMOS—A review. *IEEE Transactions on Nuclear Science* 60:1767–1790.

Fermi, E. and Teller, E. 1947. The capture of negative mesotrons in matter. *Physical Review* 72: 399–408.

Fleischer, R. 1983. Cosmic ray interactions with boron: A possible source of soft errors. *IEEE Transactions on Nuclear Science* 30(5):4013–4015.

Gadlage, M.J., Schrimpf, R.D., Benedetto, J.M., et al. 2004. Single event transient pulse widths in digital microcircuits. *IEEE Transactions on Nuclear Science* 51:3285–3290.

Gaillard, R. 2011. Single event effects: Mechanisms and classification. In M. Nicolaidis (ed.), *Soft Errors in Modern Electronic Systems*. New York: Springer.

Heijmen, T., Kruseman, B., van Veen, R., and Meijer, M. 2004. Technology scaling of critical charges in storage circuits based on cross-coupled inverter-pairs. *Proceedings of the IEEE International Reliability Physics Symposium*, 675–676. IEEE.

Karnik, T., Hazucha, P., and Patel, J. 2004. Characterization of soft errors caused by single event upsets in CMOS processes. *IEEE Transactions on Dependable and Secure Computing* 1:128–143.

Liden, P., Dahlgren, P., Johansson, R., and Karlsson, J. 1994. On latching probability of particle induced transients in combinational networks. *Proceedings of the IEEE International Symposium on Fault-Tolerant Computing*, pp. 340–349, 15–17 June, Austin, TX, IEEE.

Lindhard, J. and Scharff, M. 1961. Energy dissipation by ions in the keV region. *Physical Review* 124:128–130.

Macdonald, B., Diaz, J.A., Kaplan, S.N., and Pyle, R.V. 1965. Neutrons from negative-muon capture. *Physical Review* 139:B1253–B1263.

Massengill, L.W. 1993. SEU modeling and prediction techniques. *IEEE Nuclear and Space Radiation Effects Conference*. Short Course Text. IEEE.

Measday, D.F. 2001. The nuclear physics of muon capture. *Physics Reports* 354:243–409.

Mukhopadhyay, N.C. 1977. Nuclear muon capture. *Physics Reports* 30:1–144.

Nakamura, K., et al.; Particle Data Group. 2010. Review of particle physics. *Journal of Physics G* 37:075021. Available at: http://pdg.lbl.gov.

Narasimham, B., Bhuva, B.L., Holman, W.T., et al. 2006. The effect of negative feedback on single event transient propagation in digital circuits. *IEEE Transactions on Nuclear Science* 53:3285–3290.

Oldham, T.R., Murrill, S., and Self, C.T. 1986. Single event upset of VLSI memory circuits induced by thermal neutrons. Hardened Electronics and Radiation Technology (HEART) Conference, July, Newport Beach, RI.

Olmos, M., Gaillard, R., Van Overberghe, A., Beaucour, J., Wen, S., and Chung, S. 2006. Investigation of thermal neutron induced soft error rates in commercial SRAMs with 0.35 μm to 80 nm technologies. *Proceedings of the IEEE International Reliability Physics Symposium*, pp. 212–216, 26–30 March, San Jose, CA, IEEE.

Rajeevakumar, T.V., Lu, N.C.C., Henkels, W.H., Hwang, W., and Franch, R. 1988. A new failure mode of radiation-induced soft errors in dynamic memories. *IEEE Electron Device Letters* 9:644–646.

Rinard, P. 1991. Neutron interactions with matter. Los Alamos Technical Report. Available at: http://www.fas.org/sgp/othergov/doe/lanl/lib-www/la-pubs/00326407.pdf.

Roche, P. 2006. Year-in-review on radiation-induced soft error rate. Tutorial at IEEE International Reliability Physics Symposium, March, San Jose. IEEE.

Shivakumar, P., Kistler, M., Keckler, S.W., Burger, D., and Alvisi, L. 2002. Modeling the effect of technology trends on the soft error rate of combinational logic. In *Proceedings of the International Conference of Dependable Systems and Networks*, pp. 389–398, Los Alamitos, CA, IEEE.

Sierawski, B.D., Mendenhall, M.H., Reed, R.A., et al. 2010. Muon-induced single event upsets in deep-submicron technology. *IEEE Transactions on Nuclear Science* 57:3273–3278.

Singer, P. 1974. Emission of particles following muon capture in intermediate and heavy nuclei. *Nuclear Physics, Springer Tracts in Modern Physics* 71:39–87.

Sobottka, S.E. and Wills, E.L. 1968. Energy spectrum of charged particles emitted following muon capture in ^{28}Si. *Physical Review Letters* 20:596.

Sundelin, R.M. and Edelstein, R.M. 1973. Neutron asymmetries and energy spectra from muon capture in Si, S, and Ca. *Physical Review C* 7:1037–1060.

Sundelin, R.M., Edelstein, R.M., Suzuki, A., and Takahashi, K. 1968. Spectrum of neutrons from muon capture in silicon, sulfur, and calcium. *Physical Review Letters* 20:1198–1200.

Suzuki, T., Measday, D.F., and Roalsvig, J.P. 1987. Total nuclear capture rates for negative muons. *Physical Review C* 35:2212–2224.

Tang, H.H.K. 1996. Nuclear physics of cosmic ray interactions with semiconductor materials: Particle-induced soft errors from a physicist's perspective. *IBM Journal of Research and Development* 40:91–108.

Tavernier, S. 2010. *Experimental Techniques in Nuclear and Particle Physics*. Berlin: Springer.

Tiomno, J. and Wheeler, J.A. 1949. Charge-exchange reaction of the μ-meson with the nucleus. *Reviews of Modern Physics* 21:153–165.

Valkovic, V. 2000. *Radioactivity in the Environment*. Amsterdam: Elsevier.

Wen, S., Wong, R., Romain, M., and Tam, N. 2010. Thermal neutron soft error rate for SRAM in the 90–45 nm technology range. *Proceedings of the IEEE International Reliability Physics Symposium*, pp. 1036–1039, 2–6 May, Anaheim, CA, IEEE.

Wrobel, F. 2002. Elaboration d'une base de données de particules responsables des dysfonctionnements dans les composants électroniques exposés à des flux de protons ou neutrons. Application au calcul de taux d'erreurs dans les mémoires SRAM en environnement radiatif [Database of particles induced failure in electronic components in proton and neutron environment. Application to the calculation of soft-error rate in SRAM memories]. PhD Thesis, Université de Montpellier II [in French].

Wyttenbach, A., Baertschi, P., Bajo, S., et al. 1978. Probabilities of muon induced nuclear reactions involving charged particle emission. *Nuclear Physics A* 294:278–292.

Ziegler, J.F., Biersack, J.P., and Ziegler, M.D. 2008. *SRIM—The Stopping and Range of Ions in Matter*. Chester, MD: SRIM.

Ziegler, J.F. and Lanford, W.A. 1979. Effect of cosmic rays on computer memories. *Science* 206:776–788.

Ziegler, J.F. and Puchner, H. 2004. *SER—History, Trends and Challenges*. San Jose, CA: Cypress Semiconductor.

Ziegler, J.F., Ziegler, M.D., and Biersack, J.P. 2010. SRIM—The stopping and range of ions in matter. *Nuclear Instruments and Methods in Physics Research Section B: Beam Interactions with Materials and Atoms* 268:1818–1823.

6

Accelerated Tests

6.1 Introduction

To predict the impact of natural radiation on the behavior of electronics and to (statistically) estimate (measure) its radiation-induced soft-error rate (SER), three different experimental methods can be envisaged (Ziegler and Puchner 2004; JEDEC 2006), excluding modeling and simulation approaches, which can be used, under certain conditions (i.e., when correctly calibrated), as predictive tools (see Chapter 10).

The first one, called *field testing*, involves collecting errors from a large number of finished products already on the market. The SER value is evaluated a posteriori from the errors experienced by the consumers themselves; it generally takes several years after the introduction of the product on the market. Measurement on production circuits or systems poses significant challenges, since the measurement must not introduce any noticeable performance impact on the existing running applications (Li et al. 2007). This method is not adapted to upstream reliability studies performed during the cycle of product development and will not be considered in the following.

The second method consists in exposing a given device (or a large number of identical devices) to terrestrial radiation over a sufficiently long period (weeks or months) in order to achieve adequate statistics on the number of accumulated errors and then on the SER value. This method is called the *real-time* SER (RTSER) test (or unaccelerated testing) and will be the topic of Chapter 7. The main advantage of RTSER tests is that they provide a direct measurement of the *true* SER that does not require intense radiation sources and extrapolations to use conditions. The major drawbacks of this method concern the cost of the system (which has to be capable of monitoring a very large number of devices at the same time) and the long duration of the experiment. All these points will be developed in Chapter 7.

The third method, the subject of the present chapter, is called *accelerated* SER (ASER) testing and will be detailed here. The chapter will address the different ASER test methodologies and procedures and main facilities. The most important types of ASER experiments, using high-energy neutrons, thermal neutrons, protons, and muons, will be described, as well as accelerated tests using intense sources of alpha particles. Finally, the differences between different artificial broad-spectrum sources of atmospheric-like neutrons will be discussed and analyzed on the basis of a complete simulation study. This chapter does not pretend to be an exhaustive review of ASER studies; reference materials cited in the text are available for further details.

6.2 Methodology and Test Protocols

Accelerated tests are commonly used to estimate the SER, since they are relatively easy to perform and cheaper and faster than real-time tests. In accelerated tests, devices are exposed to various radiation sources that reproduce the natural radiation environment. The intensity of these sources is much higher than the ambient radiation levels that the device would normally meet. Thus, accelerated tests provide useful data in a fraction of the time required by real-time tests (JEDEC 2006). Only a few devices are needed, and comprehensive evaluations can often be performed in a few hours or days instead of weeks or months. Accelerated tests can be used to complete the real-time characterization, as will be shown in Section 7.5. The main disadvantages of the accelerated tests are as follows:

1. The results should be extrapolated to real conditions.
2. Several different sources of radiation should be employed to ensure that the evaluation takes into account the soft errors due to alpha particles, atmospheric neutrons (high-energy cosmic-ray neutrons and, if necessary, thermal neutrons), and, for the most recent and sensitive technologies, protons and muons.

The conditions and procedures for the accelerated tests are defined by different standard tests. We can cite the three most important and most used standards developed by the Joint Electron Device Engineering Council (JEDEC) (2006), the Japan Electronics and Information Technology Industry Association (JEITA) (2005), and the International Electrotechnical Commission (IEC) (2008a). These specification guidelines define the standard requirements and procedures for terrestrial SER testing of integrated circuits and reporting of results. Both real-time (unaccelerated) and accelerated testing procedures are described. The JEDEC JESD89A standard is relatively general and can be applied to both memory (e.g., dynamic random-access memory [DRAM] and SRAM) and logic components (e.g., flip-flops). The JEITA SER testing guideline is more specifically applied for SRAMs and DRAMs. The IEC technical specification is intended to help aerospace equipment manufacturers and designers to standardize their global approach to SEE in avionics.

In the following, we will frequently refer to the JESD89A standard, which covers in detail the three sources of soft errors: alpha particles, high-energy neutrons, and thermal neutrons. This standard is widely referenced in the literature for SER studies. The JESD89A standard defines the test plans, the equipment requirements, and the standardized methodology of results presentation for

1. Real-time SER testing
2. Accelerated alpha-particle SER testing
3. Accelerated high-energy-neutron SER testing
4. Accelerated thermal-neutron SER testing

6.2.1 SEU Cross Section

By analogy with the cross section introduced in nuclear physics to express the probability of interaction between particles, the single-event upset (SEU) cross section is a quantity that allows the sensitivity of a component, chip, or circuit to a particle species (e.g., neutron, proton, heavy ion, etc.) to be estimated. This quantity is denoted σ and is generally expressed in square centimeters. The SEU cross section characterizes the sensitive area of a given component, cell (memory bit), or circuit to irradiation flux; it depends on the particle type and particle energy and can be considered as an intrinsic parameter of the chip/circuit under test that quantifies its response to the particle species used. In general, it is also a function of the operating conditions of the irradiated chip (e.g., power-supply voltage, temperature, etc.) (JEDEC 2006).

The SEU cross section is connected to the total number of events N induced by irradiation in the chip under test by the following equation:

$$N = \sigma \times F \tag{6.1}$$

where F is the fluence of the particle beam, expressed in units of particles per square centimeter; it corresponds to the particle flux integrated in time over the entire duration of the experiment. The SEU cross section therefore provides an estimation of the number of events for a component or circuit exposed to a particle beam. For digital circuits, the SEU cross section is often given per bit or per megabit. Multiplying the cross section by the radiation fluence rate Φ (in particles $\text{cm}^{-2} \text{h}^{-1}$) results in upsets per hour for the considered circuit.

The purpose of accelerated tests is to estimate the three cross sections of interest at ground level: σ_{cosmic} for cosmic rays (high-energy neutrons), σ_{thermal} for thermal neutrons, and σ_{alpha} for alpha particles. Using these cross sections, the error rate, in failure in time (FIT), is given by the following equation:

$$\text{SER}_{\text{FIT}} = 10^9 \times \left(\sigma_{\text{cosmic}} \Phi_{\text{cosmic}} + \sigma_{\text{thermal}} \Phi_{\text{thermal}} + \sigma_{\text{alpha}} \Phi_{\text{alpha}} \right) \tag{6.2}$$

6.2.2 Test Equipment Requirements

As explained in the JESD89A standard, accelerated tests are generally conducted using automatic test equipment (ATE) capable of remotely controlling a small number of devices under test (DUTs) at a time while they are irradiated. Figure 6.1 shows the schematic of a typical ATE and the detailed view of a current high-performance tester. The equipment has been constructed in two parts: a holder (DUT board) with the DUTs to be measured and a separate electronic board corresponding to the core of the tester hardware. During a test, the DUT board is placed in the particle beam (or placed in front of an intense solid source for alpha tests), while the tester board can be shielded in the experimental room away from the beam. The cabling between them must be carefully designed and constructed to minimize the potential for errors during operations (JEDEC 2006). With a long-distance Ethernet connection, the control PC can be located in the control room and used by the operators during the experiment.

ATE characteristics and performances must fulfill very strict conditions, for both hardware and software, in order to obtain accurate accelerated SER testing results. The tester board must be able to tolerate stray radiation and must be capable of exercising the device(s) under test over the range of operating conditions, such as power-supply voltage, access-cycle speed, and temperature, that are specified in the test plan (JEDEC 2006). The ATE software must be able to create the proper conditions for the test and to identify,

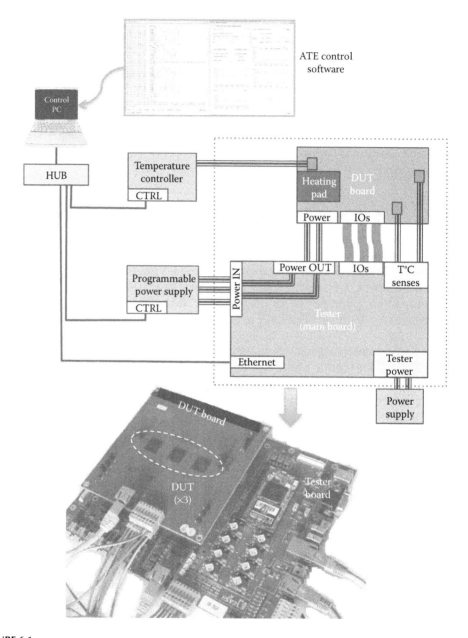

FIGURE 6.1
Schematic illustration of an automatic test equipment (ATE) used in accelerated SER testing (*top*) and detailed view (DUT and tester boards) of a modern tester (*bottom*). (Courtesy of EASII-IC, Grenoble, France.)

record, and correct errors as they are detected. As indicated in the JESD89A standard, the ATE software should be capable of

1. Controlling device initialization and rudimentary functional checks.
2. Dynamic or static device operation, as required by the test plan.
3. Resetting the DUT during irradiation or real-time testing.

4. Error detection and logging, including the time when the error was detected. It is important during error detection that new errors are not omitted or that corrections are made for system dead time.

6.2.3 Test Plan

As specified in the JESD89A standard (JEDEC 2006) and in Slayman (2010c), for all SER testing it is necessary to accurately measure

1. The particle flux at the device from the accelerated source
2. The number of circuit elements that are subject to upset
3. The number of upsets that occur in each circuit element

The test plan, developed to support each test, serves as a guide for the procedures and decisions to be made during the irradiation period. For accelerated testing, a minimum test plan should include (Slayman 2010c)

1. Setup and checking of ATE
2. Dosimetry calibration
3. Correlation of SER with previously tested device if available
4. Initial test run to confirm exposure rate
5. Data collection for each voltage, frequency, test pattern, temperature, and angle planned
6. Final test run repeating initial test to confirm repeatability and determine whether any device damage or drift has occurred

6.2.4 Test Conditions

Test conditions include several parameters that accurately define the state of the DUT(s) during the experiment. They include test pattern, supply voltage, DUT temperature, and the device's access method (operation mode).

- *Test pattern*: Basic test patterns for memory circuits include all "0," all "1," and checkerboard. For testing when there is no a priori knowledge of the device, the test pattern should balance the numbers of "0" and "1." If detailed layout information for the DUT is available, a physical checkerboard pattern should be used. In other cases, a logical checkerboard pattern, alternating by address and bit, should be considered (JEDEC 2006).
- *Supply voltage*: Because the SER and the SEU cross section are very sensitive to the power-supply voltage (the cross section increases as the supply voltage decreases), it is extremely important that the power-supply voltage be accurately measured and controlled (JEDEC 2006). It is also important to note that the internal cell voltage may be different from the power-supply voltage. Monitoring of the core voltages *in situ* is preferable.
- *Temperature*: Both the DUT internal temperature (generally measured with a dedicated test device—a p-n junction embedded on the chip) and the ambient temperature should be measured and possibly monitored during the whole experiment.

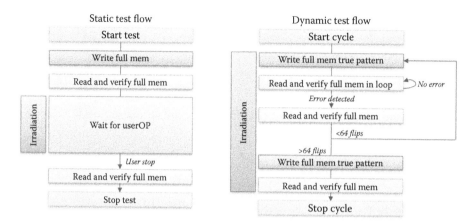

FIGURE 6.2
Flow diagrams comparing typical static and high-speed dynamic test algorithms used in ASER testing. (Courtesy of EASII-IC, Grenoble, France.)

- *Operation mode*: Static or dynamic operation of the DUT can be performed. As explained in JESD89A (JEDEC 2006) and illustrated in Figure 6.2, static tests are performed using the following steps:
  1. The DUT is initialized to a known state.
  2. The DUT is irradiated.
  3. After irradiation, the DUT state is read and compared with the initialization value to determine the number of upset events.

In dynamic tests, the steps are (JEDEC 2006)

1. DUT is initialized before irradiation to a known state.
2. Once the irradiation starts, the DUT is accessed continuously, at a specific rate in the test plan, and upset events are counted while they are detected.
3. When the irradiation stops, sufficient time is allowed to read all the DUTs to ensure that all upsets induced by irradiation were tabulated.

6.3 Experiments Using Intense Beams of Particles

6.3.1 High-Energy Neutrons

High-energy particle beams are widely used in accelerated tests of electronic devices and systems in order to characterize their sensitivity to neutron-induced single events. To study how atmospheric neutrons cause SEU in circuits, it is possible to use intense beams of particles, which provide particle fluxes with intensities several orders of magnitude higher than those encountered at the terrestrial level. There are three types of facilities that can offer this type of high-energy particle beam: (1) spallation neutron sources, (2) monoenergetic neutron sources, and (3) quasi-monoenergetic neutron sources. We describe in Sections 6.3.1.1–6.3.1.3 these different sources of high-energy neutrons.

6.3.1.1 Spallation Neutron Sources

The spallation type of neutron source is created by the interaction of a high-energy proton beam with a large, dense target, producing secondary neutrons (IEC 2008b; Zhang 2013). Figure 6.3 shows a schematic representation of the wide neutron source facility (TNF) at Tri-University Meson Facility (TRIUMF) accelerators, Vancouver, Canada (Blackmore 2009; Blackmore et al. 2003). Neutrons are produced by the spallation reaction of a 400–450 MeV proton beam (100–150 µA) on an aluminum plate absorber surrounded by a water moderator. Neutron channels transport these neutrons through the shielding surrounding the TNF. The flux of neutrons at the normal operating currents is 6×10^6 neutrons cm^{-2} s^{-1} above 1 MeV and has energies extending to about 400 MeV, as shown in Figure 6.4. Thermal neutrons are also present if desired; they can be easily removed using a cadmium shield.

Since neutrons are created in the atmosphere in the same way, spallation neutron sources provide neutrons over a wide range of energies. The shape of this spectrum is similar to the neutron terrestrial environment over a certain interval of neutron energy: broad-spectrum neutron sources duplicate the atmospheric neutron spectrum more or less closely but at a much higher intensity (see Table 6.1). Measurements of the upset rate in such beams scales directly to environmental upset rates. There are several main atmospheric-like neutron test facilities (nonexhaustive list):

1. The TNF facility at TRIUMF, as presented above
2. Los Alamos Neutron Science Center (LANSCE), New Mexico, USA (Lisowski et al. 1990)

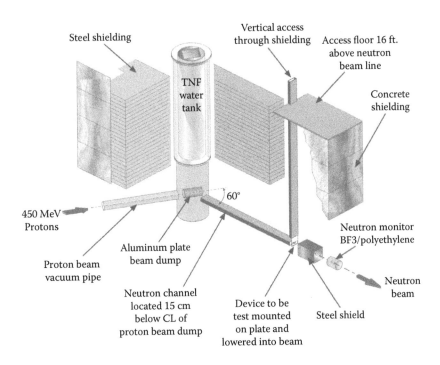

FIGURE 6.3
Illustration of the TRIUMF Neutron irradiation Facility (TNF) offering atmospheric-like neutrons with energies ranging from thermal to 400 MeV at high intensity. (Courtesy of E. Blackmore.)

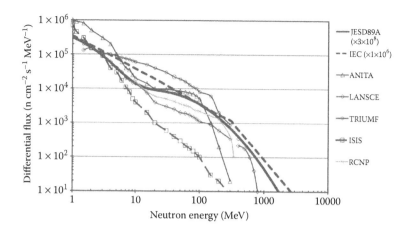

FIGURE 6.4

High-energy neutron spectra for different sources of neutrons (LANSCE, TRIUMF, ANITA, RNCP, ISIS) compared with JEDEC and IEC reference distributions. For comparison, curves have been normalized to the integral flux above 10 MeV, as in Platt, S.P., Prokofiev, A.V., and Xiao, C.X., *Proceedings of the IEEE International Reliability Physics Symposium*, pp. 411–416, 2–6 May, Anaheim, CA, IEEE, 2010. (Reprinted from Slayman, C., Theoretical correlation of broad spectrum neutron sources for accelerated soft error testing, *IEEE Trans. Nucl. Sci.* 57, 3163–3168. © (2010) IEEE. With permission.)

3. The Theodor Svedberg Laboratory (TSL) Atmospheric-like Neutrons from thIck TArget (ANITA) at Uppsala University, Sweden (Prokofiev et al. 2009)

4. Research Center for Nuclear Physics (RCNP), Osaka, Japan (Sakai et al. 1996)

5. ISIS neutron source at Rutherford Appleton Laboratory, Oxfordshire, UK (Andreani et al. 2008; Violante et al. 2007)

6. The CERN-EU high-energy Reference Field (CERF, for avionic altitudes and space), Geneva, Switzerland (Mitaroff and Silari 2002)

Figure 6.4 compares the energy distributions of the neutron flux (differential flux) for five of these sources (LANSCE, TRIUMF, ANITA, RNCP, and ISIS) together with the terrestrial spectra referenced in the JEDEC (2006) and IEC (IEC 2008a) standards. The main characteristics of these sources are summarized in Table 6.1. This table also indicates the acceleration factor (AF) for these sources with respect to both JEDEC and IEC references, as calculated by Slayman (2010a) using the following formula:

$$AF = \frac{\int_{E_{min}}^{\infty} \Phi_{acc}(E)dE}{\int_{E_{min}}^{\infty} \Phi_{jedec/iec}(E)dE} \tag{6.3}$$

where:

Φ_{acc} = flux of the considered neutron source

$\Phi_{jedec/iec}$ = reference flux

E_{min} = 10 MeV as specified in JEDEC and IEC standards

TABLE 6.1

Summary of the Main Characteristics of Several Neutron Wide-Spectrum Facilities

Facility	Type of Source	Target/ Moderator	Maximum Energy (MeV)	JEDEC Acceleration Factor (>10 MeV)	IEC Acceleration Factor (>10 MeV)
ANITA	Proton spallation	Tungsten	800	2.7×10^8	6.1×10^5
LANSCE	Proton spallation	Tungsten	800	1.3×10^8	2.9×10^5
TRIUMF	Proton spallation	Aluminum surrounded by water moderator	450	7.6×10^8	1.7×10^6
ISIS	Proton spallation	Tungsten (beam heavily moderated by water)	800	1.5×10^7	3.5×10^4
RCNP	Proton spallation	Lead	392	1.8×10^8	4.0×10^5

Source: Adapted after Slayman, C., Acceleration factors for neutron beam soft error testing in commercial and avionic environments, Workshop on Accelerated Stress Testing and Reliability (ASTR), 2010; Slayman, C., *IEEE Trans. Nucl. Sci.* 57, 3163–3168, 2010.

The values of AF for the different sources evidence the very high fluxes available during ASER tests, which allow users to perform extensive measurements in a very short time (typically in the range of minutes or hours of irradiation).

Because neutron interaction with a given target nucleus primarily depends on the neutron's energy and the target's properties, the capability of the source to replicate the natural neutron spectrum is fundamental in the estimation of the SER. A few recent works (Platt et al. 2010; Slayman 2010b,d) focus precisely on the characterization of various atmospheric-like neutron facilities to quantify the soft errors induced in such accelerated testing with respect to neutron production methods, facility design, and experimental conditions. As shown in Figure 6.4 and already highlighted by Slayman (2010d), none of these synthetic sources is able to accurately replicate the terrestrial neutron spectrum across the full range 1 MeV–1 GeV, mainly due to the limited energy of the incident proton beam used to produce by spallation any neutron flux above a few hundreds of MeV. Such deviations with respect to the natural radiation background should have direct consequences for the distributions of generated products in the silicon target (i.e., the circuit) and indirect consequences for the resulting circuit SER. A detailed simulation study (Serre et al. 2012) on precisely this point will be described in Section 6.5.

To conclude, a typical minimum test plan for wide-spectrum neutron testing, as detailed in Wilkinson (2010), should be as follows:

1. DUT setup.
2. Beam initialization.
3. Dosimetry check on beam.
4. Expose DUTs to beam and record errors and time.
5. Calculate the average cross section, in square centimeters, using

$$\overline{\sigma_{cosmic}} = \frac{N_{SEU}}{\Phi_{beam}} \tag{6.4}$$

where:

N_{SEU} = number of errors detected during the run
Φ_{beam} = total beam fluence (in the energy range of the source used) provided by the test facility (in-line dosimetry) for the run duration (in n cm^{-2})

6. Calculate the cosmic ray device error rate in failure in time (FIT), using

$$SER_{FIT} = 10^9 \times \overline{\sigma_{cosmic}} \times \Phi_{env} \tag{6.5}$$

where $\Phi_{env} = 13h^{-1} \, cm^{-2}$ represents the integrated flux for the reference flux (New York City at sea level outdoors) above 10 MeV (see Section 1.3.1.2).

6.3.1.2 Monoenergetic Neutron Sources

As explained in JESD89A (JEDEC 2006), a spallation neutron source is used to measure the SEU rate and to deduce an average SEU cross section over the energy range covered by the source. Since neutrons from a spallation source cover a wide spectrum of energies, it is not possible to extract an SEU cross section at a specific energy from these measurements. For measuring neutron-induced SEU cross sections, monoenergetic neutron beams may be used. While the spallation type is useful for testing conditions similar to the natural environment at ground level and in the lower atmosphere (troposphere), monoenergetic sources are useful for measuring the energy dependence of single-event effects (Zhang 2013).

The JESD89A (JEDEC 2006) standard gives some details concerning these sources, detailed in the following. There are three main types of truly monoenergetic neutron beams >1 MeV in which almost all the neutrons are within ±1 MeV of the peak energy. All are produced by accelerating a charged particle into a tritium (T) or deuterium (D) target. D–T reactions produce neutrons of 14 MeV, and this is the most common type of neutron generator. D–D reactions produce neutrons of 3–5 MeV, depending on the energy of the deuteron, and p–T reactions produce neutrons with energies depending on the energy of the proton. A large number of facilities produce monoenergetic neutrons: a detailed list can be found in the Compendium of International Radiation Test Facilities (Compendium 2011). We can cite, for example, the Boeing Radiation Effects Laboratory (BREL) in the United States, which provides 14 MeV neutrons, and the Atomic Weapons Establishment (AWE) in the United Kingdom, which produces 3 and 14 MeV neutrons (Zhang 2013).

The monoenergy SEU cross section $\sigma(E_n)$ can be calculated from the number of errors N_{SEU} and the total fluence $\Phi_{beam}(E_n)$ for the tested energy E_n (JEITA 2005):

$$\sigma(E_n) = \frac{N_{SEU}}{\Phi_{beam}(E_n)} \tag{6.6}$$

The total fluence $\Phi_{beam}(E_n)$ may be calculated from flux $\phi(E_n)$ and the test duration T as

$$\Phi_{beam}(E_n) = \phi(E_n) \times T \tag{6.7}$$

only when the flux is very stable, which is not the case in many facilities, so that integrated dosimetry for fluence is recommended (JEITA 2005).

Several measurements, with a judicious choice of the discrete measurement energies, are necessary to construct the $\sigma(E_n)$ discrete curve. The following four-parameter Weibull curve is often used to fit the discrete measurements and to obtain an analytical approximation of the SEU cross-section curve:

$$\sigma_{appx}(E) = \sigma_{\infty}\left[1-\exp\left\{-\left(\frac{E-E_{th}}{W}\right)^{S}\right\}\right] \tag{6.8}$$

where:

σ_{∞} = asymptotic neutron cross section
E_{th} = threshold (or cutoff) energy below which the cross section is zero
S = shape factor
W = width parameter

Figure 6.5 shows a typical example of a SEU cross section versus neutron energy for a 65 nm SRAM. Experimental points are well fitted using Equation 6.8, with the four-fitting parameter values indicated in the inset of Figure 6.5. Empirically, it has been observed that the shapes of the $\sigma(E_n)$ curves for memory devices are quite similar to each other with different threshold energy E_{th} and saturation level σ_{∞}. As recommended by JEITA (2005), neutron energy below 20 MeV is recommended for test energy to identify or estimate the threshold energy E_{th}, where $\sigma(E_n)$ shows a sharp increase. Some energy points between 20 and 70 MeV are also recommended as test energies to identify the shape of the curvature approaching the saturated cross section at higher energies. Finally, energies from 70 to over 100 MeV are recommended as test energies to identify or estimate the saturation level.

Once the analytical curve is determined from monoenergetic measurements, the SER for any given neutron spectrum can be obtained by

$$SER_{FIT} = 10^{9} \times \int_{E_{th}}^{E_{max}} \frac{d\Phi(E)}{dE} \times \sigma_{appx}(E)\, dE \tag{6.9}$$

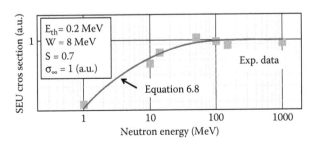

FIGURE 6.5
Typical SEU cross-section curve for a 65 nm SRAM circuit deduced from monoenergetic neutron measurements. The four-parameter Weibull fit (Equation 6.8) is also plotted, with the corresponding values of the fitting parameters indicated in the inset.

where:

$d\Phi(E)/dE$ = differential flux of the particle in the considered environment, given in units of particle number cm⁻² MeV⁻¹ s⁻¹

$\sigma_{appx}(E)$ = SEU cross section at neutron energy E (Equation 6.8), given in units of square centimeters per device or square centimeters per bit

E_{th} and E_{max} = lower and upper limits of the energy spectrum over which $\sigma_{appx}(E)$ is defined

6.3.1.3 Quasi-Monoenergetic Neutron Sources

Finally, we mention the use of quasi-monoenergetic neutron sources to measure neutron-induced SEU cross sections (Grandlund et al. 2004; JEDEC 2006). As detailed in JESD89A (JEDEC 2006), these beams are different from truly monoenergetic neutron beams, since a significant proportion of neutrons are at energies less than the maximum energy; they can be used as an alternative to spallation neutron sources. The standard beam of this type is obtained by accelerating monoenergetic protons into a lithium target, although other production mechanisms are also possible. The neutrons from this beam comprise a two-part distribution: the neutrons at peak energy (~1–2 MeV below the proton energy) and the neutrons within the so-called low-energy tail (from E_{peak} ~2 MeV down to ~0 MeV). An example is TSL at Uppsala University, Sweden (Prokofiev et al. 2006, 2008). TSL provides quasi-monoenergetic neutron beams with energies in the range of 20–180 MeV. Typically, about 40% of the neutrons in the TSL field are at the nominal energy and 60% in the low-energy tail (Zhang 2013).

6.3.2 Thermal Neutrons

The sensitivity of integrated circuits (ICs) to thermal neutrons is a recurrent reliability issue for past and current technologies. The underlying mechanism of this sensitivity, that is, the interaction of thermal neutrons with the ¹⁰B isotope of boron present in circuit materials, has been described in detail in Chapter 5, Section 5.1.4. ICs have to be tested to evaluate the impact of thermal neutrons on the SER, even if modern processes have massively (but not totally) eliminated the presence of ¹⁰B, notably in the back-end-of-line materials (borophosphosilicate glasses; see Section 5.1.4). The chip SER due to thermal neutrons can be evaluated with ASER testing using high-flux thermal neutron fluxes obtained from nuclear reactors or particle accelerators. Thermal neutrons are available from nuclear reactors operated by different centers, for example the NIST Center for Neutron Research in the United States (20 MW split-core research reactor), the Institut Laue-Langevin (ILL) Research Reactor in France (58 MW research reactor), and the Australian Nuclear Science and Technology Organisation (20 MW pool-type nuclear research reactor). As detailed in Ziegler and Puchner (2004), the neutrons are extracted using pipes leading to the reactor core and are then filtered to eliminate charged particles and some gamma rays. Thermal neutrons can alternatively be produced in particle accelerator facilities from spallation reactions of energetic protons on Li targets (JEDEC 2006). Then, the energy of the produced neutrons is lowered by passing them through low-Z materials such as polyethylene. Thermal neutrons can also be produced using 14 MeV neutron generators and, as previously, a moderating low-Z material to strongly reduce the energy of high-energy neutrons.

As discussed in JESD89A (JEDEC 2006), a critical issue for thermal neutron accelerated testing is the calibration, because the SER cross section varies significantly with small changes in neutron energy. Thus, the measurement is very sensitive to the energy distribution of the low-energy neutrons.

The test plan for thermal neutron testing is similar to that considered for high-energy neutrons (see Section 6.3.1.1). In Equations 6.4 and 6.5, the average CR cross section is replaced by the average thermal neutron cross section, and the thermal neutron device error rate, in FIT, is estimated using

$$\text{SER}_{\text{FIT}} = 10^9 \times \overline{\sigma_{\text{thermal}}} \times \Phi_{\text{env}} \tag{6.10}$$

where $\Phi_{\text{env}} = 1.8 \times 10^{-3}\,\text{cm}^{-2}\,\text{s}^{-1} = 6.5\,\text{cm}^{-2}\,\text{h}^{-1}$ is the nominal integral thermal neutron flux on the ground, centered around E 0.025 eV (JEDEC 2006). It is important to note that the thermal neutron flux is highly variable (Gordon et al. 2004). Then, the calculated error rate has significant uncertainty due to the variable nature of the environment (Wilkinson 2010).

Limitations and strengths of accelerated thermal neutron tests are discussed in Slayman (2011). The main advantage of accelerated thermal neutron testing is that it allows rapid determination of whether thermal neutrons play a role in the device SER. If this is the case, special techniques can be used to mitigate the problem (see Section 5.1.4). The inconvenience of accelerated thermal neutron tests consists in the difficulty of translating the SER from an accelerated thermal neutron test to an actual field SER at other locations. The conversion of the measured cross section to a situation of real field use requires measuring the thermal flow of neutrons at the desired location (Slayman 2011).

6.3.3 Protons

Proton interactions with matter have been described in detail in Section 5.3. Schematically, protons present two asymptotic behaviors as a function of their energy. At low energies (<1–10 MeV), they behave as charged particles and deposit their energy by direct ionization. The experimental evidence of direct ionization from low-energy protons in SRAMs and latches has been demonstrated recently in advanced complementary metal–oxide–semiconductor (CMOS) technologies; this point will be presented and discussed in Section 10.4.1. The growing importance of low-energy protons for current and future technological nodes has a direct consequence in the domain of accelerated tests; the volume of experimental work in dedicated facilities has significantly increased in recent years. At high energies (>50 MeV), protons interact with matter like neutrons and the nuclear reactions are practically the same for neutrons and protons. This similarity between protons and neutrons has been exploited in SER studies for a long time: high-energy proton (accelerator) experiments are performed to investigate the high-energy neutron response of device and circuits.

For both types of experiments, a wide variety of proton test facilities can be envisaged, as listed in Compendium (2011). These facilities are available at universities or government laboratories using accelerators that run, or have run in the past, for high-energy or nuclear physics programs. Note also that, during the last decade, a significant number of proton therapy centers have been built around the world to provide treatment for cancer. These facilities generally provide intermediate and high proton energies using different types of accelerators (Van de Graaff, cyclotron, linac, synchrotron), typically with a maximum energy of 20–35 MeV for Van de Graaff, 10–560 MeV for cyclotrons, 500–800 MeV for linacs, and >1 GeV for synchrotrons. Low-energy protons are directly produced with specific machines (Cockcroft-Walton, Van de Graaff, Tandem Van de Graaff) or from the energy degradation of low-energy protons by using different types of material (thick degraders) placed in front of the DUT to slow the beam. Aluminum and Mylar are common degrader materials for protons, though higher-Z materials like tantalum are also employed (Sierawski 2011). However, high-energy protons degraded substantially to lower energies are no longer truly monoenergetic. In Sierawski (2011), to investigate the viability of this approach and the shape of the proton energy distribution after slowing down the beam, radiation transport simulations of a 65 MeV proton beam were performed with the Monte Carlo radiation transport code MRED (see Chapter 9). In each simulation, protons passed through an aluminum slab of variable thickness, and the energy of each particle was recorded as it exited the slab. Figure 6.6 shows the resulting histograms smoothed and normalized to the particle fluence (Sierawski 2011). The curves represent a probability density function of incident kinetic energies for the degrader selection. The simulations demonstrate that degrader thicknesses of 5, 10, and 15 mm still produce reasonably shaped energy spectra for single-event upset testing with peak energies at 53, 40, and 20 MeV (Sierawski 2011). For thicker degraders producing lower-energy protons (<10 MeV), the beam becomes widely spread, and, in this case, the use of a direct production method with a dedicated low-energy accelerator is preferable.

Typical test plans for monoenergetic protons are similar to those considered for monoenergetic neutrons (see Section 6.3.1.1). Especially, high-energy proton cross-section values combined with additional data obtained with low-energy neutrons can be used to estimate the energy-dependent cross-section approximation using Equation 6.8 and then to calculate the circuit SER from Equation 6.9.

6.3.4 Muons

Similarly to low-energy protons, the evolution of CMOS technologies toward the decananometer range of integration has led to expanding the spectrum of atmospheric particles to which a circuit is sensitive. Low-energy muons are now capable of creating upsets despite their very low ionizing power. As we will detail in Section 10.4.2, the first experiments with muon beams on modern technologies have been conducted in very recent years. Only a few muon beam facilities (coupled with high-energy proton beams) exist at international level; we can cite the following most important facilities:

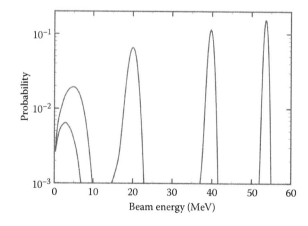

FIGURE 6.6
Simulated kinetic-energy spectra of a 65 MeV proton beam after degradation by 5, 10, 15, 17, and 17.25 mm aluminum slabs. (After Sierawski, B., The role of singly-charged particles in microelectronics reliability, PhD Dissertation, Vanderbilt University, 2011.)

1. The Swiss muon source at the Paul Scherrer Institute (PSI), Switzerland

2. The RIKEN-RAL muon facility at ISIS, Rutherford Appleton Laboratory (RAL), United Kingdom

3. The M20 beam line at TRIUMF (Marshall 1992)

4. The PHASOTRON facility at the Joint Institute for Nuclear Research (JINR, Dubna), Russia

Given the growing importance of low-energy muon reliability issues for future generations of CMOS technologies, it is very likely that muon-accelerated test studies will increase in the coming years. Future revisions of the different test standards and guidelines will certainly include a dedicated section about these tests in order to recommend standardized procedures for the radiation effect community.

6.4 Alpha-Particle Accelerated Tests Using Solid Sources

In alpha-particle accelerated tests, the chip under test is subjected to an intense flux of alpha particles with similar characteristics (energies) to those of alphas emitted by residual traces of radioactive impurities (such as uranium and thorium) or by the radioactive materials present in the circuit or packaging materials. The alpha source species should therefore be carefully selected in order to obtain an alpha-particle spectrum similar to that related to the package and the device:

- Sources that emit alpha particles with energy spectra similar to those of uranium and thorium impurities are well adapted to simulate the particular radiation environment encountered in wire-bonded components encapsulated in molding compound (JEDEC 2006). ^{238}U or ^{232}Th metal sources can be used, although they present some limitations (toxicity, mixed thick spectrum, low surface flux), and the use of these sources is restricted in certain countries. In this case, a ^{241}Am source should be used as a substitute.

- ^{241}Am sources have similar spectra to ^{210}Po and can be used to reproduce the radiation environment of components in a flip-chip arrangement with solder bumps (JEDEC 2006; Slayman 2011). Alternatively, ^{210}Po sources can be directly used, in spite of the short half-life of this isotope (138.4 days) and the resulting uncertainty in determining the source activity and effective flux.

Several types of sources exist: thin-film, metal-foil, and solid sources. Since thin-film and metal-foil sources have discrete energy spectra (Figure 6.7), they provide a representative spectrum if the real source of alpha particles that exists in the product comes from a thin layer at the die surface. Solid sources that are physically thicker broaden the alpha-particle spectrum, as illustrated in Figure 6.7. Indeed, since alpha particles can be emitted at any depth from the surface, discrete emission lines are broadened into a continuous spectrum of alpha energies. These thick solid sources emulate the alpha spectrum as it would be in a real packaged device with radioactive impurities embedded in the bulk of the packaging materials.

Alpha-particle accelerated tests can be performed either in ambient air or in primary vacuum using a dedicated setup (vacuum chamber). In all cases, the DUT's surface must

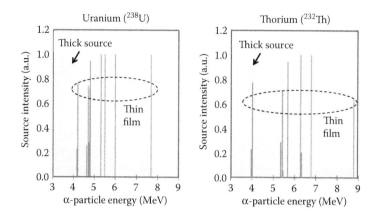

FIGURE 6.7
Comparison between the emission spectra obtained for thin-film and thick sources of ^{238}U and ^{232}Th.

be directly exposed to the source without any intervening solid material (JEDEC 2006). Ceramic dual-in-line (CERDIP) or pin-grid array (CERPGA) packages are the recommended DUT package types for such experiments. When performing the test in ambient air, the distance between the source and the DUT should be typically less than 1 mm in order to minimize energy loss of alpha particles in air. As noted by Tam (2011), important differences (between 1 and 3 mm) can be observed in this minimal distance between the source and the die surface as a function of the device configuration (direct soldered device, device mounted in debug or custom-designed socket, etc.). Figure 6.8 illustrates the placement of an intense ^{241}Am source over the DUT before its characterization using an industrial tester. The custom socket has been designed as a function of the source geometry to support it perfectly at a precisely known distance from the die surface. Wafer-level testing can also be used to perform alpha-particle testing. The advantage of wafer-level testing is evident in terms of simplification and cost reduction due to the elimination of the steps of wafer dicing and chip packaging. However, the wafer-level test system must have a special probecard configured to allow the alpha-particle source to be placed accurately and in close proximity to the die without shorting the probe pins (JEDEC 2006).

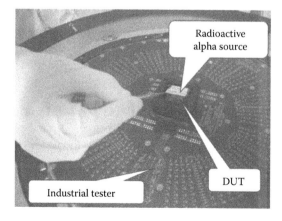

FIGURE 6.8
Placement of a radioactive alpha source (^{241}Am) over the DUT before its characterization using an industrial tester. (Courtesy of P. Roche.)

To conclude, a typical test plan for accelerated alpha particle testing should be as follows:

1. DUT setup.
2. Expose DUTs to source and record time and errors; at this level, to check the consistency of the overall test setup (hardware and software), the number of errors measured (N_{SEU}) must be perfectly proportional to the test duration (T) without any offset. An example of real α-ASER measurements is shown in Figure 6.9 for a 65 nm SRAM memory irradiated using an intense [241]Am source (activity of 3.7 MBq).
3. Calculate the average cross section, in square centimeters, using

$$\overline{\sigma_{alpha}} = \frac{N_{SEU}}{\Phi_{DUT}} \tag{6.11}$$

where Φ_{DUT} is the source fluence (α cm^{-2}) at the level of the die surface calculated from the source flux ϕ_{Source} (α cm^{-2} s^{-1}) and the test duration T (s):

$$\Phi_{DUT} = \phi_{Source} \times F_{Correction} \times T \tag{6.12}$$

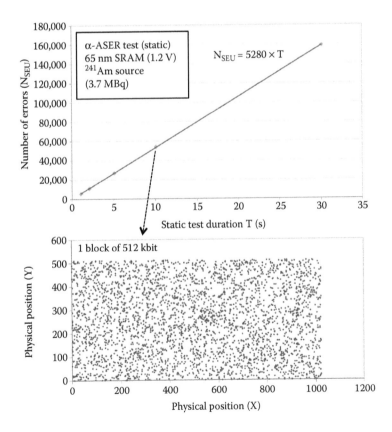

FIGURE 6.9
Illustration of an α-ASER test performed on a 65 nm SRAM circuit using an intense [241]Am source (thin foil, 3.7 MBq). *Top*: Number of detected errors in the chip as a function of test duration. *Bottom*: Error bitmap after 10 s of alpha irradiation for a memory block of 512 kbit.

In Equation 6.12, the term $F_{Correction}$ (≤ 1) is a correction factor introduced to take into account the flux reduction at the level of the die surface due to all the geometrical and shielding effects introduced by the experimental setup (source geometry and dimensions, source–device spacing, source–device alignment, etc.). Actual device and source geometries can be accurately calculated with computer modeling, accounting for typical test situations such as rectangular die and nonaxial alignments (JEDEC 2006). If the test is performed in a vacuum with a source that is very much larger than the component being tested, with a known and calibrated source flux and with a small DUT-to-alpha source spacing (<1 mm), the need for an accurate geometry calculation is minimized, since this correction factor will be close to 1 (JEDEC 2006).

4. Finally, evaluate the alpha SER of the device, in FIT, using

$$SER_{FIT} = 10^9 \times \overline{\sigma_{alpha}} \times \Phi_{package} \times \frac{FIT}{Errors/Hour} \tag{6.13}$$

where $\Phi_{package}$ is the background alpha flux from the actual product package and device impurities. Ideally, this value should be independently measured using experimental techniques as described in Chapter 4. Alternatively, the reported SER values may be extrapolated to the nominal alpha flux corresponding to the expected emissivity of circuit materials (e.g., 0.02 α cm^{-2} h^{-1} for low alpha or 0.002 α cm^{-2} h^{-1} for ultralow alpha).

6.5 Evaluation of Various Neutron Broad-Spectrum Sources from a Simulation Viewpoint

As shown in Section 6.3.1, several broad-spectrum, high-energy neutron source facilities are available at international level for ASER testing. However, none accurately replicates terrestrial neutron spectra across the full energy range above 1 MeV. A few recent works (Platt et al. 2010; Slayman 2010b,d) investigated in detail the characteristics of these atmospheric-like neutron facilities to quantify the errors introduced in accelerated soft-error measurements from these facilities. In Serre et al. (2012), we investigated by simulation the neutron–silicon (n–Si) nuclear events resulting from the interaction of neutrons produced by these different synthetic sources. Our primary objective was to understand, from possible differences in nuclear event probabilities, the differences observed in the SER of the same circuit estimated from these different sources of neutrons. This section aims to summarize this study and to highlight the most important conclusions for ASER testing.

6.5.1 Simulation Details

6.5.1.1 Wide Neutron Sources

For the purposes of this study, three different broad-spectrum, high-energy neutron sources have been compared with the terrestrial neutron spectrum at New York City referenced in the JEDEC standard (JEDEC 2006): these facilities correspond to the atmospheric-like TRIUMF, LANSCE, and ANITA neutron spallation sources already introduced in

Section 6.3.1. The neutron-flux energy distributions (differential flux) of these sources are presented in Figure 6.4. None of these synthetic sources is able to accurately replicate the terrestrial neutron spectrum above 1 MeV, mainly due to the limited energy of the incident proton beam used to produce by spallation any neutron flux above a few hundreds of MeV. In addition to the four broad-spectrum neutron sources, we have also considered five monoenergetic values (1, 10, 50, and 100 MeV and 1 GeV) for data analysis and discussion. To accurately generate neutron flux following the distributions shown in Figure 6.4, we developed dedicated source models using the General Particle Source (GPS), a specialized Geant4 class (class name: G4GeneralParticleSource) used as the particle generator for Geant4 simulation (Geant4-c).

6.5.1.2 Geant4 Simulations

To produce nuclear event databases for the mentioned neutron sources, we used Geant4.9.4p01 (Geant4-a). The list of physical processes employed in simulation was based on the standard package of physics lists QGSP_BIC_HP (Geant4-physics-list). The complete list of Geant4 classes that we considered for this general-purpose neutron simulation is summarized in Table 6.2. For each neutron source, a simulation run consists in the generation of 5×10^8 primary neutrons incident on a 20 μm thick silicon (natural isotope composition) layer perpendicular to its surface (1 cm²). The choice of the target thickness is justified by the fact that the electrical charge generated by a reaction product beyond 20 μm would not drift or diffuse to the active area and, consequently, would not play any role in the soft-error occurrence. All the resulting silicon recoil nuclei with a kinetic energy greater than 40 keV and secondary charged particles (except electrons and pions) are accumulated into the database (reaction position, kinetic energy, and momentum). Table 6.3 summarizes the main size characteristics of the different generated databases. Two size parameters (or metrics) are reported in Table 6.3 to quantify the different databases in terms of n–Si interactions: the total number of events (i.e., the total number of elastic and inelastic events for the incoming neutrons) and the total number of generated products, including all silicon recoil nuclei with a kinetic energy above 40 keV and all secondary ions (i.e., fragments) created as a result of inelastic collisions. This threshold value of 40 keV is quite arbitrary and corresponds to silicon nuclei able to deposit in silicon material a maximum of 1.8 fC

TABLE 6.2

List of Considered Geant4 Classes in Simulation Flow for Description of Neutron Interactions

Neutron Process	Energy	Geant4 Model	Dataset
Elastic	<20 MeV	G4NeutronHPElastic	G4NeutronHPElasticData
	>20 MeV	G4LElastic	—
Inelastic	<20 MeV	G4NeutronHPInelastic	G4NeutronHPInelasticData
	(20 MeV, 10 GeV)	G4BinaryCascade	—
	(10 GeV, 25 GeV)	G4LENeutronInelastic	—
	(12 GeV, 100 TeV)	QGSP	—
Fission	<20 MeV	G4NeutronHPFission	G4NeutronHPFissionData
	>20 MeV	G4LFission	—
Capture	<20 MeV	G4NeutronHPCapture	G4NeutronHPCaptureData
	>20 MeV	G4LCapture	

Source: Serre, S. et al., *IEEE Trans. Nucl. Sci.* 59, 714–722, 2012.

TABLE 6.3

Characteristics of Databases Generated from the Irradiation of a 20 μm Thick Silicon Layer with 5×10^8 Neutrons

Neutron Source	Total Number of Events[a]	Total Number of Generated Products[b]
1 MeV	86,682	86,682
10 MeV	79,228	101,016
100 MeV	49,961	90,808
1 GeV	39,731	106,493
JEDEC	66,364	93,715
LANSCE	71,054	92,935
TRIUMF	61,077	84,509
ANITA	80,039	93,549

Source: Serre, S. et al., *IEEE Trans. Nucl. Sci.* 59, 714–722, 2012.

[a] Elastic + inelastic events.

[b] Silicon recoil nuclei (E > 40 keV) + secondary ions.

of electrical charge; below this value, the deposited charge is considered to be too small to induce any upset in the SRAM test vehicle (65 nm technology) used later in this study (see Section 6.5.3). The two metrics logically show that the probability of n–Si interactions (in the range 10^{-5}–10^{-6} per micron) strongly depends on the energy/spectrum of the neutron source, and also that the nature of interactions can vary from pure elastic events (for 1 MeV) to predominantly inelastic reactions with high multiplicity (for 1 GeV, we have on average 106,493/39,731 or approximately three fragments per reaction).

6.5.2 Nuclear Event Analysis

Figures 6.10 through 6.13 graphically recapitulate the analysis of the databases in terms of secondary-ion production (Figure 6.10), energy histogram of produced ions (Figure 6.11), nuclear-reaction-induced shower multiplicity (Figure 6.12), and ratio of elastic/inelastic processes (Figure 6.13) for the different atmospheric(-like) and monoenergetic neutron sources. The analysis takes into account all secondary ions produced by both elastic and inelastic processes. As shown in Figure 6.10, the distributions of secondary-ion products are a strong function of the incident neutron energy or energy spectrum: at 10 MeV, for example, the only products that appear in the database are hydrogen, helium, magnesium, aluminum, and silicon. With neutron energy increase, further secondary products can be created, such as fluorine, beryllium, and phosphorus created by silicon (^{30}Si) capturing the incident neutron and decaying by beta emission. Concerning the broad-spectrum sources, these results highlight some differences between the JEDEC, LANSCE, TRIUMF, and ANITA databases, especially at high energies (>100 MeV): this is evidenced by the different proportions of beryllium, boron, carbon, nitrogen, and fluorine ions (Figures 6.10 and 6.11) and by a higher shower multiplicity for JEDEC and LANSCE, with respect to TRIUMF and ANITA particularly (Figure 6.12). Results for other secondary ions are comparable, demonstrating the relatively good matching of these atmospheric-like neutron distributions at intermediate energies (10–100 MeV) with respect to the JEDEC distribution. The differences in triggering some nuclear reactions with threshold energies above

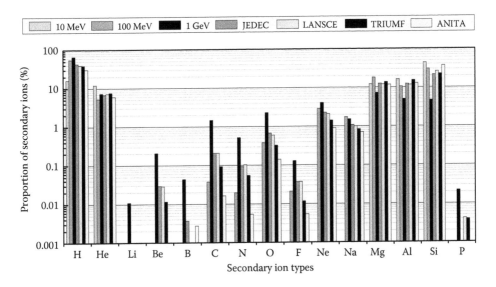

FIGURE 6.10

Proportion of secondary ions produced by n–Si interactions (elastic and inelastic processes) in the compiled databases for atmospheric(-like) and monoenergetic sources of neutrons. For each secondary-ion type, the histogram bars correspond to (from *left* to *right*) 10 MeV, 100 MeV, 1 GeV, JEDEC, LANSCE, TRIUMF, and ANITA sources, respectively. (Reprinted from Serre, S. et al., Geant4 analysis of n–Si nuclear reactions from different sources of neutrons and its implication on soft-error rate, *IEEE Trans. Nucl. Sci.* 59, 714–722. © (2012) IEEE. With permission.)

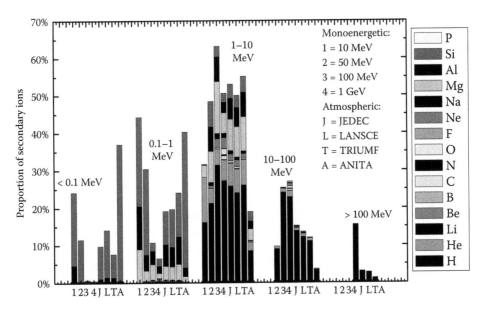

FIGURE 6.11

Energy histogram for the secondary-ion cocktails (from Figure 6.10) produced by n–Si interactions for the different neutron sources considered. (Reprinted from Serre, S. et al., Geant4 analysis of n-Si nuclear reactions from different sources of neutrons and its implication on soft-error rate, *IEEE Trans. Nucl. Sci.* 59, 714–722. © (2012) IEEE. With permission.)

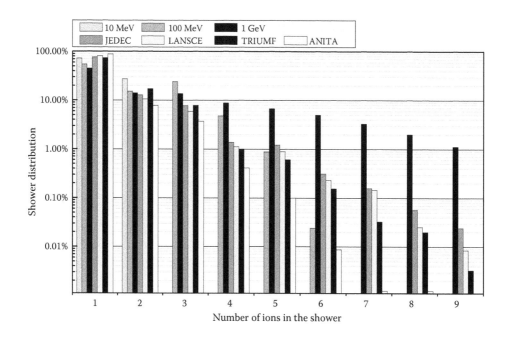

FIGURE 6.12

Shower distribution as a function of shower multiplicity (i.e., number of ions generated in n–Si interaction). Elastic recoil events are taken into account in this distribution with a unit multiplicity. For each shower multiplicity, the histogram bars correspond to (from left to right) 10 MeV, 100 MeV, 1 GeV, JEDEC, LANSCE, TRIUMF, and ANITA sources, respectively. (Reprinted from Serre, S. et al., Geant4 analysis of n-Si nuclear reactions from different sources of neutrons and its implication on soft-error rate, *IEEE Trans. Nucl. Sci.* 59, 714–722. © (2012) IEEE. With permission.)

100 MeV can be directly linked to the values of the cutoff energies, which are not the same for the facilities LANSCE (using a spallation of a heavy metal target by 800 MeV protons), TRIUMF (irradiation of a lead absorber with protons up to 500 MeV), and ANITA (spallation of a tungsten target with a 180 MeV proton beam) (Slayman 2010b). At this stage of database analysis, this suggests that the JEDEC spectrum represents a slightly more severe environment above 10 MeV than the one related to LANSCE, TRIUMF, and ANITA, if we consider a strict count of the number of by-products generated, especially above 100 MeV (Figure 6.11). On the contrary, below 1 MeV, ANITA induces the largest number of recoil nuclei and fragments as compared with the three other sources.

To conclude this analysis, we report in Figure 6.13 the ratio of the reaction types (elastic and inelastic) for the different neutron energies/sources. In a first approximation, for atmospheric-like neutron spectra, elastic interactions are the dominant mechanism because of the relatively high proportion of low-energy neutrons (well below 10 MeV), which cannot cause any inelastic nuclear reaction, since the incident neutron energy is lower than the energy threshold of the majority of nuclear reactions of this type. In more detail, the ratio of reaction types as a function of neutron energy is the direct consequence of the nuclear-interaction cross-section libraries and models used in the MC simulation. Geant4 uses, below 150 MeV, evaluated values for the nuclear-interaction cross sections given by the G4NDL neutron data library (a combination of Evaluated Nuclear Data File [ENDF] and Japanese Evaluated Nuclear Data Library [JENDL] libraries) and, above 150–200 MeV, predicted values by the binary cascade model BIC (see Geant4-b for details). The data shown in Figure 6.13 show the same trends as the variations of the cross sections with incident

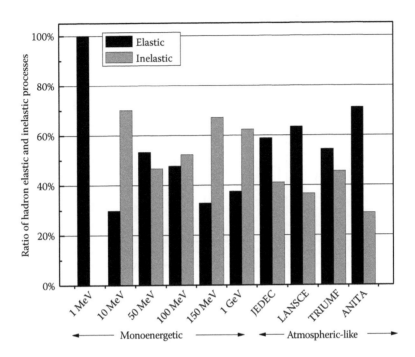

FIGURE 6.13
Ratio of elastic and inelastic processes for the different sources of neutrons. (Reprinted from Serre, S. et al., Geant4 analysis of n-Si nuclear reactions from different sources of neutrons and its implication on soft-error rate, *IEEE Trans. Nucl. Sci.* 59, 714–722. © (2012) IEEE. With permission.)

neutron energy according to the ENDF/B-VII.0 and JENDL-3.3 databases (not shown), which explains the irregular behavior of the respective contributions of elastic and inelastic processes with energy, highlighted in Figure 6.13.

6.5.3 Implications for the Soft-Error Rate

For this second part, we interfaced the different databases with our TIARA MC SER simulation code (described in detail in Section 9.3) to evaluate, for each atmospheric or monoenergetic neutron source, the SER related to a well-known and well-characterized SRAM test-vehicle circuit manufactured in a commercial CMOS 65 nm technology. The results of these simulations will be detailed in Section 9.4.3. In the following, we only consider the SER values deduced from these simulations (reported in the second column of Tables 6.5 and 6.6) for the purpose of the discussion.

6.5.3.1 SER Estimations from Monoenergetic Values

We first examine the possibility of simply estimating the SER values related to the different JEDEC, TRIUMF, LANSCE, or ANITA sources from monoenergetic SER data obtained by simulation. In other words, considering, on the one hand, the different spectra of Figure 6.4 and, on the other hand, the SER values for monoenergetic neutrons of Figure 6.14, an evident question should be: how can the SER value for a *broad-spectrum neutron source* be estimated from a few discrete simulation values obtained with monoenergetic neutrons? The principle of our methodology is as follows. Partitioning the distributions of neutron

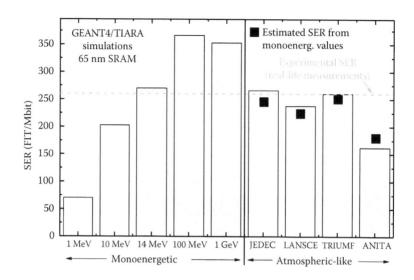

FIGURE 6.14

Comparison between SER values obtained for the 65 nm SRAM test vehicle on ASTEP (real-time testing corrected for alpha-SER contribution) and at LANSCE and TRIUMF (estimation only) facilities. (Reprinted from Serre, S. et al., Geant4 analysis of n-Si nuclear reactions from different sources of neutrons and its implication on soft-error rate, *IEEE Trans. Nucl. Sci.* 59, 714–722. © (2012) IEEE. With permission.)

flux into three segments of 1–10 MeV, 10–100 MeV, and >100 MeV, as performed by Slayman (2010b) and reported in Table 6.4, we first evaluate the average SER for a given interval from monoenergetic SER values for the bounds of this interval. Due to the logarithmic scale for neutron energy, we opt for a geometric (or logarithmic) average value, as reported in Table 6.5 for the three considered intervals. In a second step, we weight these different values by their relative fraction in the total neutron flux. We obtain

$$SER[SOURCE] = A \times SER[1-10] + B \times SER[1-100] + C \times SER[100-1000]$$

$$= A \times \sqrt{SER_1 SER_{10}} + B \times \sqrt{SER_{10} SER_{100}} + C \times \sqrt{SER_{100} SER_{1000}} \quad (6.14)$$

TABLE 6.4

Fractions (Coefficients A, B, and C) of Neutron Flux in Three Energy Segments

	Neutron Flux		
Source	1–10 MeV A	10–100 MeV B	>100 MeV C
JEDEC (New York City)	0.35	0.35	0.3
LANSCE	0.46	0.29	0.25
TRIUMF	0.24	0.54	0.21
ANITA	0.65	0.28	0.07

Source: Modified from Slayman, C., *IEEE Trans. Nucl. Sci.* 57, 3163–3168, 2010.

TABLE 6.5

SER Values for Monoenergetic Neutron Sources and Geometric Averaged SER Values on the Intervals between Two Consecutive Neutron Energies

Neutron Energy (MeV)	Monoenergetic SER (FIT/Mbit)	Averaged SER on Interval Domain (FIT/Mbit) $\sqrt{SER_i SER_{i+1}}$
1	$SER_1 = 70$	$SER[1-10] = 119.2$
10	$SER_{10} = 203$	$SER[10-100] = 272.6$
100	$SER_{100} = 366$	$SER[100-1000] = 359.4$
1000	$SER_{1000} = 353$	

Source: Serre, S. et al., *IEEE Trans. Nucl. Sci.* 59, 714–722, 2012.

where A, B, and C are the fractions of the neutron flux for the three energy segments defined in Table 6.4, and SER_1, SER_{10}, SER_{100}, and SER_{1000} the four SER values deduced from monoenergetic tests/simulation runs at 1, 10, and 100 MeV and 1 GeV, respectively (Table 6.5).

The resulting SER values are plotted in Figure 6.14 (squares) and reported in Table 6.5. We show excellent agreement for the four broad-spectrum sources, with an error always less than 10% on the SER values. Such a quantitative example illustrates the capability of this simple approximation to estimate a broad-spectrum source SER from a few monoenergetic-neutron SER values. The method should be applied and generalized for other sources, other energy intervals, and numbers of discrete values. Its main interest is not necessarily experimental (this would probably be difficult because of the number and range of energy of neutron beams to be considered and would not be economically viable for conducting accelerated tests), but, rather, computational, since it represents a gain of time (and also a simplification in the simulation input) to compute the SER values for a panel of broad-spectrum sources from a single set of monoenergetic values.

6.5.3.2 SER Estimations from the Neutron Cross Section

Another alternative method to deduce the SER value of a broad-spectrum neutron source is to consider the cross-section distribution derived from monoenergetic simulations (or measurements) and to apply the methodology proposed by Platt et al. (2010) and adopted by Slayman (2010b). To do this, we first deduce from the data of Table 6.4 the monoenergetic cross-section values for the SRAM circuit investigated. Then, we approximate this cross-section distribution with the four-parameter Weibull distribution given by Equation 6.8. Using this parameterized equation and applying the method proposed by Slayman (Slayman 2010b), it is possible to evaluate numerically the SER ratio for the different facilities (LANSCE, TRIUMF, and ANITA), taking as the reference the SER value obtained from the JEDEC spectrum. This ratio quantifies the capability for a given source to predict the SER value specified by the standard, that is, the source chosen as reference. Table 6.6, rightmost column, reports these different SER ratios for the different sources with respect to the JEDEC reference. Multiplying each ratio value by the JEDEC SER, equal to 266 FIT/Mbit, allows us to estimate the SER for the other sources: 238, 241, and 195 FIT/Mbit for the LANSCE, TRIUMF, and ANITA sources, respectively. These estimations are in good agreement with TIARA values obtained by considering the full source spectrum in the MC simulations (Table 6.6, second column), with a maximum error of about 20% for ANITA. If we remember that all these results have been derived from the combination of

TABLE 6.6

Summary of SER Values Obtained from TIARA Simulations and Estimated from Monoenergetic Values and from SER Ratio Method

	SER (FIT/Mbit)		
Source	TIARA + Full Source Spectrum	TIARA Monoenergetic + Equation 6.14	From σ(E) + Integration JEDEC SER Ratio/SER Value
JEDEC	266	245	1/–
LANSCE	238	224	0.894/238
TRIUMF	260	251	0.905/241
ANITA	162	179	0.732/195

Source: Serre, S. et al., *IEEE Trans. Nucl. Sci.* 59, 714–722, 2012.
Note: The experimental SER measured by real-time measurements is 259 FIT/Mbit.

Geant4 databases with TIARA simulations, they are in good agreement with the predictions deduced from the abacus calculated by Slayman (Slayman 2010b); in this sense, they offer a numerical verification of such a methodology for a given 65 nm SRAM technology. The generalization of the present work to other technologies should therefore be subjected to additional verification.

References

Andreani, C., Pietropaolo, A., Salsano, A., et al. 2008. Facility for fast neutron irradiation tests of electronics at the ISIS spallation neutron source. *Applied Physics Letters* 92:114101–114101-3.

Blackmore, E. 2009. Development of a large area neutron beam for system testing at TRIUMF. In *Proceedings of the IEEE Radiation Effects Data Workshop*, pp. 157–160. 20–24 July, Quebec City, QC, IEEE.

Blackmore, E., Dodd, P., and Shaneyfelt, M. 2003. Improved capabilities for proton and neutron irradiations at TRIUMF. In *Proceedings of the IEEE Radiation Effects Data Workshop*, pp. 149–155. 21–25 July, Monterey, CA, IEEE.

Compendium. 2011. Compendium of international irradiation test facilities. In S. K. Höffgen and G. Berger (eds), *RADECS 2011 Conference*, Seville, Spain. Available at: http://acdc.sav.us.es/cna/images/documentos/Irradiation%20Facilities%20Catalogue%20RADECS%202011.pdf.

Geant4-a. Geant4 9.4, released 25 February 2011 (patch-01), http://geant4.cern.ch/

Geant4-b. http://geant4.cern.ch/Geant4/UserDocumentation/UsersGuides/PhysicsReference Manual/fo/PhysicsReferenceManual.pdf.

Geant4-c. Geant4 general particle source. Geant4. http://geant4.web.cern.ch/geant4/UserDocumentation/UsersGuides/ForApplicationDeveloper/html/ch02s07.html.

Geant4-physics-list. http://geant4.cern.ch/support/proc_mod_catalog/physics_lists/referencePL.shtml.

Gordon, M.S., Goldhagen, P., Rodbell, K.P., et al. 2004. Measurement of the flux and energy spectrum of cosmic-ray induced neutrons on the ground. *IEEE Transactions on Nuclear Science* 51:3427–3434.

Grandlund, T., Granbom, B., and Olsson, N. 2004. A comparative study between two neutron facilities regarding SEU. *IEEE Transactions on Nuclear Science* 51:2922–2926.

IEC. 2008a. Semiconductor devices—Mechanical and climatic test methods—Part 38: Soft error test method for semiconductor devices with memory, IEC Std. IEC 60749-38:2008, May 2008.

IEC. 2008b. Process management for avionics—Atmospheric radiation effects—Part 2: Guidelines for single event effects testing for avionics systems, IEC Std. IEC/TS 62396-2:2008, August 2008.

JEDEC. 2006. Measurement and reporting of alpha particle and terrestrial cosmic ray-induced soft errors in semiconductor devices, JEDEC Standard JESD89A, October 2006. Available at: http://www.jedec.org/sites/default/files/docs/jesd89a.pdf.

JEITA. 2005. JEITA SER testing guideline, JEITA Std. EDR-4705. Available at: http://tsc.jeita.or.jp/tsc/standard/pdf/EDR-4705.pdf.

Li, X., Shen, K., and Huang, M.C. 2007. A memory soft error measurement on production systems. In *Proceedings of the USENIX Annual Technical Conference*, pp. 275–280, 17–22 June, Santa Clara, CA.

Lisowski, P.W., Bowman, C.D., Russell, G.J., and Wender, S.A. 1990. The Los Alamos National Laboratory spallation neutron sources. *Nuclear Science and Engineering* 106:208–218.

Marshall, G.M. 1992. Muon beams and facilities at TRIUMF. *Zeitschrift fur Physik C Particles and Fields* 56:226–231.

Mitaroff, A. and Silari, M. 2002. The CERN-EU high-energy reference field (CERF) facility for dosimetry at commercial flight altitudes and in space. *Radiation Protection Dosimetry* 102(1):7–22.

Platt, S.P., Prokofiev, A.V., and Xiao, C.X. 2010. Fidelity of energy spectra at neutron facilities for single-event effects testing. In *Proceedings of the IEEE International Reliability Physics Symposium*, pp. 411–416. 2–6 May, Anaheim, CA, IEEE.

Prokofiev, A.V., Blomgren, J., Nolte, R., Platt, S.P., Röttger, S., and Smirnov, A.N. 2008. Characterization of the ANITA neutron source for accelerated SEE testing at the Svedberg Laboratory. In *Proceedings of the European Workshop on Radiation Effects on Components and Systems*, pp. 260–267, 10–12 September, Jyväskylä, Finland.

Prokofiev, A.V., Blomgren, J., Nolte, R., Platt, S.P., Röttger, S., and Smirnov, A.N. 2009. ANITA—A new neutron facility for accelerated SEE testing at the Svedberg Laboratory. In *Proceedings of the IEEE International Reliability Physics Symposium*, pp. 929–935, 26–30 April, Montreal, Canada.

Prokofiev, A.V., Byström, O., Ekström, C., et al. 2006. A new neutron beam facility at TSL. In *Proceedings of the International Workshop on Fast Neutron Detectors and Applications (FNDA2006)*, April 2006, poS(FNDA2006)016. Available at: http://pos.sissa.it/archive/conferences/025/016/FNDA2006_016.pdf.

Sakai, H., Okamura, H., Otsu, H., et al. 1996. Facility for the (p, n) polarization transfer measurement. *Nuclear Instruments and Methods in Physics Research. Section A. Accelerators, Spectrometers, Detectors and Associated Equipment* 369:120–134.

Serre, S., Semikh, S., Uznanski, S., et al. 2012. Geant4 analysis of n-Si nuclear reactions from different sources of neutrons and its implication on soft-error rate. *IEEE Transactions on Nuclear Science* 59:714–722.

Sierawski, B. 2011. The role of singly-charged particles in microelectronics reliability. PhD Dissertation, Vanderbilt University.

Slayman, C. 2010a. Acceleration factors for neutron beam soft error testing in commercial and avionic environments. 2010 Workshop on Accelerated Stress Testing and Reliability (ASTR). Available at: http://www.ieee-astr.org/past_presentations/2010-Slayman-Acceleration%20Factors%20for%20Neutron%20Beam%20Soft%20Error%20Testing.pdf.

Slayman, C. 2010b. Accuracy of various broad spectrum neutron sources for accelerated soft error testing. In *Proceedings of the SELSE Workshop*, pp. 1–11, 23–24 March, Stanford, CA.

Slayman, C. 2010c. Impact and mitigation of soft errors. In Tutorial at the IEEE International Reliability Physics Symposium, 2–6 May, Anaheim, CA.

Slayman, C. 2010d. Theoretical correlation of broad spectrum neutron sources for accelerated soft error testing. *IEEE Transactions on Nuclear Science* 57:3163–3168.

Slayman, C. 2011. JEDEC standards on measurement and reporting of alpha particle and terrestrial cosmic ray induced soft errors. In M. Nicolaidis (ed.), *Soft Errors in Modern Electronic Systems*. New York: Springer.

Tam, N. 2011. Investigation of accelerated alpha testing with vacuum. In *Third Annual IEEE-SCV Soft Error Rate (SER) Workshop*, Santa Clara, CA. Available at: http://ewh.ieee.org/soc/cpmt/presentations/cpmt1110w-5.pdf.

Violante, M., Sterpone, L., Manuzzato, A., et al. 2007. A new hardware/software platform and a new 1/e neutron source for soft error studies: Testing FPGAs at the ISIS facility. *IEEE Transactions on Nuclear Science* 54:1184–1189.

Wilkinson, J. 2010. Standard test methods. In Tutorial at the IEEE International Reliability Physics Symposium, 2–6 May, Anaheim, CA.

Zhang, L.H. 2013. Neutron beam monitoring for single-event effects testing. PhD Thesis, University of Central Lancashire.

Ziegler, J.F. and Puchner, H. 2004. *SER—History, Trends and Challenges*. San Jose, CA: Cypress Semiconductor.

7

Real-Time (Life) Testing

7.1 Introduction

As explained in the previous chapter, different experimental approaches can be considered to estimate the SER of a given device, circuit, or system (Ziegler and Puchner 2004; JEDEC 2006): field tests, accelerated tests, and real-time (or unaccelerated) tests. Field testing consists in collecting errors from a large number of finished products already on the market. The SER value is evaluated a posteriori from the errors experienced by consumers themselves; it generally takes several years after the introduction of the product on the market. Accelerated tests have been largely described in Chapter 6. Accelerated tests use intense particle beams or sources chosen for their capability to mimic the atmospheric (neutron) spectrum or to generate alpha particles within the same energy range as the alphas emitted by radioactive contaminants. This accelerated SER (ASER) method is fast (data can be obtained in a few hours or days instead of months or years as for the other methods), is a priori easy to implement, and only requires a few functional chips to estimate the SER. This allows the manufacturer to perform such radiation tests relatively early in the production cycle. Another major and growing advantage is its capability to quantify from a very large statistic (cumulated number of events) the importance of multiple cell/multiple bit upsets in the radiation response of ICs fabricated in technological nodes typically below 65 nm. But data can be potentially tainted by experimental artifacts (more or less well controlled according to the facility, the experimental setup, or other various experimental conditions). As a direct consequence, ASER results must be extrapolated to use conditions, and several different radiation sources must be used to ensure that the estimation accounts for soft errors induced by both alpha-particle and cosmic-ray-neutron events. We will discuss these issues in Section 7.5.

Real-time SER (RTSER), or life testing, can be considered, in a way, as a middle path between the two previous approaches. As in field testing, the method considers a large population of circuits working in the natural radiation environment and, as in accelerated testing, at least for neutrons, the intensity of the natural radiation can be increased by deploying the test at altitude. However, the acceleration factor (i.e., the ratio of the neutron integrated flux at the test location divided by its reference value at New York City; see Section 1.3.1.1) has nothing to do with those reached in accelerated tests. Considering the composition of the radiation background to be equivalent at altitude and at sea level (this point will be discussed in Section 7.2), typical values between 5 and 20, as a function of the test location on Earth, can be expected. Devices must thus be tested for a long enough period of time (months or years) until enough soft errors have been accumulated to give a reasonably confident estimate of the SER.

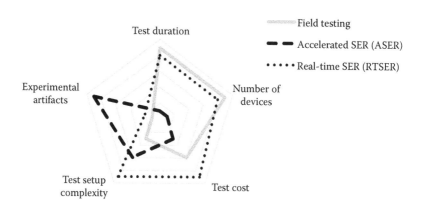

FIGURE 7.1

Radar chart illustrating the test specificities of each approach in terms of test duration, number of devices, test cost, test setup complexity, and experimental artifacts (the relative position and angle of the axes is uninformative). (Reprinted from *Microelectronics Reliability*, 54, Autran, J.L., Munteanu, D., Roche, P., and Gasiot, G., Real-time soft-error rate measurements: A review, 1455–1476, Copyright (2014), with permission from Elsevier.)

As declared by the JEDEC standard JESD89A (JEDEC 2006), the term *accelerated* should be reserved for intense radiation sources that do not occur in natural terrestrial environments. *System SER* is another term that is often used and is considered synonymous with RTSER. Real-time testing has the major advantage of being a direct measurement of the actual product SER, requiring no intense radiation sources, extrapolations to use conditions, and so on (JEDEC 2006). However, real-time testing does require an expensive system capable of monitoring hundreds or thousands of devices in parallel, for long periods of time. All these aspects will be described and discussed in details in the following sections.

Figure 7.1 summarizes the different ways to evaluate the SER in the form of a two-dimensional chart, highlighting test specificities of each approach. Of course, such information is only indicative, but this figure has the merit of clearly showing fundamental differences between RTSER and ASER test strategies, notably in terms of test duration, number of devices, test cost, test setup complexity, and experimental artifacts.

7.2 Real-Time Testing Methodology

The primary objective of a RTSER test is to obtain a well-defined estimation of the total SER for the component/circuit under test and, ideally, to determine the respective contributions of the different radiation constraints to this failure rate. As the SER is generally (extremely) low at ground level, the methodology consists in a direct observation of a (very) large number of devices working in parallel under standard operating conditions and exposed to ambient background radiation. We examine in the following several key points of the RTSER test methodology, including some instrumentation issues, the different ways to separate the SER components, and the importance of radiation background metrology for the accurate estimation of SER.

7.2.1 Instrumentation Issues

In RTSER experiments, the role of the automatic test equipment (ATE) is crucial in the detection and identification of errors related to the population of devices under test. Because soft errors can be considered as *rare events* in such a real-time approach (precisely due to the weak natural radiation constraint), the design of this ATE is technically complex in order to ensure that neither the hardware nor the software introduces any artifact or additional error during the process of detection and counting of soft errors. Moreover, the number of chips involved in the experiment must be as large as possible to reach satisfactory statistics in a reasonable duration; this also introduces additional difficulties in terms of setup complexity, power management, cost, and test operation.

Figure 7.2a shows a RTSER setup illustrating the different parts of a typical ATE (Autran et al. 2012). The circuits to test (packaged devices) are assembled on one or different IC printed boards. A modular configuration composed of several daughterboards connected to motherboards offers the possibility of replacing/isolating faulty devices during the test or reusing the setup for other circuits. Another advantage is to use the same core tester with a single daughter card for accelerated tests. The test is controlled by the test processor for lower-level functions (memory-array write/read, data comparison, current and voltage monitoring) and by the control software for higher-level functions (selection of test conditions, generation of data pattern, test flow sequencing, data processing) (Ziegler and Puchner 2004). Figure 7.2b shows a screenshot of the main page of the control software specially developed for the ATE detailed above (Autran et al. 2012). Different windows and graphs allow the user to visualize in real time the most important test parameters and control and monitoring signals, in particular card temperatures, voltage stability, and power current consumption. The PC provides network connection, time stamping, and data storage operations. The setup also frequently includes a controlled power switch (for remote control) and uninterruptible battery-backed power supplied to protect against electric microcuts and temporary power instabilities. Figure 7.3 shows other RTSER setups related to recent experiments (Autran et al. 2009; Lesea 2011; Puchner 2011; Seifert and Kirsch 2012). These different systems are capable of monitoring several hundred chips and performing all requested operations, such as writing/reading data to the chips, comparing output data with written data, and recording details on the different errors detected in chips. The design and operation of ATEs are expected to follow the guidelines of the JEDEC Standard JESD89A (JEDEC 2006). These latter have been issued to minimize the different error sources coming from external or system noise and thus to ensure that all detected errors correspond well to the signature of soft errors occurring within the devices under test. In addition, these rules recommend logging the maximum information related to the test sequence (including supply voltage, temperature, and, if possible, current monitoring of every device) and to all detected errors (type of error, written pattern and read pattern, test chip identification, logical address, etc.) for postprocessing and verification operations.

Figure 7.4 shows a typical test flow implemented in an ATE for monitoring errors in SRAM circuits (Autran et al. 2009). This sequence corresponds to *dynamic* testing, since the circuits are continuously read with access cycles at frequencies compatible with the circuit operation (Ziegler and Puchner 2004). Combined with real-time current consumption monitoring at chip level, this procedure allows discrimination between different error types (Munteanu and Autran 2008) as a function of write/read and rewrite/reread operation results, including single-bit upset (SBU), multiple-bit upset (MBU), single-event functional interrupt (SEFI), and single-event latchup (SEL) events.

(a)

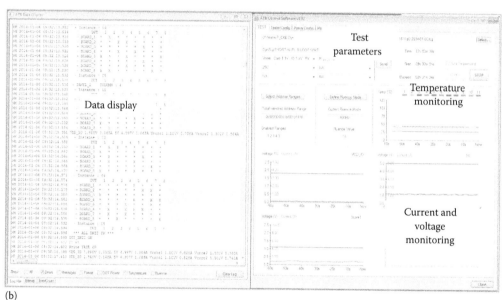

(b)

FIGURE 7.2

(a) Global view of a recent setup illustrating the different parts of a typical RTSER test equipment. This setup has been developed by EASII-IC and embeds 512 SRAM circuits manufactured in CMOS 40 nm. It has been installed since March 2011 on the ASTEP. *Inset*: Detail of one of the 64 DUT boards with eight chips in super ball grid array (SBGA) downwardly oriented cavity packages. (b) Screenshot of the main page of the control software specially developed (under Linux) for the ATE detailed above. (Courtesy of EASII-IC, Grenoble, France.) (Reprinted from *Microelectronics Reliability*, 54, Autran, J.L., Munteanu, D., Roche, P., and Gasiot, G., Real-time soft-error rate measurements: A review, 1455–1476, Copyright (2014), with permission from Elsevier.)

For each error type, the ATE software is able to estimate in real time an SER from the number of errors N_{err} observed at time T:

$$SER = \frac{N_{err}}{\Sigma} \times 10^9 \; (FIT/Mbit) \tag{7.1}$$

FIGURE 7.3
Different RTSER setups used in recent experiments: (a) from Cypress Semiconductor; (b) from Intel; (c) from ST Microelectronics; (d) from Xilinx. (Reprinted from *Microelectronics Reliability*, 54, Autran, J.L., Munteanu, D., Roche, P., and Gasiot, G., Real-time soft-error rate measurements: A review, 1455–1476, Copyright (2014), with permission from Elsevier.)

where Σ is the number of megabits $\times$ hours cumulated at time T, given by

$$\Sigma = \int_{0}^{T} N(t)\,dt \qquad (7.2)$$

where N(t) represents the number of megabits under test at time T. This quantity may be variable in the case of device automatic disconnection by the system due to abnormal current consumption or other anomaly detected at device, daughtercard, or mothercard level.

From Equation 7.1, SER estimations can be performed by considering for N_{err} different quantities related to the events to characterize, for example the total number of bit flips (bit-flip SER), the number of upset events (SEU SER), the number of single-bit upsets (SBU SER) or multiple-cell upsets (MCU SER), and so on. These different error rates can be considered as raw SER values; they take into account all SER components and are dependent on the test location. Section 7.2.2 will explain how to express the SER components separately and normalize the values with respect to reference natural radiation conditions.

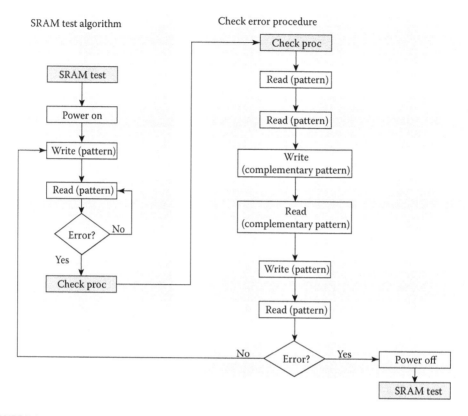

FIGURE 7.4

Typical test flow and error-check procedure for RTSER monitoring of SRAMs. (Reprinted from *Microelectronics Reliability*, 54, Autran, J.L., Munteanu, D., Roche, P., and Gasiot, G., Real-time soft-error rate measurements: A review, 1455–1476, Copyright (2014), with permission from Elsevier.)

7.2.2 Differentiation of the SER Components

As briefly stated in the introduction to the book, and neglecting sensitivity to thermal neutrons and to atmospheric charged particles (muons, protons), the SER of a given electronic component or circuit at ground level has two main components.

- The first is due to an external constraint, the atmospheric particle flux, and particularly high-energy neutrons (see Chapter 5) interacting with IC materials. This component is labeled neutron-SER or n-SER and is proportional to the intensity of the high-energy neutron flux at the test location (see below).

- The second component is an internal constraint due to the presence of traces of radioactive elements in the circuit materials (see Chapter 3), mainly U or Th contaminations at sub–parts per billion concentrations or the natural alpha-emitter isotopes of certain elements used in modern fabrication processes, Hf or Pt, for example. This second component is labeled α-SER and is, of course, independent of the test location.

Because the neutron component of the SER depends on the neutron-flux characteristics of the test location, it is necessary to express it, that is, to normalize it, with respect to a reference atmospheric spectrum and flux. As already introduced in Section 1.3.1.1, this reference corresponds to New York City (NYC) outdoors at sea level for quiet sun conditions and for neutron energies above 1 MeV. The total integral flux of this spectrum above 1 MeV is exactly $IF_{NYC} = 20.0$ neutrons $cm^{-2}\, h^{-1}$ (see Figure 1.11), the reference value used in the following for the normalization of neutron-SER.

In real-time experiment and by analogy with accelerated tests, we usually introduce a so-called acceleration factor (AF), a term introduced in Section 1.3.1.2, which defines the relative neutron flux with respect to sea level (NYC) and is given by

$$AF = \frac{IF_{test}}{IF_{NYC}} \tag{7.3}$$

where IF_{test} is the total integral flux (always above 1 MeV) of the atmospheric neutron spectrum at the test location and corresponding to the period of the test. The term *acceleration factor* suggests that, for a given exposure to atmospheric radiation, an experiment conducted at altitude will be shorter by this factor (AF) than a similar experiment carried out at sea level.

The measured global SER at a given test location is, then, the result of a weighting of the two SER components previously defined, as a function of the test location:

$$SER\big|_{test\ location} = AF \times n\text{-}SER\big|_{NYC} + \alpha\text{-}SER \tag{7.4}$$

where $n\text{-}SER\big|_{NYC}$ is the value of n-SER normalized for NYC conditions.

Because there are two unknown quantities to determine, that is, $n\text{-}SER\big|_{NYC}$ and $\alpha\text{-}SER$, a minimum of two measurements with different test environments are needed to solve the problem. Generally, the test can be conducted in series (with the same setup) or in parallel (with two identical setups) at altitude and underground or by choosing two (very) different altitudes. As we just mentioned, measurements are performed at altitude to take advantage of a higher neutron flux than at sea level, thus reducing the test duration. Underground measurements are chosen to screen the atmospheric component and to mainly detect soft errors induced by the internal chip radioactivity. In this last case, $SER\big|_{Underground} \approx \alpha\text{-}SER$, if all other sources of radiation related to the underground site can be neglected, in particular low-energy neutrons emitted from the rock or concrete cavity and alpha particles emitted by the decay of radon and/or thoron present in the ambient air.

In the practical case of combined altitude and underground experiments, Equation 7.4 immediately gives

$$SER\big|_{NYC} = \frac{SER\big|_{Altitude} - \alpha\text{-}SER}{AF} \approx \frac{SER\big|_{Altitude} - SER\big|_{Underground}}{AF} \tag{7.5}$$

where:
AF   = acceleration factor of the altitude site
$SER|_{Altitude}$ = global SER estimated from altitude test data

In the absence of an underground test, a minimum of two ground-level tests can be combined to separate neutron from alpha-particle contribution to the global SER:

$$\begin{cases} SER\big|_{Site\ 1} = AF_1 \times n\text{-}SER\big|_{NYC} + \alpha\text{-}SER \\[2mm] SER\big|_{Site\ 2} = AF_2 \times n\text{-}SER\big|_{NYC} + \alpha\text{-}SER \end{cases} \tag{7.6}$$

where:

   AF_1 and AF_2 = acceleration factors of the two ground-level test locations
   $SER|_{Site\,1}$ and $SER|_{Site\,2}$ = two corresponding measured failure rates

Of course, more than two test sites can be used. Beyond the logistic and financial aspects of such an experiment, this latter is not without interest to unambiguously confirm that Equation 7.4 is well verified for each site with a unique couple of n-$SER|_{NYC}$ and α-SER values. This point will be discussed in Section 7.3.

7.2.3 Statistics for RTSER: Typical Example

In RTSER experiments, in spite of an elevated number of devices implemented, soft errors remain events that can be qualified as rare events. The direct consequence is that the SER value evaluated using Equations 7.1 and 7.2 may be affected by a significant uncertainty interval that progressively decreases when the experiment is continued for longer. An important issue in RTSER is thus to accumulate enough errors to assert that the device under test has an SER below or above certain confidence limits. In other words, the experimenter should determine when enough errors have been accumulated to make an acceptable estimation of the SER.

From a statistical formalism point of view (JEDEC 2006), we can assume that soft errors are random in time and space (within the circuits) with a constant probability, and that their number is small relative to the number of memory elements considered in the test. In addition, soft errors are not permanent because they are eliminated when new data is written after detection, following the test flow of Figure 7.4: a RTSER test is thus equivalent to a life test with replacement (in which a failing device is replaced with a new device immediately on failure detection) (JEDEC 2006). In such conditions, soft errors can be considered to follow the Poisson distribution of probability given by

$$f(t) = N \lambda e^{-N\lambda t} \tag{7.7}$$

where N is the number of memory elements on test and λ is the mean error rate, the quantity to estimate within a given confidence interval.

The maximum likelihood estimate for λ is simply the number of detected errors divided by the time and the number of units on test. Expressing this latter quantity in megabits, for example, and time in hours, λ can be directly estimated from quantities defined in Equations 7.1 and 7.2: $\lambda = N/\Sigma$. The confidence interval for this value can then be expressed using the relationship between the cumulative distribution functions of the Poisson and chi-squared distributions. The upper and lower estimates of the SER quantity defined in Equation 7.1 are then given by

$$\frac{1}{2\Sigma}\chi^2\left(\frac{\alpha}{2}, 2(N_{err}+1)\right)\times 10^9 < SER\left[Eq.\,7.1\right] < \frac{1}{2\Sigma}\chi^2\left(1-\frac{\alpha}{2}, 2(N_{err}+1)\right)\times 10^9 \tag{7.8}$$

where:

   $\chi^2(p, n)$ is the quantile function (corresponding to a lower-tail-area p) of the chi-squared distribution with n degrees of freedom
   α is a parameter that defines the $100(1-\alpha)$ percent confidence interval
   $2(N_{err}+1)$ represents the number of degrees of freedom for the SER

Σ is the cumulated number of Mbit$\times$h

N_{err} is the number of errors detected

the factor $\times 10^9$ is to express the limits in failure in time

These calculations are now illustrated in the following example of a real RTSER test conducted at altitude during the year 2006 (Autran et al. 2007, 2010). This experiment concerned the characterization of single-port SRAMs manufactured in 130 nm technology. Figure 7.5a shows the cumulated distribution of bit flips logged during the test. A total of 72 bit flips were detected during a time interval corresponding to a cumulated number of 1.5457×10^7 Mbit$\times$h (after deducing all downtimes). The mean SER was then evaluated using Equation 7.2: $\langle SER \rangle = 72 \times 10^9 / 1.5457 \times 10^7 = 4658$ FIT/Mbit. Choosing $\alpha = 0.1$ to define a 90% confidence interval and considering $2 \times (72 + 1) = 146$ degrees of freedom, we obtain $\chi^2(0.05, 146) = 119.075$ and $\chi^2(0.95, 146) = 175.198$. The evaluation of lower and upper limits immediately gives:

- Lower limit: $119.075 \times 10^9 / (2 \times 1.5457 \times 10^7) = 3852$ FIT/Mbit
- Upper limit: $175.198 \times 10^9 / (2 \times 1.5457 \times 10^7) = 5667$ FIT/Mbit

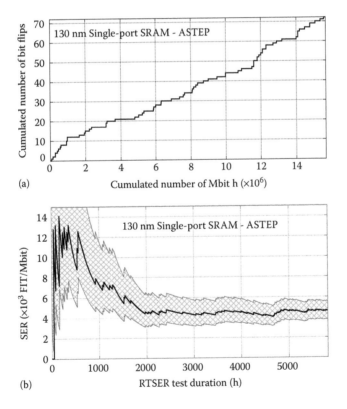

(a)

(b)

FIGURE 7.5

(a) Cumulative number of bit flips versus test duration for 130 nm single-port SRAMs detected at altitude on the ASTEP. Test was conducted under nominal conditions: $V_{DD} = 1.2$ V, room temperature, standard checkerboard test pattern. Test period was March 31, 2006–November 26, 2006 (b) Bit-flip SER (FIT/Mbit) versus test duration calculated from above data. The 90% confidence interval is also indicated (hatched area). (Reprinted from *Microelectronics Reliability*, 54, Autran, J.L., Munteanu, D., Roche, P., and Gasiot, G., Real-time soft-error rate measurements: A review, 1455–1476, Copyright (2014), with permission from Elsevier.)

Figure 7.5b shows the evolution of the SER and of the upper and lower limits of the 90% confidence interval as a function of time. This graphical representation is important to verify the convergence of the SER during the experiment toward a single asymptotic value, which precisely represents the error rate to be characterized.

7.2.4 Metrology of Atmospheric Neutron Flux

As mentioned in Section 7.2.2, the neutron flux, and thus the AF, is generally assumed to be a nominal value, based on the location of the RTSER experiment (JEDEC 2006). But this view is a first-order approximation based on a relatively long-term averaging (several years) and for midlevel solar activity. With a shorter-duration observation window, typically 1 week, 1 day, or less, the neutron flux at a given location is found to fluctuate in time, predominantly as a function of atmospheric pressure, as previously discussed in Section 1.3.2.2 and illustrated in Chapter 2. In particular, the resulting pressure dependence of the neutron flux at ground level has been illustrated in Figure 2.5 for experimental data recorded over more than 5 years with the Plateau de Bure Neutron Monitor (PdBNM) (Semikh et al. 2012). As previously mentioned (Section 2.2.2), the lowest pressure values are observed during winter periods, whereas the highest pressures characterize summer periods. In complement to Figure 2.5, Figure 7.6 shows the corresponding values of AF averaged over a 3-month period (winters and summers) from 2008 to 2013. The mean AF value during this 5-year period is 6.28, but up to 15% variation is observed around this value, evidencing that AF values are season dependent. Implications for RTSER experiments are significant, considering the importance of the AF in the expression of the SER (Section 7.2.2). This totally justifies the JEDEC recommendation (JEDEC 2006) cited here: "a real-time experiment should benefit from onsite monitoring of the atmospheric particle flux, to reevaluate in time its acceleration factor, in the same manner that the flux of a beam accelerator is monitored throughout the test duration."

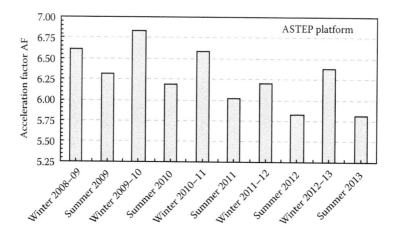

FIGURE 7.6

Averaged acceleration factor (over a 3 month period) for the ASTEP in winter and in summer deduced from neutron monitor data presented in Figure 2.5. (Reprinted from *Microelectronics Reliability*, 54, Autran, J.L., Munteanu, D., Roche, P., and Gasiot, G., Real-time soft-error rate measurements: A review, 1455–1476, Copyright (2014), with permission from Elsevier.)

7.3 Survey of a Few Recent RTSER Experiments

This section reviews some recent RTSER experiments reported in the literature, not including our own work, which is reported in Section 7.4. Our objective was not to be exhaustive, but to show and to discuss, through several representative works, some important SER-characterization results and related test issues. In particular, it is undeniable that several semiconductor manufacturers and integrated circuit suppliers have performed RTSER experiments for private (internal) use only. The present survey is based on published results or public documents. The studies are reported as a function of the main contributor groups (manufacturers, IC suppliers, research centers, and laboratories) in the domain, considering that most of these groups have continuously worked for many years on this RTSER issue. For confidentiality reasons, most of the numerical values reported for the SER are expressed in arbitrary units.

7.3.1 IBM

IBM researchers played a pioneering role in the real-time characterization of soft errors in electronics. Since the 1980s, IBM has conducted numerous RTSER experiments on product-like test vehicles to determine both alpha and cosmic-ray SER contributions. In particular, IBM performed the first field test of the cosmic-ray SER of chips, demonstrating that natural atmospheric radiation does cause soft fails at ground level (Ziegler et al. 1996). The 1996 September issue of the *IBM Journal of Research and Development* reported several key contributions from Ziegler, O'Gorman (1994), and O'Gorman et al. (1996) in RTSER tests, also called *system life testing* or *field testing* by these authors.

Focusing on more recent results, Cannon et al. (2004) and Gordon et al. (2008) reported RTSER results obtained with SRAM test vehicles representative of different bulk and SOI technologies manufactured in 180, 130, and 90 nm nodes. These tests were performed to validate product SER estimates based on accelerated tests and Monte Carlo numerical simulations. To differentiate neutron and alpha-particle components of the SER, they performed tests at three locations: Burlington (Vermont) to provide a reference SER value near sea level (altitude 60 m, acceleration factor $AF \approx 1$); Leadville (Colorado) to provide a high-altitude (3100 m, $AF \approx 11$) environment dominated by terrestrial cosmic rays; and Kansas City (Missouri) in a quarry (mineshaft) in order to isolate the alpha-particle component in an environment with greatly reduced cosmic flux.

Table 7.1 summarizes real-time results for the 180 and 130 nm nodes (Cannon et al. 2004). The life-test vehicles were all designed without solder bumps over the memory arrays under test to reduce the level of the alpha flux from the packaging materials. In general, low-alpha packaging materials were used to build the modules under test; the only exception was the 130 nm bulk SRAM array, which used regular (nonlow-alpha) packaging materials with an alpha flux reported by the supplier on the order of 18 counts cm^{-2} kh^{-1}. In contrast, typical alpha-emission levels for low-alpha materials are a few counts cm^{-2} kh^{-1}. The 180 nm bulk hardware includes ^{10}B in the borophosphosilicate glass (BPSG) layer, and thus is sensitive to SER events from neutron-induced ^{10}B fission. The process has been changed to eliminate ^{10}B from this node: IBM's 130 and 90 nm bulk and SOI technologies (and also all the following) are thus not sensitive to ^{10}B-induced soft errors. In complement to real-time data, Table 7.1 also reports Monte Carlo model predictions obtained with the IBM's SEMM-2 simulation tool (Tang 2008; Tang and Cannon 2004). These predictions are found to be in excellent agreement with the life-test results: the estimated SER values are

TABLE 7.1

RTSER Results and SER Estimates from Monte Carlo Modeling (SEMM-2)

Technology/Location	Measured SER per Bit (A.U.) with 90% Binomial Confidence	Estimated SER per Bit (A.U.) from Monte Carlo Model
180 nm bulk SRAM with ^{10}B (8 Mbit, $V_{DD}=1.6$ V)		
Burlington, VT	1.0 (0.6–1.6)	1.0
Kansas City, MO (underground)	0.6 (0.4–0.7)	0.7
Leadville, CO	5.5 (4.5–6.6)	6.4
130 nm bulk SRAM (14 Mbit, $V_{DD}=1.5$ V) with α-emissivity of 18 counts cm^{-2} kh^{-1}		
Burlington, VT	5.1 (4.4–5.8)	4.7
130 nm SOI SRAM (18 Mbit, $V_{DD}=1.2$ V)		
Burlington, VT	0.4 (0.3–0.5)	0.4
Leadville, CO	1.3 (0.9–1.7)	0.9

Source: After Cannon, E.H. et al., *Proceedings of the IEEE International Reliability Physics Symposium*, pp. 300–304. 25–29 April, IEEE, 2004.

consistently within the 90% confidence intervals of the experimental results across three technologies (two bulk, one SOI). Finally, results shown in Table 7.1 indicate that, for identical radiation environments, per-bit SER sensitivities tend to decrease with scaling, with SOI technologies having significantly lower SER values than bulk technologies at the same lithographic node (130 nm).

7.3.2 Intel

In their famous paper published in 1978, May and Woods evidenced for the first time the occurrence of soft errors in dynamic memories at sea level induced by energetic particles emitted from traces of radioactive contaminants in the memory-packaging materials. In order to understand and characterize the SER of different kinds of devices, they performed the very first real-time alpha-SER test, monitoring with an error-logging system test a total of more than 25 million device hours with various data patterns in a read-only mode (May and Woods 1978).

More than three decades after this pioneering work, Seifert and Kirsch (2012) recently published an RTSER experiment concerning one of Intel's latest bulk CMOS technologies, incorporating high-κ gate dielectrics and metal gate at the 45 nm node. The objective was to verify SER projections based on accelerated alpha-particle and neutron testing as well as alpha-particle flux estimates of the tested component under ambient conditions. In this work, they used the SRAM real-time setup shown in Figure 7.3b, which consists of a rack with 15 vertically stacked boards (containing a total of 120 SRAM chips), a logic tester, and a PC that controlled the experiment. The total number of bits exposed in the experiment was in the range of several gigabits. Other details about the setup can be found in Seifert and Kirsch (2012). The testing equipment was set up and run for about 5000 h (May 2007 until April 2008) on the top floor of one of Intel's engineering buildings in Hillsboro (Oregon) at an altitude of about 61 m (which corresponds to *sea-level* conditions, the deviation from NYC neutron flux being approximately 3%), and then for about 6000 h (January 2010 until March 2011) 655 m below ground, at the Waste Isolation Pilot Plant (WIPP) facility in Carlsbad (New Mexico) (www.wipp.energy.gov). WIPP was selected because of its well-characterized low-radiation environment and very low radon concentrations (<7 Bq m^{-3}). Neutron and muon fluxes were measured at the levels of 1.4×10^{-3} n cm^{-2} h^{-1} and 7×10^{-4} μ cm^{-2} h^{-1}, respectively, which is several orders of magnitude below typical sea-level fluxes (Balbes et al. 1997; Esch et al. 2005). At the end of the two RTSER runs, a total of 22 SBUs and 5 MCUs were detected for the sea-level experiment and only 1 SBU for the underground test. All detected MCUs were double-bit-flip errors (two bit upsets, all adjacent except for one pair), representing a total of $5 \times 2 = 10$ bit flips. Their fraction represents $10/32 = 31\%$ of the total number of bit flips detected during the sea-level experiment, and $5/27 = 18.5\%$ of the detected events. This MCU rate is on the low side compared with results reported for other bulk CMOS technologies (see Section 7.4). Figure 7.7a shows the bit-flip SER versus the cumulated number of upsets, demonstrating the convergence of the SER as a function of time. These results also show excellent agreement between this real-time SER and the SER value projected from accelerated beam testing (using the Los Alamos Neutron Science Center [LANSCE] facility [http://lansce.lanl.gov]), which is found within the 90% confidence interval computed from RTSER data. In complement to this figure, sea-level and underground real-time data are plotted in terms of time-to-failure (TTF) distribution in Figure 7.7b. The corresponding Weibull fitting lines are also drawn, highlighting the fact that the alpha-particle-induced SER of the tested SRAMs is on average 22 times smaller (eight times at 90% confidence) than the sea-level total bit-flip SER. In other words, this result demonstrates that devices manufactured with modern fabrication materials (in particular with high-κ dielectrics and metal gate), as used in Intel 45-nm technology, are characterized by practically negligible alpha-particle-induced upset rates when compared with high-energy-neutron-induced rates.

7.3.3 Sony

Kobayashi et al. (2002, 2004, 2009) from Sony Corporation published RTSER studies with original test protocols for investigating the impact of thermal neutrons and the alpha contamination of packaging materials on the SER. In Kobayashi et al. (2002, 2004), they performed RTSER tests at sea level (Oita, Japan, AF = 0.4), at high altitude (Lake Tahoe, AF = 4.8), and underground (Oto Cosmo Observatory of the Research Center for Nuclear Physics of Osaka University, 476 m below ground) to estimate the SER induced independently by high-energy neutrons, thermal neutrons, and alpha particles in 0.25 and 0.18 μm SRAM devices. For the 0.18 μm technology using a BPSG dielectric layer, the high-energy-neutron-induced SER and the thermal-neutron-induced SER have been differentiated in high-altitude real-time tests using additional thermal neutron shielding (boron nitride–containing silicone rubber sheets with a thickness of 4 mm) applied to the circuits under test. ^{10}B in the sheets captures thermal neutrons, and the thermal neutron flux is reduced to less than 1/100 using these sheets. They thus demonstrated that three quarters of the SER was induced by thermal neutrons, and one quarter was induced by high-energy neutrons, alpha-particle-induced SER measured underground being less than 1% for this 0.18 μm technology.

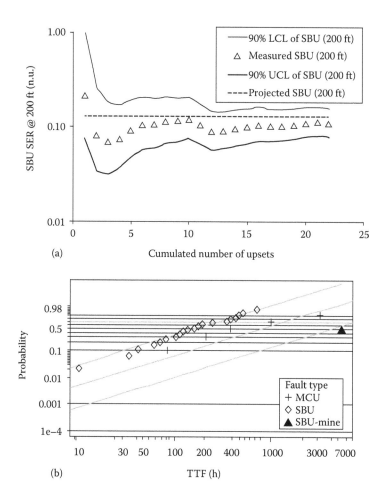

FIGURE 7.7

(a) Normalized, measured bit-flip SER at 61 m above sea level for Intel 45 nm SRAMs. Projected upset rates denote SER extrapolated from accelerated neutron testing. LCL and UCL refer to lower and upper confidence limits, respectively. (b) TTF distributions of all detected fails: SBU at sea level (open diamonds), MCU at sea level (crosses), and SBU underground (filled triangle). Lines denote Weibull fits with beta equal to 1 (i.e., exponential distributions). (Reprinted from Seifert, N. and Kirsch, M., Real-time soft-error testing results of 45-nm, high-κ metal gate, bulk CMOS SRAMs, *IEEE Trans. Nucl. Sci.* 59, 2818–2823. © (2012) IEEE. With permission.)

Kobayashi et al. (2009) focused on underground real-time tests (performed in the same underground laboratory in Japan) to estimate the α-SER with high accuracy for different generations of SRAMs assembled in a wire-bonded structure, and compared the α-SER with the n-SER obtained from accelerated tests in order to estimate the maximum permissive rate of alpha emission from package resin. Several hundred SRAMs manufactured in 180, 130, and 90 nm technologies were tested for several months. In order to obtain better statistical accuracy, package resins with high alpha-emission rates were used intentionally for 130 and 90 nm SRAMs to accelerate tests (the usual package resin was used for 180 nm SRAMs). This approach represents a very interesting way to accelerate testing in underground real-time tests. The AFs for 130 and 90 nm SRAMs with respect to 180 nm SRAMs were estimated as 59 and 45, respectively. Table 7.2 reports experimental results obtained for this underground test, as well as the estimated alpha-emission rates

TABLE 7.2

Summary of Underground Real-Time α-SER Tests

Technology (nm)	180	130	90
Duration (h)	4194	4854	5059
^{238}U concentration (ppb)	0.25	17	8.7
^{232}Th concentration (ppb)	0.22	7.3	17
Alpha-emission rate from package resin ($cm^{-2}h^{-1}$)	1.7×10^{-4}	1.0×10^{-2}	7.6×10^{-3}
Acceleration factor	1	59	45
# of errors	1	142	244
Statistical error (σ)	100%	8.3%	6.4%

Source: After Kobayashi, H. et al., *Proceedings of the IEEE International Reliability Physics Symposium*, 206–211, 26–30 April, Montreal, QC, IEEE, 2009.

Note: Conducted by Kobayashi et al. at the Oto Cosmo Observatory of the Research Center for Nuclear Physics of Osaka University (476 m below ground).

Package resins with high alpha-emission rates were used intentionally for 130 and 90 nm devices to accelerate tests for better statistical accuracy.

measured by inductively coupled plasma mass spectrometry (ICP-MS; Thomas 2013) considering the concentrations of ^{238}U and ^{232}Th elements. In more than 4000 h for the test on 180 nm SRAM, only one error was detected. For the 130 and 90 nm devices, a large number of errors were collected, demonstrating the impact of the package's high emissivity on the α-SER. In the same study, the authors also investigated the alpha-emission rates of numerous package resins bought from various vendors. Furthermore, they performed computer simulations to estimate the scaling trends of both the α-SER and the n-SER up to 45 nm technologies.

7.3.4 Tohoku University, Hitachi, and Renesas Electronics

Nakamura and Baba from Tohoku University, Ibe and Yahagi from Hitachi, and Kameyama from Renesas Electronics coauthored in 2008 the book *Terrestrial Neutron-Induced Soft Errors in Advanced Memory Devices* (Nakamura et al. 2008), in which they surveyed several key issues for RTSER testing, including a reference chapter on terrestrial-neutron dosimetry and spectrometry (Chapter 2), a review of the methods and equipment used to characterize neutron soft errors in the terrestrial field (Chapter 3), and an in-depth discussion on the comparison of real-time data with accelerated tests, modeling, and simulation results (Chapters 6 and 7).

In Chapter 3, the authors reported a multistage RTSER experiment conducted near sea level and at various altitudes to characterize neutron-induced SER. Different locations both in Japan and in the United States were chosen for this RTSER testing.

In a first stage, they verified the effect of concrete shielding by comparing the predictions of Equation 1.16 (see Chapter 1) with the ratio of the SER values evaluated both inside and outside a concrete building. RTSER measurements were conducted in Tokyo using two sets of 180 nm CMOS SRAM devices. The SER were measured for each set by using more than 1000 pieces for up to 1000 h of operation in test locations both outside and inside (1 m concrete thickness). The two SER attenuation rates obtained for the two series of devices were 0.26 and 0.32, respectively, in very good agreement with Equation 1.16, with a concrete density of 2.54 g cm^{-3} and an attenuation length of 216 g cm^{-2}, giving $\phi/\phi_0 = \exp(-1.17 \times 1) = 0.31$ for 1 m of concrete.

In a second stage, Kameyama et al. (2007) verified the altitude dependence of the SER. They considered three locations (86 m, 766 m, and 1988 m high, at 26° of geomagnetic latitude) in Japan and two locations (0 m and 1700 m high, at 44° of geomagnetic latitude) in the United States. These RTSER tests conducted in Japan used more than 1000 pieces of low-power 180 nm SRAM devices; similarly, tests conducted in the United States used more than 500 pieces of high-speed 130 nm SRAM devices. Figure 7.8a shows the experimental SER data for both the 130 nm SRAM device and the 180 nm SRAM device as a function of the neutron dose rates measured using a rem counter. These SER values show a strong dependence on neutron dose rate that increases along with altitude. This dependence is true as long as the effects of alpha particles and thermal neutrons are negligibly

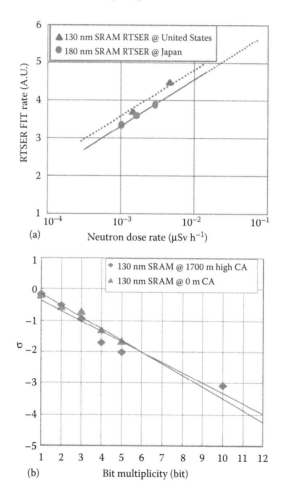

FIGURE 7.8

(a) Real-time SER versus neutron dose rate for both series of 180 and 130 nm SRAMs at three locations in Japan and two locations in the United States, respectively. (b) Comparison data for multiple-cell upset (MCU) ratio of RTSER at different altitudes in the United States (ground level and 1700 m high). (Reprinted from Kameyama, H. et al., A quantitative analysis of neutron-induced multi cell upset in deep sub micron SRAMs and of the impact due to anomalous noise, in *Proceedings of the IEEE International Reliability Physics Symposium*, pp. 678–679. © (2007) IEEE. With permission.)

small for such SRAM devices. These RTSER results show that the altitude dependence of 180 nm SER for neutron doses in the range of 10^{-3}-10^{-2} µSv h^{-1} is the same as that of 130 nm SRAM SER. The difference in increasing SER error rates for 130 nm devices and 180 nm devices might be due to the susceptibility of the devices or the influence of geomagnetic difference between Japan and the United States. In addition, Figure 7.8b shows that there is no altitude dependence for the MCU/SEU ratio between the data at sea level and at 1700 m. This result agrees with the trend of neutron-lethargy studies shown in the literature (Kameyama et al. 2007).

Finally, Kameyama et al. (2007) and Nakamura et al. (2008) reported and discussed the impact of an anomalous neutron dose rate over a limited period on the real-time SER. This peculiar phenomenon has an unknown origin but may suggest some signs related to a geophysical mechanism (a disturbance of the mesosphere, a meteorological effect, etc.). Figure 7.9a shows the important changes recorded during the monitoring of the neutron dose rate around the test site (in the United States at 1700 m of altitude). The cumulative and differential SERs are shown in Figure 7.9b, before and after this period of time. After a statistical analysis of RTSER data and neutron-monitoring data (see Nakamura et al. [2008] for details of this part), the authors concluded that these abnormal phenomena of the neutron dose rate were well synchronized with increases in both cumulative soft errors and differential SER, reporting for the first time such a correlation between a neutron-dose-rate change at ground level and an increased occurrence of soft error.

7.3.5 Cypress

Cypress has much experience in real-time testing and in 2004, edited the book *SER— History, Trends and Challenges* with Ziegler and Puchner as principal coauthors (Ziegler and Puchner 2004). Chapter 7 of this book is dedicated to "System SER Measurements"; it provides a reference text in the domain of RTSER tests with an in-depth discussion about test-system setups, test algorithms, and SER data analysis. A number of real-time experiments conducted by Cypress until 2004 are described in detail in this chapter. They concern the characterization of SRAM memories manufactured down to the 0.15 µm technological node. Measurements were performed at the Cypress Philippines manufacturing site (sea level) and at the Cypress Colorado Design Center (altitude 1830 m, AF = 4.3); the AF at these two sites differs by a factor of four, which makes it possible to differentiate n-SER and α-SER following Equation 7.4 and to compare these results with accelerated tests (see chapter 7 of Ziegler and Puchner [2004] for details).

Puchner (2011) reported more recent experiments performed on 90 nm SRAMs. For these real-time tests, Cypress operated two SRAM system SER (SSER) setups, installed at the Cypress Colorado Design Center and at Mauna Kea, Hawaii (AF = 9.22). For the first setup, installed in Colorado, 386 chips of 36 Mbit SRAM were monitored (accessed serially) from September 2007 to March 2008, representing more than 1.24 million device × hours. For the second setup, installed at the summit of Mauna Kea, a total of 269 chips of 72 Mbit SRAM were also monitored from June to December 2008 (1.06 million device × hours). Figure 7.10a and b shows the cumulated distribution of bit flips as well as the evolution of the SER (and the lower limit of the 90% confidence interval) versus the cumulated number of Mbit × h. From the two SER values measured on site and considering Equation 7.6, it is possible to separate neutron contributions from alpha-particle contributions to the global SER. The solving of this system, as reported by Puchner (2011), is shown in Table 7.3 for the total SER (SBU + MCU). Experimental SER component values deduced from accelerated tests

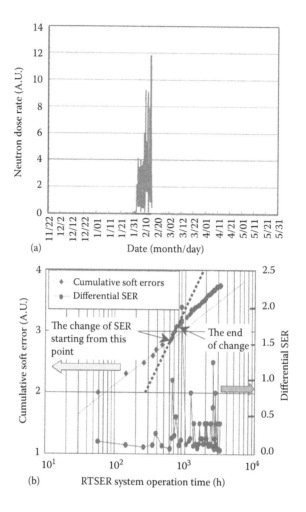

(a)

(b)

FIGURE 7.9
(a) Extraordinary changes shown in linear plot while monitoring the neutron dose rate around the test site located in California at an altitude of 1700 m. (b) Cumulative soft errors and differential SER for 130 nm SRAMs during the period of changes in the neutron dose rate for the California test site. (Reprinted from Kameyama, H. et al., A quantitative analysis of neutron-induced multi cell upset in deep sub micron SRAMs and of the impact due to anomalous noise, in *Proceedings of the IEEE International Reliability Physics Symposium*, pp. 678–679. © (2007) IEEE. With permission.)

are also indicated. These results evidence a significant contribution from alpha particles to the total SER for this 90 nm technology. They also show that RTSER tests measure a higher α-SER than accelerated tests using an intense alpha-particle source, the two sets of data being in relatively good agreement within ±25% uncertainty margins.

7.3.6 Xilinx

In 2002, Xilinx launched an ambitious test and qualification program, called the Rosetta Program (in reference to the famous Rosetta Stone, which had enabled the decryption of Egyptian hieroglyphs) (Lesea et al. 2005). This program was created when Xilinx discovered that an alpha source had contaminated a lot of solder bumps in a number of flip-chip packages (Lesea et al. 2008). This contamination issue resulted in a financial loss and a

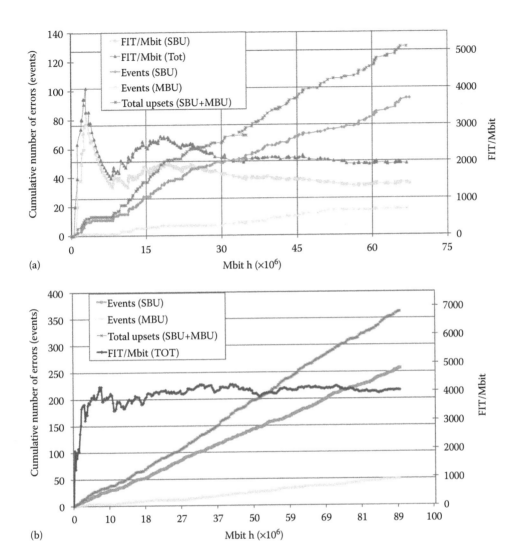

(a)

(b)

FIGURE 7.10
Cumulative distribution of bit flips and convergence of bit-flip SER versus time for RTSER experiments conducted in Colorado (a) and at the summit of Mauna Kea (b). (Courtesy of H. Puchner.)

recall of product. It was decided to create a viable test and qualification program to verify that new packaging solutions had solved the contamination issue. Plans began immediately to address prevention of a reoccurrence, as well as a plan for qualification of new substrates.

The Rosetta test program consists of multiple sets of hundreds of Xilinx's largest field-programmable gate arrays (FPGAs) of differing technologies exposed to natural radiation at different altitudes (see an example of a Xilinx setup in Figure 7.3d). Four test sites were considered when the program started in 2002: San Jose, California (sea-level, AF ≈ 1); Albuquerque, New Mexico (altitude of 1570 m, AF = 3.3); White Mountain Research Center, California (altitude of 3801 m, AF ≈ 15); and Mauna Kea Observatory, Hawaii (altitude of 4023 m, AF ≈ 9). In 2006, three additional sites in France were involved in the program: the Plateau de Bure (Altitude SEE Test European Platform [ASTEP], altitude 2552 m, AF = 6.3),

TABLE 7.3

SER Values Measured at Two Different Test Sites (with Different Acceleration Factors) and Extraction of the n-SER and α-SER Components from These Values

Technology (nm)	Test Site	Measured SER at Test Site (FIT/ Mbit)	AF	Solar Activity	α-SER (FIT/ Mbit)	n-SER at Test Site (FIT/ Mbit)	n-SER(NYC) (FIT/Mbit)
90	Mauna Kea	3922	9.22	115%	308	3614	340
90	Colorado	1928	4.13	115%	308	1620	340
90	Cypress	N/A	N/A	N/A	225	N/A	N/A
90	TSL	N/A	N/A	N/A	N/A	N/A	360

Source: After Puchner, H., *Proceedings of the IEEE 3rd Annual SER Workshop*, San Jose, CA, 2011.

Note: The two last lines indicate SER values deduced from accelerated tests performed at Cypress and at The Svedberg Laboratory (TSL) facility.

Marseille (altitude 124 m, AF ≈ 1), and Rustrel (low-noise underground laboratory, 500 m below the surface). Currently, Xilinx is considering 10 different test locations, as reported in Lesea (2011). All tested circuits have been fabricated by UMC or Toshiba in their 300 mm submicron fabrication lines using standard-logic CMOS processes and the new Triple-Gate-Oxide CMOS process. The devices currently under test concern the Virtex and Spartan product families, manufacturing in technologies ranging from 150 nm (Virtex II) to 40 nm (Virtex 6) (Lesea 2011). Memory cells that are used for configuration (configuration RAM) and in block RAM (BRAM) are monitored during the RTSER tests. Table 7.4 recapitulates the evolution of the real-time SER values as a function of the technological node for the Virtex FPGA. This table summarizes all the efforts by Xilinx to reduce the SER in these products, notably since the 130 nm technological node. Note that Xilinx regularly updates the values of Table 7.4 in a public document called Device Reliability Report

TABLE 7.4

SER Values Compiled from the Rosetta Experiment for the Xilinx Virtex Family of Products

Technological Node (nm)	Product Family	Real-Time Soft-Error Rate (FIT/Mbit)		Error (90% Confidence Interval)
		Configuration Memory	Block RAM	
250	Virtex	160	160	±20%
180	Virtex-E	181	181	±20%
150	Virtex-II	405	478	±20%
130	Virtex-II Pro	437	770	±20%
90	Virtex-4	263	484	±20%
65	Virtex-5	165	692	±20%
40	Virtex-6	101	232	±20%
28	7 series FPGA	85	77	±20%

Source: Data from Xilinx, Xilinx device reliability report, first quarter 2014, UG116 (v10.0), May 2014.

(Xilinx 2014). The report summarizes the atmospheric and beam-test results for all Xilinx FPGA device technologies.

7.3.7 NXP

Heijmen and Verwijst from NXP reported at the RADECS 2009 conference a real-time SER test performed on embedded SRAMs in a 130 nm process technology (Heijmen and Verwijst 2009). Five hundred and eleven samples of a product IC (~4 Mbit SRAM cores) have been tested in two phases. First, a *mountain test* was performed for 3 months at the Jungfraujoch (Switzerland, altitude of 3450 m, AF = 12.8). The second part of this RTSER characterization, a *cave test*, took place at the underground laboratory of Modane (LSM, Fréjus Tunnel, 1700 m below the surface) for 11 months. The results of the mountain and cave tests as a function of the number of device hours are shown in Figure 7.11, together with the 90% confidence intervals. During the mountain test, a total of 34 soft errors were observed. One of the events was an SEL and the others were SEUs. Similarly, a total of 16 soft errors (bit flips) were registered during the cave test. The resulting

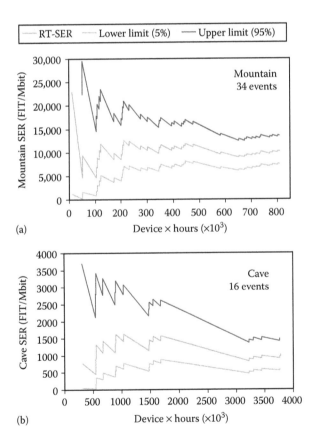

FIGURE 7.11
Real-time SER versus time (expressed in devices × hours) measured in the mountain test (a) and during the cave experiment (b). (Reprinted from Heijmen, T. and Verwijst, J., Altitude and underground real-time SER tests of embedded SRAM, in *Proceedings of the European Conference on Radiation and Its Effects on Components and Systems*, pp. 342–347. © (2009) IEEE. With permission.)

neutron-SER (717 FIT/Mbit) is in very good agreement with data from accelerated SER tests (617 FIT/Mbit). Also, the real-time alpha SER (960 FIT/Mbit) agrees within 90% confidence limits with the accelerated SER result (1425 FIT/Mbit). However, the deviation in the case of alpha-SER is larger than for neutron-SER, which can be attributed to uncertainties in the alpha-particle flux observed by the product die.

In addition to these results, Heijmen and Verwijst (2009) report an interesting analysis and discussion concerning RTSER data. Radiation-induced soft errors are generally assumed to be random external events, resulting in a constant upset rate. From a statistical point of view, this corresponds to an exponential distribution (McPherson 2013). In the present work, the authors applied statistical analysis to the measured real-time SER data. It is shown that the exponential distribution does indeed fit the mountain-SER data very well. In contrast, the cave-SER data are fitted rather poorly by the exponential distribution. From the author's point of view, the fact that the exponential distribution fits the cave-SER test data rather poorly indicates that possibly multiple sources of alpha particles are causing soft errors under these conditions. This could explain the apparent bimodal distribution observed for these data (Heijmen and Verwijst 2009). However, the quantity of data seems too low to draw conclusions, and the authors conclude that a cave test in which more than 30 events are detected would be needed to investigate whether the distribution is multimodal.

7.4 RTSER Experiments Conducted at ASTEP and LSM

This part of the chapter surveys the RTSER experiments conducted by the authors since the creation of the ASTEP in 2004 and its extension to LSM in 2007. After briefly presenting these two RTSER-dedicated installations, we will survey the main results concerning the soft-error characterization of CMOS technologies ranging from 130 down to 40 nm technological nodes. To protect the confidentiality of such proprietary information, all numerical SER values reported in this section have been normalized by a common arbitrary scaling factor (set lower than 3x), not affecting the order of magnitude of the SER and data interpretation.

7.4.1 ASTEP and LSM Test Platforms

ASTEP (www.astep.eu) is a permanent mountain laboratory and a dual academic–industry research platform created by Aix-Marseille University, Centre national de la recherche scientifique (CNRS), and STMicroelectronics in 2004. The current platform, operated by IM2NP Laboratory (www.im2np.fr) and fully operational since March 2006, is dedicated to the problem of SEEs induced by terrestrial radiation in electronic components, circuits, and systems. Located in the French Alps on the Plateau de Bure desert at an altitude of 2552 m (Dévoluy mountains), the platform is hosted by the Institut de Radioastronomie Millimétrique (IRAM, Grenoble, France) Observatory (www.iram.fr) in a one-floor metal-walled building. Figure 7.12a shows a general external view of the building. The ground-level floor hosts long-term RTSER setups installed in different dedicated experimental rooms. As specified in Section 2.2, the building also permanently hosts a neutron monitor (Semikh et al. 2012) installed in the cupola (first floor). A muon monitor was installed in 2011 to complete the panel of background characterization instruments (www.astep.eu). The AF for the site is about 6 (see Figure 7.6).

(a)   (b)

FIGURE 7.12
(a) External view of the ASTEP building located at an altitude of 2552 m on the Plateau de Bure (August 2012). Both experimental RTSER rooms and muon monitor are at ground level; the neutron monitor is installed in the cupola at first floor. (b) View of the main LSM experimental hall. RTSER experiments are located in a closed dedicated room (white walls) located at the left of the stairway. (Reprinted from *Microelectronics Reliability*, 54, Autran, J.L., Munteanu, D., Roche, P., and Gasiot, G., Real-time soft-error rate measurements: A review, 1455–1476, Copyright (2014), with permission from Elsevier.)

In October 2007, a complementary test platform was created at the Underground Laboratory of Modane (LSM) to perform measurements in order to separate the contribution of atmospheric particles to SER from the contribution of internal alpha-particle emitters. This high-energy physics laboratory is located about 1700 m under the top of the Fréjus mountain (4800 m water equivalent), near the middle of the Fréjus highway tunnel connecting France and Italy (www-lsm.in2p3.fr). Figure 7.12b shows a general view of the main experimental hall in LSM. RTSER experiments have been installed in a dedicated room (an old clean room) located in the lower level of the hall. Due to the depth of the LSM, the average particle flux inside the laboratory is greatly reduced (Chazal et al. 1998; Yakushev 2006): about 4 muons m^{-2} day^{-1}, corresponding to a reduction factor of 2 million compared with the flux at sea level, and a few thousand fast neutrons m^{-2} day^{-1} (depending on the neutron energy and the measurement location in the laboratory) emitted by natural radioactivity from the rock, the neutron component of cosmic rays being totally eliminated at this depth. In addition, the radon in the laboratory is maintained at a very low rate of ~20 Bq m^{-3} by an air purification system that totally renews the volume of the air inside the laboratory twice an hour. With such an extremely reduced radiation background, SEEs detected at LSM during RTSER experiments can be attributed with high confidence to internal chip mechanisms, primarily internal radioactivity due to ultratraces of alpha-particle emitters.

7.4.2 RTSER Experiments

Table 7.5 summarizes the different RTSER experiments conducted on the ASTEP and at LSM during the period 2006–2014. Two types of measurements have been conducted: a series of classical RTSER tests, using a large collection of packaged SRAMs controlled by

TABLE 7.5

Summary of RTSER Experiments Conducted on the ASTEP platform and at LSM during the Period 2006–2014

Platform	Circuit and Technology	Test Temperature	Initial Mbit under Test	Start Date	Stop Date	Effective Cumulated Hours of Test
ASTEP	130 nm	RT	3,664	03/31/2006	11/26/2006	5,200
LSM	SRAM		3,472	10/16/2007	10/15/2010	24,747
ASTEP	65 nm	RT	3,216	01/21/2008	05/07/2009	11,278
	SRAM	85°C	2,884	06/26/2009	01/24/2013	28,482
LSM		RT	3,226	04/11/2008	Ongoing	>48,000
ASTEP	40 nm SRAM	RT	7,168	11/03/2011	Ongoing	>23,000
ASTEP	90 nm Flash (wafer level)	RT	~50,000	12/07/2011	Ongoing	>8,000

an ATE, and a new type of experiment for nonvolatile memories (NOR flash memories), using fully processed wafers exposed to natural radiation and periodically read with an industrial tester. We will only review results obtained for SRAM devices; the other (wafer-level) approach for flash memories is not, strictly speaking, a *real-time* method (Just et al. 2013). This will be covered in detail in Chapter 11.

Three technological generations of single-port SRAMs manufactured by STMicroelectronics have been characterized, successively, in CMOS 130, 65, and 40 nm. The different fabrication processes are based on a BPSG-free BEOL that eliminates the major source of ^{10}B in the circuits and drastically reduces the possible interaction between ^{10}B and very-low-(thermal)-energy neutrons (Baumann 2005). The test chips for the three technologies contain, respectively, 4 Mbit of SRAM for the 130 nm technology (bitcell area of 2.5 μm^2), 8.5 Mbit (bitcell area of 0.525 μm^2) for the 65 nm, and $7 + 7 = 14$ Mbit (bitcell area of 0.299 and 0.374 μm^2) for the 40 nm. The core voltage is 1.2 V for 130 and 65 nm nodes and 1.1 V for 40 nm. No deep N-well was used in these different technologies.

For the purposes of the different RTSER experiments, three distinct pieces of automatic test equipment, fully compliant with all the specifications of the JEDEC standard JESD89A (JEDEC 2006), have been successively designed and constructed. Details about these set-ups can be found in Autran et al. (2007, 2009, 2012). Only the 65 nm experiment has been duplicated to run in parallel at ASTEP (Figure 7.3b) and LSM. The 130 nm setup has been deployed at ASTEP and then transported to Modane. For the 40 nm experiment, the setup (Figure 7.2) has so far only been installed at ASTEP.

Figures 7.5 and 7.13 through 7.16 show in *extenso* the raw RTSER data, that is, the time distributions of the cumulated number of bit flips for the different experiments. Each figure also presents the convergence of the SER as a function of experiment duration. Data for 65 and 40 nm experiments, currently in progress, were updated on January 6, 2014. Data for 65 nm high-temperature tests have not been previously published. Figures 7.5 and 7.13 show a remarkable convergence of the SRAM for both 130 nm experiments, at altitude and underground, with very regular staircase distributions of bit flips versus time. A stable SER value was reached after approximately 2000 h of experiment at altitude and after a longer period (~10,000 h) underground. MCU event frequency versus multiplicity is reported in Figure 7.17. For the 130 nm technology, in contrast to the other two, only a few MCUs of double multiplicity (adjacent bitcells) have been detected. Figure 7.18 shows the extraction of the two SER components from the previous data.

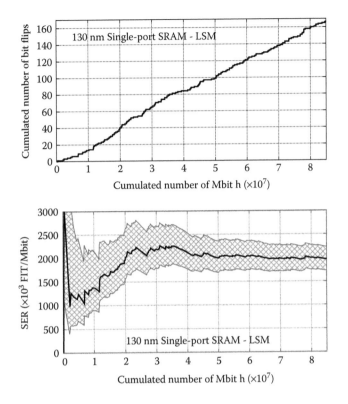

FIGURE 7.13
As Figure 7.5, for 130 nm single-port SRAMs characterized underground at LSM. Test period was October 16, 2007–October 15, 2010. (Reprinted from *Microelectronics Reliability*, 54, Autran, J.L., Munteanu, D., Roche, P., and Gasiot, G., Real-time soft-error rate measurements: A review, 1455–1476, Copyright (2014), with permission from Elsevier.)

Figures 7.14 and 7.15 show the same types of data for the 65 nm technology. These figures demonstrate that strictly the same response is obtained for these SRAMs exposed to natural radiation at two different temperatures (ambient and 85°C). In particular, it must be noted that no SEL has been detected for any of the characterized technologies, including during this test at high temperature. We therefore note a sudden increase of the bit-flip occurrence at the very end of the 85°C experiment (>25,000 h), certainly due to the emergence of a degradation mechanism activated by temperature. Negative-bias temperature instability (NBTI) (Bagatin et al. 2010; Kauppila et al. 2012) is suspected to be at the origin of such behavior after more than 3 years of continuous operation of circuits at 85°C. For underground measurements (performed in parallel to the altitude test with a second duplicated setup), a relatively good convergence of the SER was observed until 8×10^7 Mbit $\times$ h (25,000 h), followed by a second period during which the SER slightly increased until the end of the experiment at 48,000 h. The origin of this slight increase is also not clearly understood. A possible explanation (Gedion et al. 2011) could be an increase of the alpha activity of circuit materials induced by a disequilibrium state of the uranium decay chain. Such an effect has already been demonstrated in Gordon et al. (2010) using lead-plated wafers: the alpha emissivity has been found to increase for several years after plating before reaching secular equilibrium.

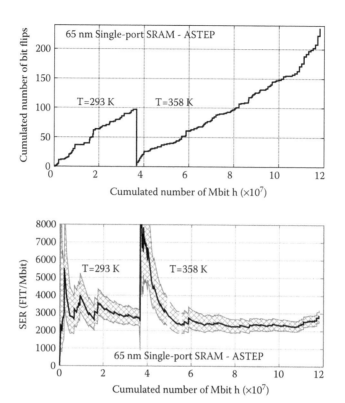

FIGURE 7.14

As Figure 7.5, for 65 nm single-port SRAMs. Devices were subjected successively to two different test temperatures: room temperature (20°C, i.e., 293 K) during the period (January 21, 2008–July 5, 2009) and high temperature (85°C, i.e., 358 K) during the period (June 26, 2009–January 24, 2013). For the high-temperature test, circuit boards were placed in a PC-controlled oven. (Reprinted from *Microelectronics Reliability*, 54, Autran, J.L., Munteanu, D., Roche, P., and Gasiot, G., Real-time soft-error rate measurements: A review, 1455–1476, Copyright (2014), with permission from Elsevier.)

RTSER altitude results for the latest technology, that is, 40 nm SRAM, are reported in Figure 7.16. The staircase bit-flip distribution is less regular than the previous ones, due to the occurrence of large MCU events, as illustrated in Figure 7.17. MCU multiplicities up to 21 have been detected experimentally for the 40 nm SRAM, against 8 for the 65 nm and only 2 for the 130 nm, demonstrating the increasing importance of the carrier channeling into narrow wells, resulting in the propagation of the transient current, which turns on the NMOS-based parasitic bipolar transistors, as explained in Munteanu and Autran (2013).

Finally, Figure 7.18 shows the evolution of the neutron- and alpha-SER, respectively, for the three generations of circuits, deduced from these different RTSER experiments. For the 40 nm alpha-SER value only, the figure shows the result of combined irradiation tests using an intense solid-state alpha-particle source (^{241}Am) and alpha-emissivity measurements at wafer level performed with an ultralow-background alpha-particle counter (Autran et al. 2012). Such a combination of techniques, associated with modeling and simulation work as described in Martinie et al. (2011a), allows us to be very confident of this extracted value (as previously demonstrated in Martinie et al. 2011b, 2012), even if no underground test has already been performed for the present 40 nm test chip.

Figure 7.18 highlights a global trend for the neutron-SER, which is found to be minimal for the 65 nm node and to increase again by about a factor of two for the 40 nm technology.

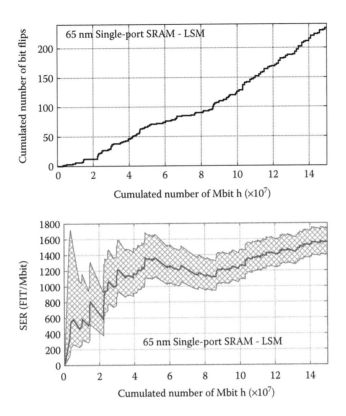

FIGURE 7.15
As Figure 7.5, for 65 nm single-port SRAMs characterized underground at LSM. Test period was November 4, 2010–January 6, 2014. (Reprinted from *Microelectronics Reliability*, 54, Autran, J.L., Munteanu, D., Roche, P., and Gasiot, G., Real-time soft-error rate measurements: A review, 1455–1476, Copyright (2014), with permission from Elsevier.)

At the same time, Figure 7.18 demonstrates constantly decreasing evolution of the alpha SER, which has been reduced by a factor of approximately four between the 130 and the 40 nm nodes. This tendency is well correlated with the values of alpha-particle emissivity measured at wafer level for these technologies (not shown), evidencing a strong decrease from typical values of $2.3 \times 10^{-3} \alpha \, cm^{-2} \, h^{-1}$ for the 130 nm technology to $0.92 \times 10^{-3} \alpha \, cm^{-2} \, h^{-1}$ for the 40 nm one. From a manufacturer's point of view, this illustrates the important effort conducted over the last 10 years to reduce the alpha-emission rate for semiconductor processing and packaging materials (drastic selection of the materials and chemical precursors used, in-line characterization, etc.). This effort should continue for future technological nodes (Roche et al. 2013).

7.5 Comparison with Accelerated Tests

In complement to Section 6.5, which investigated by simulation the differences (and their implications for the SER) between the energy distribution of various artificial neutron sources used for ASER testing and the natural environmental spectrum at ground level,

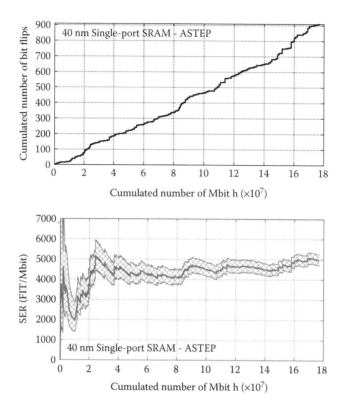

FIGURE 7.16

As Figure 7.5, for 40 nm single-port SRAMs. Test period was March 11, 2014–January 6, 2014. The nominal voltage core for this technology is 1.1 V. (Reprinted from *Microelectronics Reliability*, 54, Autran, J.L., Munteanu, D., Roche, P., and Gasiot, G., Real-time soft-error rate measurements: A review, 1455–1476, Copyright (2014), with permission from Elsevier.)

this last section concludes the chapter with some recent comparisons between RTSER and ASER tests. One of the main objectives is to discuss an important question in the field of SER characterization related to the *equivalence* between these two approaches. Concerning this crucial aspect, it must be noted that contradictory results have been reported in the literature (for the essential points, see the works cited in Section 7.3), ranging from a very good matching of ASER results with RTSER data (Cannon et al. 2004; Heijmen and Verwijst 2009; Lesea 2011; Puchner 2011; Seifert and Kirsch 2012) to more or less important discrepancies, up to 50% or more in certain studies (Autran et al. 2007, 2009, 2010, 2012; Kobayashi et al. 2004; Nakamura et al. 2008). But, due to the diversity of circuits and technologies under test, the wide range of accelerated facilities considered, and the experimental protocols used, it is difficult to conclude on a common origin for all observed discrepancies. Multiple factors are susceptible to impacting ASER and RTSER tests; we already discussed in Section 7.2 some important issues for RTSER tests. In comparison with RTSER, ASER allows useful data to be obtained in a fraction of the time required by unaccelerated testing with only a few circuits (JEDEC 2006). ASER results must therefore be extrapolated to natural backgrounds for both internal (alpha-particle emitters) and external (atmospheric particles) radiation constraints; this point requires knowledge of the particle flux and spectra delivered by the accelerated source and, of course, its exact quantification with respect to the same physical quantity related to the natural environment. ASER tests must be also

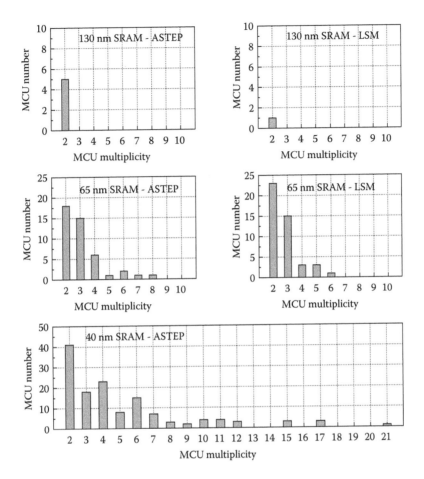

FIGURE 7.17
Experimental distributions of multiple-cell upsets (MCUs) as a function of event multiplicity for 130, 65, and 40 nm single-port SRAMs. Test conditions are defined in previous figure captions for the different technologies. (Reprinted from *Microelectronics Reliability*, 54, Autran, J.L., Munteanu, D., Roche, P., and Gasiot, G., Real-time soft-error rate measurements: A review, 1455–1476, Copyright (2014), with permission from Elsevier.)

conducted with extreme precautions to avoid or limit potential sources of inaccuracy during the measurements, for example, beam uniformity and its fluctuations in time, uncertainties in particle flux and dosimetry, chip misalignment with respect to the beam, and so on.

In their work published in 2004, Kobayashi et al. reported a large discrepancy, of a factor of 2.6, between RTSER and ASER of 0.18 μm SRAMs. They investigated the possible causes of this discrepancy, including accuracy of the neutron flux, high flux effect, dependence on the angle of the incident neutron, temperature dependence, and time variation of the cosmic rays. Table 7.6 proposes an inventory of such possible causes of discrepancy between RTSER and ASER tests, also indicating a numerical estimation of the degree of discrepancy cause per cause. The experimental accuracies are relatively satisfactory, and the degree to which any of these causes, except the neutron flux at sea level, contributes to the discrepancy is negligible compared with the factor of 2.6 observed. The authors concluded that the discrepancy might originate from the value assumed for the integrated neutron flux at sea level. In the meantime, Gordon et al. (2004) have reported more recent

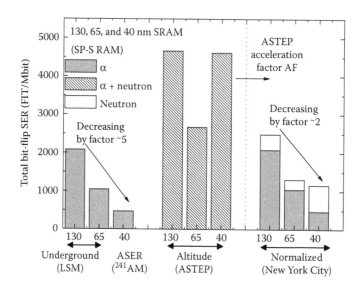

FIGURE 7.18
Synthesis of experimental RTSER values obtained for 130, 65, and 40 nm single-port SRAM from altitude and underground experiments and normalization of the SER for New York City reference conditions. (Reprinted from Autran, J.L. et al., Real-time soft-error testing of 40 nm SRAMs, in *Proceedings of the IEEE International Reliability Physics Symposium*, pp. 3C.5.1–3C.5.9. © (2012) IEEE. With permission.)

TABLE 7.6

Causes of Discrepancy between RTSER and ASER Tests and Estimation of Degree of Discrepancy

Causes of Discrepancy between RTSER and ASER	Degree of Discrepancy
Accuracy of neutron flux in ASER test	<5%
Difference in neutron energy spectra between that used in ASER test and cosmic neutrons at sea level	<25%
Thermal-neutron contamination in ASER test	<3%
Thermal-neutron generation in devices by high-energy neutrons in ASER test	<3%
Malfunction of devices induced by high flux beam in ASER test	<10%
Angular dependence	<20%
Temperature dependence	<10%
Statistical error in RTSER	<30%
Time variation of RTSER related to solar activity	<20%
Accuracy of neutron flux at sea level	>2×

Source: Kobayashi, H. et al., *Proceedings of the IEEE International Reliability Physics Symposium*, 288–293, 25–29 April, IEEE, 2004.

and accurate values for this flux: the conclusion formulated in 2004 by Kobayashi et al. should possibly be revised for later studies.

Finally, and in the light of the recent simulation study reported in Section 6.5, it should be noted that the direct comparison between ASER and RTSER underlines not only possible experimental causes of errors in ASER and RTSER test procedures but also intrinsic differences in the SER due to the fact that the natural background and the artificial source do not have the same energy distribution of particles. The example of the SER calculation for a 65 nm SRAM test circuit considering different broad-spectrum or monoenergetic sources, reported in Section 6.5, is significant. It demonstrates the fact that ASER and RTSER values can differ, and that this difference has its origin in the convolution of the device cross section with the natural or artificial broad spectrum. The question of ASER versus RTSER is, thus, not trivial and cannot be limited to a simple comparison of SER values.

References

Autran, J.L., Munteanu, D., Roche, P., et al. 2010. Soft-errors induced by terrestrial neutrons and natural alpha-particle emitters in advanced memory circuits at ground level. *Microelectronics Reliability* 50:1822–1831.

Autran, J.L., Munteanu, D., Roche, P., and Gasiot, G. 2014. Real-time soft-error rate measurements: A review. *Microelectronics Reliability* 54 :1455–1636.

Autran, J.L., Roche, P., Borel, J., et al. 2007. Altitude SEE Test European Platform (ASTEP) and first results in CMOS 130 nm SRAM. *IEEE Transactions on Nuclear Science* 54:1002–1009.

Autran, J.L., Roche, P., Sauze, S., et al. 2009. Altitude and underground real-time SER characterization of CMOS 65 nm SRAM. *IEEE Transactions on Nuclear Science* 56:2258–2266.

Autran, J.L., Serre, S., Munteanu, D., et al. 2012. Real-time soft-error testing of 40 nm SRAMs. In *Proceedings of the IEEE International Reliability Physics Symposium*, pp. 3C.5.1–3C.5.9. 15–19 April, Anaheim, CA, IEEE.

Bagatin, M., Gerardin, S., Paccagnella, A., and Faccio, F. 2010. Impact of NBTI aging on the single-event upset of SRAM cells. *IEEE Transactions on Nuclear Science* 57:3245–3250.

Balbes, M.J., Boyd, R.N., Kalen, J.D., et al. 1997. Evaluation of the WIPP site for the supernova neutrino burst observatory. *Nuclear Instruments and Methods in Physics Research Section A* 399:269–274.

Baumann, R.C. 2005. Radiation-induced soft errors in advanced semiconductor technologies. *IEEE Transactions on Device and Materials Reliability* 5:305–316.

Cannon, E.H., Reinhardt, D.D., Gordon, M.S., and Makowenskyj, P.S. 2004. SRAM SER in 90, 130 and 180 nm bulk and SOI technologies. In *Proceedings of the IEEE International Reliability Physics Symposium*, pp. 300–304. 25–29 April, Phoenix, AZ, IEEE.

Chazal, V., Brissot, R., Cavaignac, J.F., et al. 1998. Neutron background measurements in the underground laboratory of Modane. *Astroparticle Physics* 9:163–172.

Esch, E.I., Bowles, T.J., Hime, A., Pichlmaier, A., Reifarth, R., and Wollnik, H. 2005. The cosmic ray muon flux at WIPP. *Nuclear Instruments and Methods in Physics Research Section A* 538:516–525.

Gedion, M., Wrobel, F., Saigne, F., Portier, M., Touboul, A.D., and Schrimpf, R.D. 2011. Effect of the uranium decay chain disequilibrium on alpha disintegration rate. *IEEE Transactions on Nuclear Science* 58:2793–2797.

Gordon, M.S., Goldhagen, P., Rodbell, K.P., et al. 2004. Measurement of the flux and energy spectrum of cosmic-ray induced neutrons on the ground. *IEEE Transactions on Nuclear Science* 51:3427–3434.

Gordon, M.S., Rodbell, K.P., Heidel, D.F., Cabral, C., Cannon, E.H., and Reinhardt, D.D. 2008. Single-event-upset and alpha-particle emission rate measurement techniques. *IBM Journal of Research and Development* 52:265–273.

Gordon, M.S., Rodbell, K.P., Heidel, D.F., et al. 2010. Alpha-particle emission energy spectra from materials used for solder bumps. *IEEE Transactions on Nuclear Science* 57:3251–3256.

Heijmen, T. and Verwijst, J. 2009. Altitude and underground real-time SER tests of embedded SRAM. In *Proceedings of the European Conference on Radiation and Its Effects on Components and Systems*, pp. 342–347. 14–18 September, Bruges, IEEE.

JEDEC. 2006. Measurement and reporting of alpha particle and terrestrial cosmic ray-induced soft errors in semiconductor devices, JEDEC Standard JESD89A, October 2006. Available at: http://www.jedec.org/sites/default/files/docs/jesd89a.pdf.

Just, G., Autran, J.L., Serre, S., et al. 2013. Soft errors induced by natural radiation at ground level in floating gate flash memories. In *Proceedings of the IEEE International Reliability Physics Symposium*, pp. 3D.4.1–3D.4.8. 14–18 April, Anaheim, CA, IEEE.

Kameyama, H., Yahagi, Y., and Ibe, E. 2007. A quantitative analysis of neutron-induced multi-cell upset in deep submicron SRAMs and of the impact due to anomalous noise. In *Proceedings of the IEEE International Reliability Physics Symposium*, pp. 678–679. 15–19 April, Phoenix, AZ, IEEE.

Kauppila, A.V., Bhuva, B.L., Loveless, T.D., et al. 2012. Effect of negative bias temperature instability on the single event upset response of 40 nm flip-flops. *IEEE Transactions on Nuclear Science* 59:2651–2657.

Kobayashi, H., Kawamoto, N., Kase, J., and Shiraish, K. 2009. Alpha particle and neutron-induced soft error rates and scaling trends in SRAM. In *Proceedings of the IEEE International Reliability Physics Symposium*, pp. 206–211. 26–30 April, Montreal, IEEE.

Kobayashi, H., Shiraishi, K., Tsuchiya, H., et al. 2002. Soft errors in SRAM devices induced by high energy neutrons, thermal neutrons and alpha particles. In *Proceedings of the International Electron Devices Meeting Technical Digest*, pp. 337–340. 8–11 December, San Francisco, CA, IEEE.

Kobayashi, H., Usuki, H., Shiraishi, K., et al. 2004. Comparison between neutron-induced system-SER and accelerated-SER in SRAMs. In *Proceedings of the IEEE International Reliability Physics Symposium*, pp. 288–293. 25–29 April, Phoenix, AZ, IEEE.

Lesea, A. 2011. Continuing experiments of atmospheric neutron effects on deep submicron integrated circuits. Xilinx white paper, 2011. Available at: http://www.xilinx.com/support/documentation/white_papers/wp286.pdf.

Lesea, A., Castellani, K., Waysand, G., Mauff, J.L., and Sudre, C. 2008. Qualification methodology of sub-micron ICs in the low noise underground laboratory of Rustrel. *IEEE Transactions on Nuclear Science* 55:2148–2153.

Lesea, A., Drimer, S., Fabula, J., Carmichael, C., and Alfke, P. 2005. The Rosetta experiment: Atmospheric soft error rate testing in differing technology FPGAs. *IEEE Transactions on Device and Materials Reliability* 5:317–328.

Martinie, S., Autran, J.L., Munteanu, D., Wrobel, F., Gedion, M., and Saigné, F. 2011a. Analytical modeling of alpha-particle emission rate at wafer-level. *IEEE Transactions on Nuclear Science* 58:2798–2803.

Martinie, S., Autran, J.L., Sauze, S., et al. 2012. Underground characterization and modeling of alpha-particle induced soft-error rate in CMOS 65 nm SRAM. *IEEE Transactions on Nuclear Science* 59:1048–1053.

Martinie, S., Autran, J.L., Uznanski, S., et al. 2011b. Alpha-particle induced soft-error rate in CMOS 130 nm SRAM. *IEEE Transactions on Nuclear Science* 58:1086–1092.

May, T.C. and Woods, M.H. 1978. A new physical mechanism for soft errors in dynamic memories. In *Proceedings of the 16th Annual IEEE International Reliability Physics Symposium*, pp. 33–40. 18–20 April, San Diego, CA, IEEE.

McPherson, J.W. 2013. *Reliability Physics and Engineering—Time-to-Failure Modeling*, 2nd edn. New York: Springer.

Munteanu, D. and Autran, J.L. 2008. Modeling and simulation of single-event effects in digital devices and ICs. *IEEE Transactions on Nuclear Science* 55:1854–1878.

Munteanu, D. and Autran, J.L. 2013. Understanding evolving risks of single event effects. European Conference on Radiation and Its Effects on Components and Systems, Oxford, UK, 2013. Short Course Text.

Nakamura, T., Baba, M., Ibe, E., Yahagi, Y., and Kameyama, H. 2008. *Terrestrial Neutron-Induced Soft Errors in Advanced Memory Devices*. River Edge, NJ: World Scientific.

O'Gorman, T.J. 1994. The effect of cosmic rays on the soft error rate of a DRAM at ground level. *IEEE Transactions on Electron Devices* 41:553–557.

O'Gorman, T.J., Ross, J.M., Taber, A.H., et al. 1996. Field testing for cosmic ray soft errors in semiconductor memories. *IBM Journal of Research and Development* 40:41–50.

Puchner, H. 2011. Correlation of life testing to accelerated soft error. In *Proceedings of the IEEE 3rd Annual SER Workshop*, 27 October, San Jose, CA.

Roche, P., Autran, J.L., Gasiot, G., and Munteanu, D. 2013. Technology downscaling worsening radiation effects in bulk: SOI to the rescue. In *Proceedings of the IEEE International Electron Devices Meeting Technical Digest*, pp. 766–769, 9–11 December, Washington, DC, IEEE.

Seifert, N. and Kirsch, M. 2012. Real-time soft-error testing results of 45-nm, high-κ metal gate, bulk CMOS SRAMs. *IEEE Transactions on Nuclear Science* 59:2818–2823.

Semikh, S., Serre, S., Autran, J.L., et al. 2012. The Plateau de Bure neutron monitor: Design, operation and Monte Carlo simulation. *IEEE Transactions on Nuclear Science* 59:303–313.

Tang, H.H.K. 2008. SEMM-2: A new generation of single-event-effect modeling tools. *IBM Journal of Research and Development* 52:233–244.

Tang, H.H.K. and Cannon, E.H. 2004. SEMM-2: A modeling system for single event upset analysis. *IEEE Transactions on Nuclear Science* 51:3342–3348.

Thomas, R. 2013. *Practical Guide to ICP-MS: A Tutorial for Beginners*, 3rd edn. Boca Raton, FL: CRC.

Xilinx. 2014. Xilinx device reliability report, first quarter 2014, UG116 (v10.0), May 2014. Available at: http://www.xilinx.com/support/documentation/user_guides/ug116.pdf.

Yakushev, E. 2006. Neutron flux measurements in the LSM. LSM internal report. Available at: http://www-lsm.in2p3.fr.

Ziegler, J.F., Curtis, H.W., Muhlfeld, H.P., et al. 1996. IBM experiments in soft fails in computer electronics (1978–1994). *IBM Journal of Research and Development* 40(1):3–18.

Ziegler, J.F. and Puchner, H. 2004. *SER—History, Trends and Challenges*. San Jose, CA: Cypress Semiconductor.

Section III

Soft Errors
Modeling and Simulation Issues

8

Modeling and Simulation of Single-Event Effects in Devices and Circuits

8.1 Interest in Modeling and Simulation

Modeling and simulation have always been used in microelectronics to study the behavior of MOS devices and very-large-scale-integration (VLSI) circuits, which have become the dominant technology in the semiconductors industry. Today, due to the reduction of metal-oxide-semiconductor (MOS) transistor dimensions, VLSI circuits commonly contain several billions of transistors with gate lengths down to 22 nm. The design of such circuit architectures is impossible without the assistance of simulation tools to predict the circuit operation before its manufacturing (Arora 2007). Computer-aided design (CAD) systems are thus essential tools for circuit design and verification operations.

Interest in simulation has also increased considerably in the last decades due to the growing complexity of CMOS manufacturing processes and to the emergence of new devices based on alternative architectures (such as multigate transistors, ultrathin film, or silicon-nanowire-based MOSFETs) requiring very specific fabrication steps that increase the manufacturing costs of prototype circuits. It is clear that exploratory studies for device optimization based only on systematic experimental tests are impossible today, because they are at the same time too long, too complex, and too expensive. Consequently, in recent years simulation has become the ideal solution to significantly reduce the cost (and duration) of the circuit-optimization manufacturing cycle. This was possible in particular thanks to spectacular progress in computer performance, which is able to offer today very cheap and extensive (i.e., quasi-unlimited) calculation resources. Simulation offers considerable advantages to circuit designers due to its predictive capabilities, thereby considerably decreasing the number of experimental tests. In addition, thanks to simulation, it is now possible to simulate hypothetical devices that are not (or that cannot be) fabricated using current technological processes.

Performances achieved in circuit design are completely dependent on the capability and efficiency of existing CAD tools. One of the tools most used for IC design is the electrical circuit simulator, which enables designers to improve their understanding of circuit operation (Galup-Montoro and Schneider 2007). Nevertheless, interest in this kind of simulation depends on its capacity to describe circuit behavior with sufficient accuracy (Arora 2007). Models used in simulation must describe as accurately as possible the physical phenomena associated with the MOSFET operation. This latter is not trivial, since it exploits complex physical phenomena. To use these devices wisely at circuit level and to obtain the benefits of all transistor facets, it is necessary to perfectly understand all these underlying physical mechanisms and to model them with the best accuracy possible. In addition,

FIGURE 8.1
Flowchart of the simulation chain considered in microelectronics. Radiation effects can be considered at both device and circuit levels.

transistor dimensions are currently entering the decananometer range; this trend reinforces the complexity of MOSFET physics with the emergence of physical phenomena specific to such low-dimensionality systems (quantum-mechanical carrier confinement, quasi-ballistic and ballistic transports, parameter fluctuations at atomic scale, etc.). Note that this increase in model complexity not only concerns electrical device-level simulation but also impacts the other simulation approaches, that is, at process and circuit levels. These three distinct (but strongly dependent) simulation domains constitute the main parts of the simulation chain used in microelectronics (Figure 8.1). In the following, we will mainly focus on simulation of electrical and radiation effects at device and circuit levels. Note simply that process simulation deals with all the technological processes used in device fabrication, such as ion implantation, oxidation, materials growth, chemical vapor deposition, annealing, etching, cleaning, metal plating, and so on.

8.1.1 Main Approaches of Electrical Simulation at Device Level

Electrical simulation at device level consists in reproducing device electrical operation as a function of the different biases applied on the device terminals and as a function of various parameters, such as temperature, for example. For MOSFETs, this consists specifically in calculating the source-to-drain current characteristics as a function of the bias applied to the gate, drain, source, and substrate electrodes. Two different types of electrical simulation can be performed to reproduce the electrical operation of a given device (Figure 8.2): device numerical simulation and simulation using compact models. Table 8.1 shows the various features of these two approaches and recapitulates their main advantages and drawbacks.

Device numerical simulations are performed using numerical codes that solve a certain number of differential equations accurately describing the device physics (e.g., Poisson's equation, Schrödinger's equation, and transport equations). The code usually solves these equations coupled together, that is, self-consistently. This kind of simulation provides

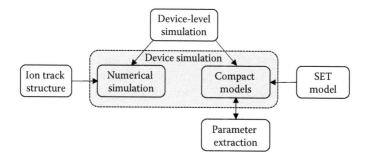

FIGURE 8.2
Different approaches to device simulation. The corresponding inputs to take into account the effects of ionizing particles are also indicated.

TABLE 8.1

Main Features of Numerical Simulation and Simulation Using Compact Models at Device Level

	Numerical Simulation	**Compact Models**
Equations	Numerical solving of different fundamental physics equations (Poisson, Schrodinger, current continuity, etc.)	Analytical simple equations
Accuracy	Very accurate	Accurate (with careful parameter extraction procedure)
Complexity	Complex, difficult to implement	Simple
CPU time	Significant CPU time	Very rapid (suitable for simulation of circuits with large number of transistors)

very good accuracy, but its major drawback is the long computing time. Both device and process numerical simulations are part of the so-called technology computer-aided design (TCAD) simulation platforms (Silvaco 2004; Synopsys 2014), described in detail in Section 8.2.3.

Device simulation with compact models consists in developing a simplified set of equations that mathematically describe the transistor operation and the associated physical phenomena. These equations are usually obtained starting from physical equations (e.g., Poisson, continuity equations) and making a number of simplifying assumptions. This allows a computation time significantly shorter than for device numerical simulation, while the accuracy of the results can be very satisfactory (see Section 8.2.4 for more details). Due to their advantages in terms of reduced computational cost, compact models are extensively used for circuit simulation considering a (very) large number of individual devices. This is the reason why compact modeling is considered as an essential bridge between circuit design and the manufacturing process (Arora 2007).

As previously stated, the accuracy of the computed device characteristics depends on the exactitude of models used in simulation. Compact models frequently contain semiempirical analytical mathematical expressions to take into account a certain number of physical phenomena that are difficult to model. These expressions include fitting parameters that very often have no clear physical meaning. It is, then, very important to extract the values of these model parameters from the characteristics of real device data in order to maintain their physical meaning (Arora 2007). Note that this parameter extraction can be performed using experimental data measured on real devices or, eventually, considering data from device numerical simulation.

8.1.2 Main Simulation Approaches at Circuit Level

Circuit-level simulation is used to investigate the electrical operation of a given circuit and to eventually optimize this latter by direct feedback on the circuit design. It is based on the collective simulation of individual devices that are interconnected and constitute the different bricks of the circuit. Depending on the number of devices, several approaches exist.

- Full numerical simulation of the entire circuit in the three-dimensional (3D) domain, in practice restricted to circuits with a small number of transistors (such as SRAM memory cells, individual flip-flops, etc.) for computational reasons.

- Partial numerical simulation of the circuit delimited to the region of interest (e.g., the device impacted by an ionizing particle), the rest of the circuit being simulated using compact models; this approach is often called *mixed-mode simulation*.
- Full-circuit simulation using compact models, also called SPICE-like circuit simulation. For designs with a large number of transistors, it is clear that a full numerical approach is impossible due to the prohibitive computation time. Compact models are therefore essential for the simulation and design of these circuits, thanks to their very short computation times.

These different approaches to circuit-level simulation will be detailed and illustrated in Section 8.3.

Concerning more particularly the effects of ionizing radiation, simulation has long been used for better understanding radiation effects on the operation of devices and circuits. In the last two decades, due to substantial progress in simulation codes and computer performance that has reduced computational times, simulation has attracted significantly increased interest. In particular due to its increasing predictive capabilities, the simulation of radiation effects offers the possibility to substantively reduce radiation experiments and to test hypothetical devices or conditions that are not feasible (or not easy) to measure by experiment. Physically based numerical simulation at device level is currently becoming an indispensable tool for the study of radiation effects in new device architectures (such as multiple-gate, silicon-nanowire MOSFET), for which experimental investigation is still limited (Munteanu and Autran 2008). In these cases, numerical simulation is an ideal investigative tool for providing physical insights and predicting the operation of future devices expected from long-term projections of the microelectronics roadmap. Last but not least, understanding of the soft-error mechanisms in such devices and prediction of their occurrence in a given radiation environment are of fundamental importance for certain applications requiring a very high level of reliability and dependability (Ziegler and Puchner 2004).

8.2 Device-Level Simulation

Simulation of radiation effects at device level aims to describe both the device properties (geometric architecture, electrical operation) and its response when it is subjected to a given radiation. Two main methods can be used for this purpose, as previously introduced: on the one hand, device numerical simulation, and, on the other hand, the use of compact models. We begin this section by introducing the transport models used in device simulation (drift-diffusion, hydrodynamic, Monte Carlo, and quantum models). Next, the emerging physical phenomena in ultrashort MOSFETs (quantum confinement, ballistic transport, tunneling) are described in detail, and the methods envisaged for taking them into account in simulation at device level are briefly presented. Finally, we describe the two simulation methods at device level (numerical simulation and compact modeling), highlighting the advantages and drawbacks of each approach. Two other important issues of SEE simulation are discussed: (1) the calibration of the device simulator and (2) the ion-track structure to be used as simulation input.

8.2.1 Transport Models

Advanced transport models describing carrier transport in semiconductors are generally derived from the solutions of the semi classical Boltzmann Transport Equation (BTE). The drift-diffusion (DD) approximation is the simplest solution of the BTE; it corresponds to one of the most widely used models in device simulation. This model assumes that the physical phenomena that occur in microelectronics devices are dependent on the electric field, although these phenomena depend on the carrier energy (Selberrer 1984). This conventional description is thus suitable only for long-channel devices and loses its validity in short channels, where the electron transport significantly differs from conventional transport. Therefore, to accurately simulate the transport in these short-channel transistors, advanced transport models, also derived from the BTE, became mandatory. These different approaches are the hydrodynamic model and the Monte Carlo solving of the BTE. Note that more advanced versions of these different approaches are now able to take into account the quantum-mechanical confinement of carriers, generally in one or two dimensions perpendicular to the direction of transport. These approaches are generally qualified as *quantum* derivations of the initial models, for instance *quantum drift-diffusion* or *quantum hydrodynamics*.

With the drastic reduction of device dimensions, average parameters such as carrier mobility or diffusion coefficient become inadequate to describe the electronic transport in devices with channel lengths as small as several nanometers. Other physical concepts, combining quantum mechanics and nonequilibrium statistical physics, are needed to model the electronic transport, referred to now as *quantum transport* and built on a formalism fundamentally different from the one considered in semiclassical approaches. Quantum transport includes model families based, for example, on the Wigner approach (Querlioz et al. 2006) or on the nonequilibrium Green's function (NEGF) formalism (Datta 2000; Ren et al. 2001, 2003; Autran and Munteanu 2008).

Figure 8.3 summarizes the different levels of approximation used to describe the carrier transport in MOS transistors. In this figure, models are divided into two categories: semiclassical and quantum approaches. Transport models are classified according to increasing accuracy, from the simplest drift-diffusion approach to the most complete quantum numerical model. As shown in this figure, the more the accuracy of the models increases, the more their complexity, their implementation difficulty, and particularly their computational time increase (Vasileska and Goodnick 2002). Table 8.2 also gives some additional information concerning several key features of all these models. In the continuation of this section, we briefly describe transport models usually used in studies of radiation effects on electronic devices, including the drift-diffusion and hydrodynamic models. Very general principles for Monte Carlo simulation are also recalled. Of course, this section does not pretend to be an exhaustive review of transport models; reference articles are specified each time, offering a more complete description of all these models.

8.2.1.1 *Drift-Diffusion Model*

The drift-diffusion model was for many years the standard level in the modeling of microelectronic devices, mainly due to its conceptual simplicity and very short simulation time. This model assumes that the carrier energy is constantly in balance with the electric field, and therefore transport depends only on the electric field. In the drift-diffusion model, the carrier transport in the MOS transistor is mainly due to electrostatic potential gradients,

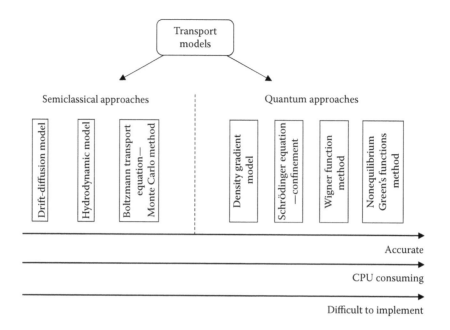

FIGURE 8.3

Schematic representation of different transport models used in device-level simulation. Models are classified according to increasing accuracy from the simplest approach (drift-diffusion model) to the more complex model (NEGF method).

TABLE 8.2

Key Features of Various Transport Models Used in Device-Level Simulation

Approach	Model	Key Features
Semiclassical	Drift-diffusion	Fast and simple to implement; extensively used in SEE simulation; appropriate for devices down to 0.2 μm; include the mobility dependence on the electric field
	Hydrodynamic	Include nonstationary effects (velocity overshoot) and carrier diffusion by electronic temperature gradient
	Boltzmann transport equation—Monte Carlo method	Accurate up to classical limit; long CPU time; difficult to implement
Quantum	Density gradient	Drift-diffusion or hydrodynamic equations + quantum corrections
	Schrödinger equation—confinement	Drift-diffusion or hydrodynamic equations + solved self-consistently with the Poisson–Schrödinger system of equations; include accurate quantum confinement
	Wigner function method	Quantum kinetic approach (determinist/Monte Carlo)
	Non Equilibrium Green Functions (NEGF) method	Mode-space, real-space, and time-domain approaches; difficult to implement; very high CPU consumption

carrier concentration gradients, or both (Selberrer 1984). The current densities of electrons and holes in a biased device are then usually modeled as follows:

$$J_n = -q\mu_n n \nabla\varphi + qD_n \nabla n \tag{8.1}$$

$$J_p = q\mu_p p \nabla\varphi - qD_p \nabla p \tag{8.2}$$

where:
μ_n = electron mobility
μ_p = hole mobility
D_n = thermal diffusion coefficient of electrons
D_p = thermal diffusion coefficient of holes
φ = electric potential
n = density of electrons
p = density of holes

$D_{n,p}$ and $\mu_{n,p}$ depend on the material and electric field and are connected by the Einstein relation:

$$D_n = \mu_n \frac{kT_L}{q} \tag{8.3}$$

where T_L is the lattice temperature. In Equations 8.1 and 8.2, therefore, current densities are given by the sum of the conduction or drift component (the first term of the right side of Equations 8.1 and 8.2) and the diffusion component (the second term).

The drift-diffusion model is generally suitable for long devices operating under moderate electric fields. This model assumes that the carriers reach the maximum energy at the same time as the electric field (Lundstrom 2000). Therefore, nonstationary effects (velocity overshoot and carrier transport by thermal diffusion process associated with electronic temperature gradients) specific to ultrashort devices are neglected. This model also considers that the carrier energy never exceeds the thermal energy and the carrier mobility is only a function of the local electric field (mobility does not depend on carrier energy). These assumptions are acceptable only if the electric field varies slowly in the active region, as is the case for long-channel devices. However, the drift-diffusion model can take into account the quantum confinement of carriers, when the Poisson equation is coupled with Schrödinger's equation.

8.2.1.2 Hydrodynamic Model

When the device dimensions are reduced, electron transport is qualitatively different from conventional transport. Contrary to one of the simplifying assumptions of the drift-diffusion model, carrier energy does not respond instantaneously to changes in the electric field. The mobility and the diffusion coefficient are now described as tensors that depend on several other parameters in addition to the electric field (Khanna 2004). In reality, the average carrier velocity does not depend on the local electric field, but is a function of the carrier energy, which depends on temporal and spatial variations of the electric field. In short devices, very abrupt variations in the electric field can occur in the active region of the device. Nonstationary phenomena, such as carrier velocity overshoot (Baccarani and

Wordeman 1985; Chou et al. 1985; Jacoboni and Reggiani 1983; Shahidi et al. 1988), appear consecutively to these spatial and temporal variations of the electric field. In short devices, nonstationary phenomena play an important role and can dominate device operation. To account for these phenomena specific to short transistors, new models of advanced transport become necessary (Apanovich et al. 1994; Stratton 1962; Blotekjaer 1970; Ieong and Tang 1997; Stettler et al. 1993; Thoma et al. 1991), such as the hydrodynamic model.

Another issue concerns the phenomenon of impact ionization, which is particularly relevant for the operation of silicon-on-insulator (SOI) devices under radiation. Impact ionization is an energy-threshold phenomenon that is directly dependent on carrier energy. The physical mechanism of impact ionization can be summarized as follows. Electron–hole pairs are generated in the regions of the device where a strong electric field occurs, such as in the vicinity of the drain region. An electron with sufficient energy in the conduction band transfers its energy to an electron from the valence band, which therefore results in a multiplication of carriers in the device. The energy threshold required for this phenomenon to occur is approximately equal to the energy of the semiconductor bandgap. In the case of MOS devices, impact ionization becomes important for devices that operate at high drain voltages. Electrons generated by impact ionization are collected into the channel and amplify the drain current. Holes are pushed toward the substrate and may or may not be rapidly evacuated, depending on device type. In bulk MOS transistors, the substrate electrode collects holes. In partially depleted SOI MOSFETs, the existence of the buried oxide prevents holes from being evacuated through the substrate electrode, and they accumulate in the neutral region. This accumulation results in an increase of the potential of the neutral region, leading to a *kink* phenomenon in the drain current. Modeling approaches of impact ionization based on the electric field (as in the traditional drift-diffusion model) lead to significant quantitative and qualitative errors (Quade et al. 1991). In particular, an overestimation of impact-ionization rates is observed even for long-channel devices. The use of models based on the carrier energy (such as the hydrodynamic model) is necessary for an accurate modeling of impact ionization (McMacken and Chamberlain 1991).

The hydrodynamic model is a macroscopic approximation of the BTE, taking into account the relaxation effects of energy and momentum. This model was developed from the first three moments of the BTE. In this model, the electron propagation in a semiconductor is treated as the flow of a charged, thermally conducting gas subjected to an electric field. This model eliminates several restrictive assumptions of the drift-diffusion model: carrier energy can exceed thermal energy, and all physical parameters are dependent on carrier energy. In the hydrodynamic model, the current density and the energy flow are modeled by the following equations, shown here for electrons as they are implemented in the Atlas device simulator (Silvaco 2004):

$$\vec{J}_n = q\mu_n \left[-n\vec{\nabla}\varphi + \frac{kT_n}{q}\vec{\nabla}n + \frac{k}{q}(1+\xi_n)n\vec{\nabla}T_n \right] \tag{8.4}$$

$$\text{div}\vec{S}_n = -\vec{J}_n\vec{\nabla}\varphi - \frac{3k}{2}\frac{\partial}{\partial t}(nT_n) - W_n \tag{8.5}$$

$$\vec{S}_n = -K_n\vec{\nabla}T_n - \frac{k\Delta_n}{q}\vec{J}_nT_n \tag{8.6}$$

$$\Delta_n = \frac{5}{2} + \xi_n \tag{8.7}$$

where:

T_n = electron temperature
ξ_n = model coefficient
S_n = the energy flow
W_n = energy density loss rate
K_n = thermal conductivity

The energy density loss W_n is given by

$$W_n = \frac{3}{2} n \frac{k(T_n - T_L)}{\tau_{rel}} + \frac{3}{2} k T_n R_{SRH} + E_g \left(G_n - R_n^A \right) \tag{8.8}$$

where:

τ_{rel} = energy relaxation time
R_{SRH} = Shockley–Read–Hall (SRH) recombination rate
G_n = impact ionization rate
R_n^A = Auger recombination rate
E_g = silicon bandgap

Similar equations are used for holes. Usually, the mobility μ_n is modeled as a decreasing function of energy (because the scattering rate increases with carrier energy).

To develop Equations 8.4 through 8.8, the hydrodynamic model uses the following simplifying assumptions (Munteanu and Autran 2008):

1. The closure condition of the hierarchy of moments is given by the following relation:

$$Q = -K_n \nabla T_n \tag{8.9}$$

This equation links the heat flux Q of the electron gas to the electron temperature. The thermal conductivity K_n is given by the Wiedmann–Franz law:

$$K_n = q n \mu_n \left(\frac{k}{q} \right)^2 \Delta_n T_n \tag{8.10}$$

2. The temperature tensor reduces to a single scalar.
3. The relaxation-time approximation is used for modeling the effects of collisions on the momentum and energy of the electrons.

To understand the superiority of the hydrodynamic model over the drift-diffusion model, it is necessary to analyze in detail Equation 8.5, also called the *energy-balance equation*. The left side of this equation represents the spatial variation of the energy flow. In the right side, the three terms are described as follows:

1. The first term is related to the energy absorbed by the electrons due to the electric field.
2. The second term is the time derivative of the energy density and is a representation of the energy supplied by the lattice by electron collisions with optical phonons.
3. The third term gives the energy loss through recombination processes.

Thus, the equation implies that the spatial variation of energy flow equals the sum total of heat flow and transported energy. The applicability of the hydrodynamic model in MOSFETs with short channels is justified since, in *hot* areas where the electron temperature is high, it predicts a greater diffusion than the drift-diffusion model, due to the finite value of the energy relaxation time (Khanna 2004). Therefore, the average energy and the electron temperature are higher than their equilibrium values in regions with strong electric fields. In addition, the thermoelectric field, given by the term $\nabla(kT_n/q)$ in Equation 8.4, produces a force that enables electron transport from hot to cold areas. This phenomenon corresponds to electron transport by a thermal diffusion process associated with electronic temperature gradients.

Carrier velocity overshoot is the immediate consequence of the finite timespan needed before the carrier energy reaches equilibrium with the electric field. This is mainly due to the nonequivalence of the electron momentum and energy relaxation times. The hydrodynamic model is able to correctly predict the velocity overshoot, which is not the case with the drift-diffusion model, as indicated above. The velocity overshoot can be easily evidenced from the hydrodynamic model (Khanna 2004). For this purpose, we consider the one-dimensional case in which the electric field increases in the direction of carrier motion. With this condition, the energy-balance equation shows that the average energy is less than the energy value corresponding to the local electric field under homogeneous conditions. Given that mobility is a decreasing function of energy, this means that the velocity given by the hydrodynamic model is higher than the velocity obtained by the drift-diffusion model, which is based on a mobility that depends only on the local electric field.

8.2.1.3 Direct Solution of the Boltzmann Transport Equation by the Monte Carlo Method

The most accurate approach used in physics and engineering for the description of carrier transport within a semiclassical description is the Monte Carlo method (Fischetti and Laux 1988; Jacoboni and Reggiani 1983; Ravaioli et al. 2000; Vasileska and Goodnick 2002).

The principle of this method is to simulate the free movement of particles (also called free flight) completed by random and instantaneous scattering events in the semiconductor. The simulation algorithm is briefly explained as follows. Carriers are considered as particles with a mass and an electric charge. In the first simulation step, free flight times are randomly generated for each particle, and all physical quantities characterizing the particle (drift velocity, energy, position) are calculated. In a second simulation step, a scattering mechanism is randomly chosen according to the scattering probabilities of all the possible scattering mechanisms. This scattering event changes the energy and the momentum of the particle. At the end of this second step, the new quantities related to each particle are recalculated. The simulation continues by repeating this procedure: the particles are subjected to the same succession of physical phenomena (free flight time ended by a scattering mechanism). Sampling the particle motion at various simulation times allows several interesting physical quantities to be statistically estimated, such as the single particle distribution function, the average drift velocity in the presence of electric field, the average energy of particles, and so on. The simulation of a set of particles, representative of the considered physical system, allows the nonstationary time-dependent variation of electron and hole distributions under the influence of a time-dependent driving force to be simulated.

Since the Monte Carlo method does not require simplifying assumptions, it provides more accurate results than energy-transport models, such as the hydrodynamic model. Monte Carlo simulation has been widely used to study the operation of MOS transistors with ultrashort channels. Monte Carlo simulation has been used in the past for simulating radiation-induced charge collection (Sai-Halasz and Wordeman 1980; Sai-Halasz et al. 1982) and SEU in short-channel SOI MOSFETs (Brisset et al. 1994). Although this method provides the most accurate semiclassical simulation results for ultrashort MOS transistors, its use is limited due to significant computational time. This drawback prevents the Monte Carlo method from being systematically used for studies of radiation effects in MOS devices.

8.2.2 Emerging Physical Effects

As explained in the introduction to this chapter, due to the reduction of MOSFET dimensions new physical phenomena occur; the most important are quantum confinement of carriers at the silicon/gate-oxide interface and ballistic transport in the channel. In this section, we describe these two emerging phenomena, which are becoming essential in the operation of current and future ultrashort devices. Existing models and approaches to take these phenomena into account in device simulation are also briefly recounted.

8.2.2.1 Quantum-Mechanical Confinement of Carriers

To contain the significant increase of parasitic short-channel effects when the gate length is reduced, the procedure for bulk MOS transistors is usually to increase channel doping and, at the same time, to reduce gate-oxide thickness. Since the supply voltage does not shrink at the same rate as the geometric dimensions, this scaling rule is at the origin of an emerging phenomenon called *quantum confinement of carriers*, explained as follows.

From the CMOS technological node, typically 0.25 μm, the increase of the electric field at the silicon/gate-oxide interface was indeed sufficient to create a deep potential well at this interface in which inversion charge carriers could be confined (Spinelli et al. 1998). Figure 8.4a schematically shows an energy-band diagram (bottom of the conduction band and top of the valence band) in a vertical cross section through the transistor, illustrating the formation of a potential well (created by the silicon conduction band bending at the interface and the silicon/gate-oxide conduction band offset) when an n-channel transistor (p-type silicon substrate) is biased in strong inversion regime. If the potential well profile is sufficiently steep (characteristic size comparable to the electron wavelength), a quasi-two-dimensional (2D) electron gas is formed in this well: carriers that occupy the lowest energy levels are free to move parallel to the interface, but are tightly confined in the direction perpendicular to the interface. Carriers located at higher energies, which are not confined in the potential well, remain free in the three directions, as illustrated in Figure 8.4a. When the electric field at the surface increases, the system becomes more quantified and more carriers are confined in the quantum potential well.

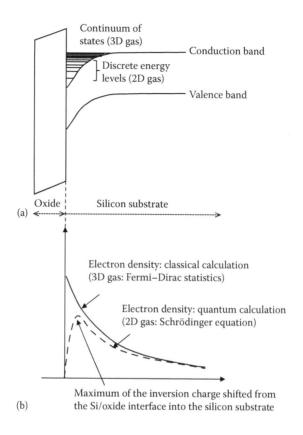

FIGURE 8.4
(a) Schematic illustration of the energy-band diagram (bottom of the conduction band and top of the valence band) in a vertical cross section through an MOS structure biased in the strong inversion regime. The splitting of energy levels into subbands due to carrier quantum confinement in the narrow potential well is also shown. (b) Schematic representation of the electron-density profiles calculated using the Fermi–Dirac statistics for the classical case and Schrödinger's equation for the quantum case.

The confinement of carriers in the well leads to quantized energy levels, with a conduction band split into subbands (Hareland et al. 1998; Taur and Ning 1998). In such a 2D system, the lowest energy level allowed for electrons in the quantum well does not coincide with the bottom of the conduction band. Moreover, the total density of states in a 2D system is lower than that in a 3D (or conventional) system, and the spatial distribution of carriers becomes controlled by the electron wave functions in the direction of the confinement. As a result, the inversion charge distribution resulting from carrier confinement at the transistor interface is qualitatively very different from that obtained by the classical theory: in addition to a reduction of the inversion charge amount (depending on the *intensity* of the confinement), the charge centroid is no longer located at the silicon/gate-oxide interface (as in the conventional case), but moves deep into the substrate (Lopez-Villanueva et al. 1997), as illustrated in Figure 8.4b. Since the source-to-drain conduction in a MOS transistor occurs at the gate-oxide/silicon interface, this double alteration of the inversion charge (amount and spatial distribution) has an immediate impact on carrier transport in the channel and on the various electrical parameters of the MOS structure (Janik and Majkusiak 1998).

- The threshold voltage of the devices increases. Due to the lower density of states in the 2D system, the total population of carriers will be smaller for the same Fermi level as in the corresponding 3D (or classical) case. This affects the net charge of carriers in the inversion layer, which requires a larger gate voltage to fill a 2D inversion layer with the same number of carriers as in the corresponding 3D system. The increase of the threshold voltage is an important issue, especially because supply voltages are reduced when the MOSFET dimensions shrink (thus requiring a reduction of the threshold voltage).
- The gate capacitance and channel transconductance decrease.
- The carrier mobility in the channel is modified with respect to the classical case.

These considerations suggest that the wave nature of electrons and holes cannot be neglected in ultrashort devices. Therefore, it becomes necessary to take these *quantum effects* into account in device simulation for current and future technological generations. Quantum confinement is also important for the response of the device to radiation. A detailed study of the impact of quantum-confinement effects on immunity to single events in single-gate SOI and multiple-gate devices will be presented in Chapter 12.

Numerous approaches have been proposed to account for quantum-confinement effects in device numerical simulation. The most accurate physical solution is obviously to include the Schrödinger equation in the system of equations to be solved (self-consistently) in order to reproduce the device characteristics (Stern 1972). This solution is fully compatible with conventional electrical simulators based on drift-diffusion and hydrodynamic models. However, this approach is rarely used because solving the Schrödinger equation is very time consuming. Different approximation methods have been proposed, especially those based on quantum corrections made to conventional models such as the drift-diffusion or hydrodynamic model. Two widely used models are the van Dort model and Hansch models.

The van Dort model (van Dort et al. 1994) expresses the impact of quantum confinement by an apparent shift of the conduction band bottom, which is a function of the electric field. This model is based on the calculation of the lowest eigenenergies of a particle confined into a triangular potential well. The device electrical characteristics obtained with the van Dort model are in very good agreement with those calculated by coupling the electrostatics and transport problems with the Schrödinger equation. However, its drawback is that it does not reproduce the correct inversion charge distribution in the device.

The Hansch model (Hansch et al. 1989) is based on a quantum correction of the density of states as a function of the depth relative to the silicon/gate-oxide interface. With this model, the charge profile in the device is better reproduced than with the van Dort model. Nevertheless, the Hansch model strongly overestimates the impact of quantum confinement on the drain-current characteristics.

Another solution to take quantum confinement into account is the Density Gradient model (Ancona and Tiersten 1987; Ancona and Iafrate 1989; Ancona et al. 1999; Gardner 1994; Grubin et al. 1993; Wettstein et al. 2002), coupled with the drift-diffusion or hydrodynamic equations. This model falls into the category of quantum corrections, since it is based on an artificial modification of the charge density in order to take quantum confinement into account. Thus, the Density Gradient model considers a modified charge-density equation taking into account an additional term that depends on the gradient of the charge density. The quantum confinement effects are reproduced by the introduction of an additional quantity Λ in the expressions of the charge density (written below for electrons):

$$n = N_C \exp\left(\frac{E_{Fn} - E_C - \Lambda}{kT} \right)$$

(8.11)

where:
  n   = electron density
  T   = carrier temperature
  k   = Boltzmann constant
  N_C = conduction band effective density-of-states
  E_C = conduction band energy
  E_{Fn} = electron Fermi level

The impact of quantum confinement on the carrier density in the device can be taken into account by properly modeling the quantity Λ. A similar equation to Equation 8.11 is applied to the holes. These new equations for the density of electrons and holes are then used in the self-consistent solving of the Poisson equation with the current continuity equations (in which the type of transport, drift-diffusion or hydrodynamic, is also included).

In the Density Gradient model, Λ is usually given in terms of a partial differential equation:

$$\Lambda = -\frac{\gamma \hbar}{6m} \frac{\nabla^2 \sqrt{n}}{\sqrt{n}}$$

(8.12)

where:
  $\hbar = h/2\pi$ is the reduced Planck constant
  m   is the electron density of states mass
  γ is a fitting factor

It has been shown in the literature that the Density Gradient model can accurately reproduce the quantum confinement of carriers in SOI and double-gate MOSFETs if the parameter γ is properly calibrated (Wettstein et al. 2002). An example of calibration of the parameter γ will be presented in Chapter 12; in this example, the exact solution of the self-consistent Schrödinger–Poisson system of equations is used to calibrate the Density Gradient model.

8.2.2.2 Quasi-Ballistic and Ballistic Transport

Ballistic transport has been, in the last decade, a particularly important issue, given that ultimate MOS devices are supposed (expected) to work near the ballistic limit. As we have already mentioned, carrier transport can be qualified as conventional as long as carriers, moving from source to drain region, are subjected to a sufficient number of scattering events (with optical and/or acoustic phonons, ionized impurities, interface, etc.) that statistically define macroscopic parameters such as carrier mobility and diffusion coefficients. This situation is encountered for typical channel lengths above a few tens of nanometers; in such devices, the transport is well described by models such as drift-diffusion or hydrodynamic models. When the transistor gate length is sufficiently reduced and the device size becomes typically of the same order of magnitude as, or less than, the carrier mean free path (around

10 nm for electrons in silicon at 300 K), the transport becomes ballistic (Lundstrom 2000). In this case, carriers traverse the channel without undergoing scattering events and the drain current reaches its maximum value. For longer gate lengths, scattering may affect the transport, which can therefore no longer be considered as purely ballistic. In other terms, carrier transport makes a transition from the ballistic to quasi-ballistic or drift-diffusive regime for channel lengths immediately longer than the carrier mean free path. Then, the drain-current value decreases compared with the maximum (ballistic) value because of scattering events.

Figure 8.5 schematically describes the different mechanisms that take place in the channel in the case of purely ballistic transport. For this purpose, the energy-barrier profile from source to drain is depicted in the case of a device biased in the on state. Naturally, both the amplitude and the width of the barrier are modulated by the gate and drain voltages. The physical mechanisms that take place are explained below.

1. Carriers from the source reservoir with an energy larger than the maximum value of the energy barrier (E_{max}) are transmitted from the source to the drain by a thermionic emission process.

2. Carriers from the source reservoir with an energy lower than E_{max} can therefore cross the channel with a certain probability by a quantum-mechanical tunneling effect through the energy barrier. Similar mechanisms (thermionic emission and quantum tunneling) take place for electrons located in the drain reservoir.

3. Quantum-mechanical reflections of carriers can also occur in the two reservoirs (source and drain) and in the channel (due to the potential energy drop induced by the drain bias), as shown in Figure 8.5. These reflections have a nonnegligible effect on both thermionic and tunneling components of pure ballistic transport.

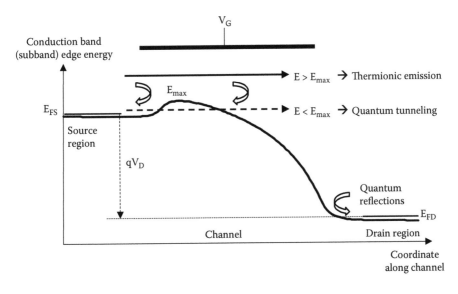

FIGURE 8.5
Schematic representation of the source-to-drain energy barrier profile in a device biased in the on state; illustration of the different mechanisms taking place in the device channel in the case of purely ballistic transport. E_{max} is the maximum value of the energy barrier. Carriers having energy larger than E_{max} are transmitted from source to drain by a thermionic emission phenomenon. Carriers having energy lower than E_{max} can cross the channel only by quantum mechanical tunneling effect through the energy barrier. Quantum reflections of carriers can also occur in the two reservoirs (source and drain) and in the channel.

Because of the emergence of these physical mechanisms, models developed so far for conventional devices (classical transport) no longer apply to ultrascaled devices. Models such as drift-diffusion or hydrodynamic fail to describe ballistic transport accurately (Banoo and Lundstrom 2000). Therefore, the understanding and modeling of ballistic transport are of primary importance for the optimal design of CMOS devices and circuits of future technological generations. Many approaches to simulating ballistic and quasi-ballistic transport have been developed (Rahman and Lundstrom 2002; Jiménez et al. 2003; Natori 1994; Autran et al. 2005; Martinie et al. 2009). Some models take into account only the thermionic emission of carriers (Natori 1994); others consider in addition the current component corresponding to the carrier flow by direct tunneling through the source–channel–drain potential barrier (Munteanu and Autran 2003; Autran et al. 2005). This phenomenon is now recognized as a fundamental limitation to the reduction of MOSFET dimensions below a few nanometers of gate length. Other models take quasi-ballistic transport into account, such as the McKelvey flux theory, which covers all types of transport, from ballistic transport to the diffusive regime (Rahman and Lundstrom 2002; Lundstrom 1997; Martinie et al. 2008). Finally, 2D and 3D numerical codes that take into account quantum transport, optionally including scattering effects, have been widely developed using the NEGF formalism or Monte Carlo/Wigner approaches. Although significant effort has been made on this topic during the last decade, ballistic transport is still not included in 2014 in commercial simulation platforms. In addition, an open question is how ballistic (and quasi-ballistic) transport will affect the immunity to radiation of future devices. This problem, which has not yet been addressed in research studies related to radiation-induced effects, will certainly require the development of new specific models and simulation tools.

8.2.3 TCAD Simulation

TCAD simulation at the device level, also known as numerical modeling (i.e., based on the numerical solving of a set of physics equations), is the most physical approach, whereby it is possible to "see" the internal behavior of a given device (Silvaco 2004; Synopsys 2014). This type of device simulation does not correspond to a large-circuit approach (it is restricted in practice to a single device, a circuit cell, or a portion of a given circuit limited to a few units/tens of transistors), but it is an essential step in IC process development. TCAD numerical modeling aims to quantify the understanding of the underlying technology and abstract this knowledge for use in circuit design. It consists of two distinct parts: simulation of the manufacturing process and simulation of device electrical operation (device electrical simulator). As explained in the introduction to this chapter, the process simulator models the various stages of device fabrication, such as ion implantation, deposition, etching, annealing, and oxidation. Device electrical simulation models the electrical behavior of a device created by process simulation, taking into account its geometry, materials, and doping profiles. The effects of radiation on device operation are also simulated at the device-electrical-simulation level. For this reason, we detail in the following the general principles of electrical device simulation, common to a large majority of electrical simulation codes (integrated in commercial simulation platforms or implemented in homemade codes).

The starting point of electrical device simulation is the device of interest, which has been previously created by process simulation. The device is represented as a meshed structure in which each node has specific associated properties, such as type of material, dopant concentration, and so on. An example of a 3D meshed device, corresponding to an SOI MOSFET, is shown in Figure 8.6. This mesh is used for solving the main differential equations, such as the Poisson equation and the transport and continuity equations.

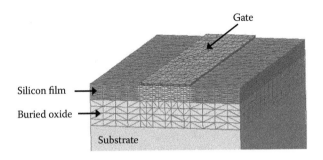

FIGURE 8.6
Illustration of the 3D mesh in an SOI MOSFET. Each mesh node has specific properties in terms of material, doping, permittivity, etc. Physical equations (Poisson, continuity equations) are discretized on the mesh structure.

The general flowchart of electrical simulation is presented in Figure 8.7. The simulation starts with different initialization steps (device geometry, mesh properties, charge density, and biases applied to the different transistor electrodes). All TCAD device-simulation approaches (Silvaco 2004; Synopsys 2014) are based on the numerical solving of a fundamental set of equations, as summarized in Figure 8.8: the Poisson equation for the electrostatic problem, the continuity equations for the carrier dynamic balance, and one or several equations to treat the transport problem. The physical description of the device electrostatics requires solving the Poisson equation, which can be written in its general form as

$$\nabla\left(\varepsilon\,\vec{\nabla}\Phi\right) = -\rho \tag{8.13}$$

where:
Φ = electrostatic potential (V)
ε = material-dependent permittivity (F m^{-1})
ρ = net charge density (C m^{-3})

The net charge density ρ is given by Taur and Ning (1998):

$$\rho = q\left[N_{DOPING} - n + p\right] \tag{8.14}$$

where:
q = absolute value of the electron charge
N_{DOPING} = net doping atom concentrations in the silicon film
n and p = electron and hole densities, respectively

The electron and hole densities are calculated using the Fermi–Dirac statistics. The Poisson equation is solved self-consistently (coupled) with continuity equations, given by

$$\nabla\vec{J}_n = q\left(G - R\right) + q\frac{\partial n}{\partial t} \tag{8.15}$$

$$\nabla\vec{J}_p = -q\left(G - R\right) - q\frac{\partial p}{\partial t} \tag{8.16}$$

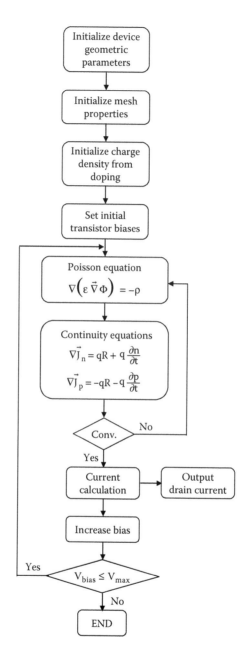

FIGURE 8.7
General flowchart of a typical numerical simulation code at device level illustrating the different equations to be solved for the classical calculation of the drain current (i.e., without quantum confinement effects).

where:

G = generation rate

R = recombination rate

The effect of a particle strike is included in R as a supplementary generation rate (depending on several irradiation parameters), representing an external generation

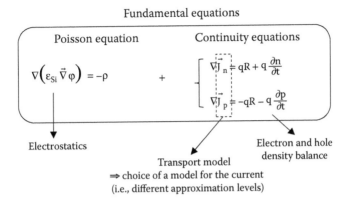

FIGURE 8.8
Schematic description of the basic set of equations to be solved in numerical simulation at device level. Device electrostatics is obtained from the Poisson equation. Continuity equations are mandatory in order to guarantee current continuity. The transport model is included via the current densities J_n and J_p. Different approximation levels can be used for the transport model (from drift-diffusion to quantum approaches) as indicated in Figure 8.3. Whatever the transport model, Poisson equation and continuity equations must be solved.

source of carriers (as explained in Section 8.2.3.2). In Equations 8.15 and 8.16, the current density is expressed from a given model, such as the drift-diffusion model or the hydrodynamic model. Nevertheless, whatever the model used for the calculation of the carrier density, the condition of current continuity should be guaranteed. Indeed, the continuity equations describe the evolution of carriers in the silicon domain (source, channel, drain, substrate) in order to maintain a constant current along the channel in which there is no charge accumulation. If the drift-diffusion model is used as the transport model, electron and hole current densities are given by Equations 8.1 and 8.2. If the hydrodynamic model is used, Poisson and continuity equation have to be coupled with the set of Equations 8.4 through 8.10.

The electrical simulation program numerically solves these main differential physical equations. For this purpose, finite element discretization of each equation is performed on the considered mesh. The calculation scheme is explained as follows. After the initialization steps, the discretized Poisson equation is solved and the electrostatic potential in the device is calculated. The potential is then injected directly into the solving of the continuity equation. The output of this module is used to calculate a new carrier density to be used in the Poisson equation. In this way, a new potential is found and is injected into the continuity equation, and so on. The loop stops when the convergence criterion is reached (Figure 8.7). The device simulation described here corresponds to a classical calculation, which means that quantum-mechanical confinement of carriers is neglected. For taking quantum effects into account, Schrödinger's equation can be added to the system of equations described in Figure 8.8, or quantum corrections can be used, as explained in Section 8.2.2.1.

The main drawback of this type of simulation is the computational time, which can be very important depending on the mesh number and the nature of the solved equations (drift-diffusion, hydrodynamic). But the significant advantage of this full numerical approach is the ability offered to the user to access internal quantities of the simulation (which cannot be measured), which substantially facilitates the thorough understanding of the physical and electrical mechanisms taking place in the device.

In the following, we will address two other important issues for the simulation of radiation effects in microelectronic devices: the calibration of the device simulator on experimental data obtained from real device characterization, and the ion-track structure used as simulation input.

8.2.3.1 Calibration of the Device Simulator

As stated before, numerical device simulation provides satisfactory results because it solves the physical equations in the device. However, using the drift-diffusion or hydrodynamic transport model signifies that a certain number of approximations have been made to simplify the equations. In addition, macroscopic parameters are used in these models, such as carrier mobility, that include the effects of various scattering mechanisms (lattice vibrations or phonons, impurity ions, other carriers, scattering on surfaces, etc.). These parameters have to be carefully calibrated on experimental data in order to accurately simulate the device characteristics. Then, the calibration of the device simulator before its use is a key step of device-level simulation. The calibration step is generally performed with respect to the drain-current characteristics (versus the drain voltage and the gate voltage) $I_D(V_G)$ and $I_D(V_D)$, in order to obtain the correct threshold voltage and the simulated characteristics to match those measured on real devices (Munteanu and Autran 2008). This implies calibration of the main physical models of the device simulator, especially the mobility model, as explained above. This mandatory calibration step can be delicate and time consuming. An example of calibration will be given in Chapter 12 (Figure 12.8); in that study, the TCAD simulation was calibrated on experimental drain-current characteristics measured on 50 nm gate-length FD SOI devices in order to accurately reproduce the correct off-state drain current, which is a key parameter for the device transient response under radiation constraint.

8.2.3.2 Ion-Track Structure

A major issue for studies of radiation effects on device operation is the ion-track structure to be used as input in simulation. In early models, the ion track was represented by a simple generation of cylindrical charge with a uniform charge distribution and a constant LET along the track. However, the real structure of the ion track is radial and varies when the ionizing particle passes through matter. When an ionizing particle hits a device, very energetic primary electrons (called δ rays) are released. They generate a very high density of electron–hole pairs in a very short time and in a very small volume around the ion path, also called the ion track. Carriers created by the ionizing-particle strike then undergo several physical mechanisms: (a) they are collected by drift and diffusion mechanisms in the device, and (b) they are recombined by different mechanisms of direct recombination (radiative Auger) in the dense track core. The combination of these mechanisms induces a significant reduction of the peak carrier concentration and, at the same time, the spatial and temporal distribution of the ion-track ion is considerably affected. As the particle passes through matter, it loses energy and δ rays become less energetic. Therefore, electron–hole pairs are generated closer to the ion trajectory. The incident-particle track shrinks gradually as the particle penetrates more deeply into the matter and generates a characteristic cone-shaped charge plasma in the device (Dodd 2005).

The Monte Carlo method was used for calculating the real ion-track structure (Hamm et al. 1979; Martin et al. 1987; Oldiges et al. 2000). Monte Carlo simulations revealed significant differences between the track structure of low-energy and high-energy particles, even if the LET is the same. More details can be found in Dodd (2005) and Dodd et al.

(1998). High-energy particles are representative of ions existing in the real space environment, but they are not available in typical laboratory SEU measurements (Dodd 1996). The simulation study of the effects of high-energy particles thus represents an interesting opportunity, which may be difficult to perform experimentally.

The ion-track structure was also modeled by analytical models, which were then implemented in simulation codes. One of the most interesting models proposed in the literature is the *nonuniform power law* track model, based on the theory of Katz (Kobetich and Katz 1968) and developed by Stapor (Stapor and McDonald 1988). In that model, the trace of the ion has a radial distribution of excess carriers expressed by a power-law distribution. In addition, the charge density may vary along the ion track (the LET is not constant along the trace) (Dussault et al. 1993). In other analytical models, the ion track is nonuniform with constant radius or with a Gaussian distribution.

As explained at the beginning of Section 8.2.3, the most commonly used commercial TCAD simulators can take into account the effects of an ionizing-particle strike on device operation. The particle impact is modeled as an external source of carrier generation. The generation of electron–hole pairs caused by the particle is taken into account directly in the continuity equations (Equations 8.15 and 8.16) through an additional generation rate (which is added to conventional generation rates corresponding to various direct generation–recombination mechanisms occurring in the device). This new rate is called the radiation-induced generation rate and can be connected to various irradiation parameters, such as the particle LET. The particle LET is defined as the energy deposited per unit length, that is, $-dE/dl$. The particle LET can be converted into an equivalent number of electron–hole pairs by unit of length using the mean energy necessary to create an electron–hole pair (E_{ehp}) (Roche 1999):

$$\frac{dN_{ehp}}{dl} = \frac{1}{E_{ehp}} \frac{dE}{dl} \tag{8.17}$$

where N_{ehp} is the number of electron–hole pairs created by the particle strike. To calculate the radiation-induced generation rate, one must consider two functions, R and T, describing, respectively, the spatial and temporal distributions of the electron–hole pair creation. Using Equation 8.17, the generation rate induced by radiation can be expressed as

$$G(w,l,t) = \frac{dN_{ehp}}{dl}(l) \cdot R(w) \cdot T(t) \tag{8.18}$$

This equation takes into account the following simplifying assumption: the spatial distribution function R(w) depends only on the distance traversed by the particle in the material, and the generation of electron–hole pairs on the ion path follows the same temporal distribution function at each point. It is important to note that the radiation-induced generation rate given by Equation 8.18 depends on the position of the ion-strike impact, the distance traversed by the particle in the matter, and the time. The number of electron–hole pairs created by the passage of the particle is finally included in the continuity equations through the generation rate given by Equation 8.18.

The radiation-induced generation rate must fulfill the condition

$$\int_{w=0}^{\infty} \int_{\theta=0}^{2\pi} \int_{t=-\infty}^{\infty} G\, w\, dw\, d\theta\, dt = \frac{N_{ehp}}{dl} \tag{8.19}$$

Then, functions R(w) and T(t) are submitted to the following normalization conditions:

$$2\pi \int_{w=0}^{\infty} R(w)\,w\,dw = 1 \tag{8.20}$$

$$\int_{t=-\infty}^{\infty} T(t)\,dt = 1 \tag{8.21}$$

Ion-track models implemented in commercial TCAD simulators usually propose a Gaussian function for the temporal distribution function T(t):

$$T(t) = \frac{e^{-(t/t_C)^2}}{t_C\sqrt{\pi}} \tag{8.22}$$

where t_C is the characteristic time of the Gaussian function. The spatial distribution function is usually modeled by an exponential function or by a Gaussian function:

$$R(w) = \frac{e^{-(w/r_C)^2}}{\pi\,r_C^2} \tag{8.23}$$

where r_C is the characteristic radius of the Gaussian function. The characteristic time and the characteristic radius are two key parameters allowing the user to adjust, respectively, the radiation-induced pulse duration and the ion-track width.

The validity of some approximations or models used to model the ion-track structure (such as the approximation of a constant LET along the track or the generation model used for the radial track) has been tested in the literature. Concerning the variation of LET, considering a constant LET along the ion track induces nonnegligible errors in the transient response of the device, since the real LET is not constant with depth along the track. Several main parameters of the transient event (maximum current, time to peak, and collected charge) show up to 20% difference when the LET varies with depth compared with the case of a constant LET (Dussault et al. 1993). However, LET variation with depth has no effect on the transient response of the current SOI devices with very thin silicon films, because the distance traversed by the particle in the device is limited to the film thickness. Concerning the various charge-generation distributions that can be used for radial ion trace, they affect the transient response of the device, but the variation is limited to 5% of the ion strikes on bulk p-n diodes (Dodd 2005; Dussault et al. 1993). The validity of Equations 8.17 through 8.23 to model the track structure of an ionizing particle, in particular with respect to ultrascaled devices and circuits, will also be discussed in Chapter 12.

8.2.4 Analytical and Compact Model Approaches

As explained in the introduction to this chapter, the second approach to device-level modeling is the use of analytical or compact models (Figure 8.2). The numerical modeling approach presented above is computer intensive because it involves detailed spatial and temporal solutions of coupled partial differential equations on 3D meshes inside the device. Although this numerical modeling of the device is intended to be very accurate, it is not fast enough for high-level circuit simulators (e.g., SPICE).

To evaluate the performances of circuits containing a large number of devices (sometimes several thousands), faster transistor models are needed. In order to be used in tools for circuit design, these MOSFET models must be compact, that is, they must provide simple, fast, and accurate algorithms for calculating charges, currents, and their derivatives. These compact models, also called circuit models, must be simple enough to be integrated into circuit simulators and sufficiently precise to make the simulation results useful to circuit designers. They are generally based on analytical expressions that describe the static and dynamic electrical behavior of the elementary devices constituting the circuit. The compact model is, then, the key element that allows one to establish the link between the device (which is itself closely related to technology) and the circuit design. In order to adapt to new nanotransistor architectures, these compact models must evolve to steadily integrate more physics without altering their flexibility, in terms of both computational time and ability to converge numerically.

There are several types of compact models (Arora 2007; Galup-Montoro and Schneider 2007):

1. Physical compact models that are based on the device physics, having as a starting point the physical device equations (Poisson, Schrödinger, transport equations...). The development of this kind of model can be very long and difficult, but the benefits of a mature physical compact model are significant: the model is predictable (it takes into account effects of the device geometry, technological parameters, temperature), and all parameters have physical meanings. Very often, a compact physical model can be applied to several generations of technology by simply changing the parameters.

2. Empirical models, in which the characteristics of the device are modeled by empirical equations (without physical basis) that match experimental data.

3. Table lookup models, which are tables containing the values of various electrical parameters of the transistor (e.g., drain current) according to electrode biases.

Of course, the latter two categories are easier to set up than a physical compact model, but they are not predictive.

It is important to note that transistor models must satisfy very strict conditions in order to be used in a circuit simulator, as explained in Arora (2007):

1. Models must be precise enough to simulate the real transistor behavior over all operating regimes. For circuit design, it is sufficient to have an accuracy of about 5% (Arora 2007) between the model and experimental data of the real device (current and capacities).

2. To avoid nonconvergence problems in the simulator, the mathematical equations of the device model must be continuous, with continuous first derivatives (Arora 2007).

3. During the transient analysis, the transistor current is computed thousands of times, so it is mandatory that compact models be both efficient and accurate (Arora 2007). Thus, models should be not only accurate, but also very simple. Generally, the more accurate a model is, the more its complexity increases, and it becomes less efficient. A compromise is necessary between the model accuracy needed to

reproduce device behavior over the operating range of interest and the computational efficiency required for the simulation of large circuits (Arora 2007). The choice of the compact model is now becoming increasingly complicated as the size and complexity of circuits increase. To help designers choose the most appropriate model for their specific application, most simulator systems provide a hierarchy of models of different accuracy levels (Arora 2007).

8.3 Circuit-Level Simulation Approaches

As explained in the introduction to this chapter, three main modeling approaches are used for the simulation of single-event effects at circuit level: SPICE-like circuit simulation, mixed-mode, and numerical simulation of the full circuit in the 3D domain. These approaches are schematically illustrated in Figure 8.9 in the case of an SRAM cell simulation; several key features, advantages, and drawbacks are also summarized in Table 8.3. In the following, we briefly describe these different circuit-level modeling approaches.

8.3.1 SPICE-Like Circuit Simulation

The most commonly used approach in the industry for integrated circuit design and optimization is the simulation using standard simulation codes, such as the popular SPICE

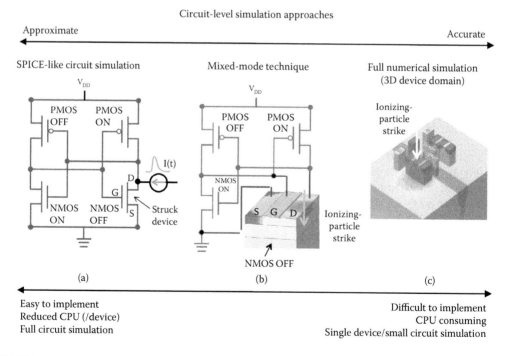

FIGURE 8.9
Schematic illustration of the simulation approaches that can be used to investigate single-event effects at circuit level (e.g., on an SRAM cell): (a) SPICE-like circuit simulation; (b) mixed-mode simulation; (c) numerical simulation of the full circuit in the 3D domain.

TABLE 8.3

Key Features, Advantages, and Drawbacks of the Three Main Approaches That Can Be Used for Simulation of Radiation Effects at Circuit Level

Approach	Key Features	Advantages	Drawbacks
SPICE-like circuit simulation	Uses compact models that describe the static/dynamic electrical behavior of the different elementary devices Single-event-induced transient usually modeled as a current source connected at the struck node of the circuit	Rapidity	Possible poor accuracy of compact model and of the transient used as input stimulus
Mixed-mode technique	Only struck device described by physically based numerical models; other devices described using compact models	Current transient directly computed by numerical simulation Considerably reduced simulation time compared with the numerical simulation of the full circuit in the 3D domain	Increased CPU time compared with a SPICE-like simulation
Numerical simulation of the full circuit in the 3D domain	Possible only recently (past decade), due to the enhancement of computer performance	The most accurate solution	Very long CPU time

codes (Berkeley), SmartSPICE (Silvaco), HPSICE (Synopsys), PSpice (OrCAD), ELDO (Mentor Graphics), and so on. Figure 8.9a shows the construction of a memory cell in a SPICE-like simulation. Circuit simulators solve systems of equations that describe the electrical operation of circuits (e.g., Kirchoff's laws). The basic components of these codes are compact models (see Section 8.2.4) describing the operation of all elementary devices (transistors, diodes, resistors, etc.) constituting the circuit. This approach is appropriate for many applications, but presents some limitations. First, compact models may introduce some numerical error, which is directly related to the accuracy of the models. Second, it is necessary that satisfactory compact models already exist. This is not always the case, especially concerning several emerging devices, for which compact models are currently under development. Finally, models that are adequate for digital-circuit simulation may be insufficient for other applications.

For simulating single-event effects at circuit level, the single-event-induced transient is usually modeled as a current source connected at the struck node of the circuit (as shown in Figure 8.9a). In order to reproduce the effect of the ionizing particle in the circuit-level simulation, a pulse-current transient obtained by numerically or analytically simulating the single-device response to radiation is usually considered. One of the most commonly used expressions is the double-exponential current transient pulse:

$$i_{inj}(t) = \frac{Q}{\tau_f - \tau_r}\left(e^{-t/\tau_f} - e^{-t/\tau_r}\right)$$

$$(8.24)$$

where:
Q = total charge deposited by the current pulse
τ_r = rising time constant
τ_f = falling time constant

A particle-induced current pulse typically has a short risetime (~10 ps) and a longer fall time (~200 ps).

An alternative solution for the modeling of the pulse-current transient is to directly import, in the circuit simulation, a numerical data file describing current intensity versus time. Data can be obtained separately using TCAD or other approaches to isolated devices. Obviously, the precision of the transient current used as input stimulus can significantly affect the accuracy of the circuit simulation. A typical example is the use of a current transient resulting from the numerical simulation of an unloaded device in the SPICE-like circuit simulation of the radiation response of a circuit. In Song et al. (1988), the response of a memory cell to a single event is simulated at the circuit level with SPICE. In that work, the stimulus is the transient response of the unloaded device obtained by 2D numerical simulation with PISCES, a device simulation code for silicon (developed at Stanford University). In this case, the circuit simulation inherits the inaccuracy of the improperly loaded device simulation (Dodd 2005).

8.3.2 Mixed-Mode Approach

Some limitations of the SPICE-like circuit simulation previously discussed can be overcome by using a complete numerical simulation of certain devices or circuit parts. This type of approach is called *mixed-mode* or *mixed-level* simulation, and it can be used in two different ways. The first possibility is to numerically simulate only a single transistor of the circuit, all others being simulated using compact models. This approach is the most commonly used in circuit simulation, for applications needing to reproduce very accurately the operation of a particular device. An example of this approach is the simulation of radiation effects in memory cells, illustrated in Figure 8.9b. In this case, the particular device that needs to be numerically simulated is the device struck by the ionizing particle. As shown in Figure 8.9b, only the struck NMOS transistor is modeled in the 3D device domain, the other transistors of the memory cell being simulated using compact models. The current transient resulting from the ion strike on the struck device is directly computed by device-domain simulation. Thus, the inaccuracy of the SPICE-like circuit simulation concerning the current transient used as stimulus can be eliminated. A second possibility is to numerically simulate all individual transistors of the integrated circuit and to connect them through the mixed-mode interface, which describes the operation of the total system. For reasons of computational resources and software limitations, it is clear that this type of simulation can be applied to very simple circuits containing only a few transistors—at most a few tens (such as inverters, ring oscillators with few stages, or individual memory cells).

In the mixed-mode approach, the two simulation domains (device and circuit) are connected by boundary conditions at contacts, and the solution of both systems of equations is implemented in a single matrix (Mayaram et al. 1993; Rollins and Choma 1988). The mixed-mode approach is implemented in all major commercial device simulators (Silvaco 2004; Synopsys 2014). This approach provides several interesting advantages. Errors inherent in compact models or inaccuracies of input stimulus can be avoided by full-device numerical simulation. It is also possible to access internal physical quantities of the numerically simulated device (such as potential, electric field, or densities of carriers) at any time during the mixed-mode simulation. Furthermore, the mixed-mode approach may be used to simulate the operation of small circuits made on emerging devices (such as ultrascaled multiple-gate or silicon-nanowire-based architectures) or to take into account new physical phenomena (e.g., quantum confinement or quasi-ballistic transport) for which compact models do not exist or are still not satisfactory in terms of accuracy. In this case, all transistors contained in the small circuit can be simulated in the 3D device domain. More particularly, the mixed-mode approach is successfully used to simulate ionizing-radiation impact on these new devices and associated small circuits. Several examples of mixed-mode simulations of SRAM cells based on SOI, double-gate, and junctionless devices will be presented in detail in Chapter 12.

The main drawback of the mixed-mode approach is the increased computational time compared with SPICE-like circuit simulation. In addition, mixed-mode simulation is not feasible for complex circuits. On the other hand, in the case of SRAM cells or CMOS inverters, 3D mixed-mode simulation significantly reduces computation time compared with circuit numerical simulation fully described in the 3D domain (see Section 8.3.3). Finally, it is important to note that mixed-mode simulation is accurate only for small circuits and when there is no coupling effect between devices (Dodd 2005). Given that the spacing between devices decreases when device dimensions are reduced, it is expected that coupling effects will become increasingly important; in this case, numerical simulation of the full circuit in the 3D domain may become mandatory (Roche et al. 1998, 1999; Dodd et al. 2001).

8.3.3 Full Numerical Simulation in the 3D Device Domain

The most accurate solution for the study of small circuits is to numerically simulate the entire circuit in the 3D domain. This has been possible only recently, typically in the two last decades, because

1. Computer performances have been significantly improved (in terms of processor clock speed, memory resources), which significantly reduces the computational time.
2. Numerical methods implemented in commercial simulators have been improved.
3. New mesh generators are used, which integrate new effective meshing strategies.
4. Existing linear iterative solvers have been significantly improved.
5. Developers have begun to intensively use code parallelization strategies; thereby a device simulation may be run on multiple processors or parallel machines.

Pioneering works of Roche et al. (1998, 1999) and Dodd et al. (2001) have demonstrated the capability of commercial codes to build and numerically simulate SEE on complete 3D SRAM cells or flip-flop stages. An example is shown in Figure 8.9c, where an SRAM cell is fully described in the 3D domain. Although the time required for the simulation of the entire cell has been greatly reduced, it is still significant compared with the time required to simulate the same circuit with mixed-mode or SPICE-like approaches. The recent emergence of parallel computers (PC clusters) with hundreds of processors and large memory resources is certainly a very promising solution for the future development of such numerical simulations of small circuits and complete portions of circuits.

References

Ancona, M.G. and Iafrate, G.J. 1989. Quantum correction to the equation of state of an electron gas in a semiconductor. *Physical Review B* 39:9536–9540.

Ancona, M.G. and Tiersten, H.F. 1987. Macroscopic physics of the silicon inversion layer. *Physical Review B* 35:7959–7965.

Ancona, M.G., Yu, Z., Dutton, R.W., Van de Voorde, P.J., Cao, M., and Vook, D. 1999. Density-gradient analysis of tunneling in MOS structures with ultra-thin oxides. In *Proceedings of the International Conference on Simulation of Semiconductor Processes and Devices*, pp. 235–238. 6–8 September, Kyoto, IEEE.

Apanovich, Y., Lyumkis, E., Polski, B., Shur, A., and Blakey, P. 1994. Steady-state and transient analysis of submicron devices using energy balance and simplified hydrodynamic models. *IEEE Transactions on Computer-Aided Design of Integrated Circuits and Systems* 13:702–711.

Arora, N. 2007. *MOSFET Modeling for VLSI Simulation—Theory and Practice.* New Jersey, London: World Scientific.

Autran, J.L. and Munteanu, D. 2008. Simulation of electron transport in nanoscale independent-gate double-gate devices using a full 2D Green's function approach. *Journal of Computational and Theoretical Nanoscience* 5:1120–1127.

Autran, J.L., Munteanu, D., Tintori, O., Decarre, E., and Ionescu, A.M. 2005. An analytical subthreshold current model for ballistic quantum-wire double gate MOS transistors. *Molecular Simulation* 31:179–183.

Baccarani, G. and Wordeman, M.R. 1985. An investigation of steady-state velocity overshoot in silicon. *Solid State Electronics* 28:407–416.

Banoo, K. and Lundstrom, M.S. 2000. Electron transport in a model silicon transistor. *Solid State Electronics* 44:1689–1695.

Blotekjaer, K. 1970. Transport equations for electron in two-valley semiconductors. *IEEE Transactions on Electron Devices* ED-17:38–47.

Brisset, C., Dollfus, P., Hesto, P., and Musseau, O. 1994. Monte Carlo simulation of the dynamic behavior of a CMOS inverter struck by a heavy ion. *IEEE Transactions on Nuclear Science* 41:619–624.

Chou, S.Y., Antoniadis, D.A., and Smith, H.I. 1985. Observation of electron velocity overshoot in sub-100-nm-channel MOSFETs in silicon. *IEEE Electron Device Letters* EDL-6:665–667.

Datta, S. 2000. Nanoscale device modeling: The Green's function method. *Superlattice Microstructure* 28:253–278.

Dodd, P.E. 1996. Device simulation of charge collection and single-event upset. *IEEE Transactions on Nuclear Science* 43:561–575.

Dodd, P.E. 2005. Physics-based simulation of single-event effects. *IEEE Transactions on Device Materials and Reliability* 5:343–357.

Dodd, P.E., Musseau, O., Shaneyfelt, M.R., et al. 1998. Impact of ion energy on single-event upset. *IEEE Transactions on Nuclear Science* 45:2483–2491.

Dodd, P.E., Shaneyfelt, M.R., Horn, K.M., et al. 2001. SEU-sensitive volumes in bulk and SOI SRAMs from first-principles calculations and experiments. *IEEE Transactions on Nuclear Science* 48:1893–1903.

Dussault, H., Howard Jr, J.W., Block, R.C., Pinto, M.R., Stapor, W.J., and Knudson, A.R. 1993. Numerical simulation of heavy ion charge generation and collection dynamics. *Transactions on Nuclear Science* 40:1926–1934.

Fischetti, M. and Laux, S. 1988. Monte Carlo analysis of electron transport in small semiconductor devices including band-structure and space-charge effects. *Physical Review B* 38:9721–9745.

Galup-Montoro, C. and Schneider, M.C. 2007. *MOSFET Modeling for Circuit Analysis and Design.* Singapore: World Scientific.

Gardner, C.L. 1994. The quantum hydrodynamic model for semiconductor devices. *SIAM Journal on Applied Mathematics* 54:409–427.

Grubin, H.L., Govindan, T.R., Kreskovsky, J.P., and Stroscio, M.A. 1993. Transport via the Liouville equation and moments of quantum distribution functions. *Solid-State Electronics* 36:1697–1709.

Hamm, R.N., Turner, J.E., Wright, H.A., and Ritchie, R.H. 1979. Heavy ion track structure in Silicon. *IEEE Transactions in Nuclear Science* 26:4892–4895.

Hansch, W., Vogelsang, T., Kirchner, R., and Orlowski, M. 1989. Carrier transport near the Si/SiO$_2$ interface of a MOSFET. *Solid State Electronics* 32:839–849.

Hareland, S.A., Jallepalli, S., Shih, W.-K., et al. 1998. A physically-based model for quantization effects in hole inversion layers. *IEEE Transactions on Electron Devices* 45:179–186.

Ieong, M. and Tang, T. 1997. Influence of hydrodynamic models on the prediction of submicrometer devices characteristics. *IEEE Transactions on Electron Devices* 44:2242–2251.

Jacoboni, C. and Reggiani, L. 1983. The Monte Carlo method for the solution of charge transport in semiconductors with applications to covalent materials. *Reviews of Modern Physics* 55:645–705.

Janik, T. and Majkusiak, B. 1998. Analysis of the MOSFET based on the self-consistent solution to the Schrödinger and Poisson equations and on the local mobility model. *IEEE Transactions on Electron Devices* 45:1263–1271.

Jiménez, D., Sáenz, J.J., Iñiguez, B., Suñé, J., Marsal, L.F., and Pallarès, J. 2003. A unified compact model for the ballistic quantum wire and quantum well MOSFET. *Journal of Applied Physics* 94:1061–1068.

Khanna, V.K. 2004. Physics of carrier-transport mechanisms and ultra-small scale phenomena for theoretical modelling of nanometer MOS transistors from diffusive to ballistic regimes of operation. *Physics Reports* 398:67–131.

Kobetich, E.J. and Katz, R. 1968. Energy deposition by electron beams and δ rays. *Physical Review* 170:391–396.

Lopez-Villanueva, J.A., Cartujo-Casinello, P., Banqueri, J., et al. 1997. Effects of the inversion layer centroid on MOSFET behavior. *IEEE Transactions on Electron Devices* 44:1915–1922.

Lundstrom, M.S. 1997. Elementary scattering theory of the Si MOSFET. *IEEE Electron Device Letters* 18:361–363.

Lundstrom, M.S. 2000. *Fundamentals of Carrier Transport*. 2nd edn. Cambridge: Cambridge University Press.

Martin, R.C., Ghoniem, N.M., Song, Y., and Cable, J.S. 1987. The size effect of ion charge tracks on single event multiple-bit upset. *IEEE Transactions on Nuclear Science* 34:1305–1309.

Martinie, S., Le Carval, G., Munteanu, D., Soliveres, S., and Autran, J.L. 2008. Impact of ballistic and quasi-ballistic transport on performances of double-gate MOSFET-based circuits. *IEEE Transactions on Electron Devices* 55:2443–2453.

Martinie, S., Munteanu, D., Le Carval, G., and Autran, J.L. 2009. Physics-based analytical modeling of quasi-ballistic transport in double-gate MOSFETs: From device to circuit operation. *IEEE Transactions on Electron Devices* 56:2692–2702.

Mayaram, K., Chern, J.H., and Yang, P. 1993. Algorithms for transient three dimensional mixed-level circuit and device simulation. *IEEE Transactions on Computer-Aided Design of Integrated Circuits and Systems* 12:1726–1733.

McMacken, J.R.F. and Chamberlain, S.G. 1991. An impact ionization model for two-carrier energy-momentum simulation. *Simulation of Semiconductor Devices and Processes* 4:499–504.

Munteanu, D. and Autran, J.L. 2003. Two-dimensional modeling of quantum ballistic transport in ultimate double-gate SOI devices. *Solid State Electronics* 47:1219–1225.

Munteanu, D. and Autran, J.L. 2008. Modeling of digital devices and ICs submitted to transient irradiations. *IEEE Transactions on Nuclear Science* 55:1854–1878.

Natori, K. 1994. Ballistic metal-oxide-semiconductor field effect transistor. *Journal of Applied Physics* 76:4879–4890.

Oldiges, P., Dennard, R., Heidel, D., Klaasen, B., Assaderaghi, R., and Ieong, M. 2000. Theoretical determination of the temporal and spatial structure of α-particle induced electron-hole pair generation in silicon. *IEEE Transactions on Nuclear Science* 47:2575–2579.

Quade, W., Rudan, M., and Scholl, E. 1991. Hydrodynamic simulation of impact-ionisation effects in P-N junctions. *IEEE Transactions on Computer Aided Design* 10:1287–1294.

Querlioz, D., Saint-Martin, J., Do, V.-N., Bournel, A., and Dollfus, P. 2006. A study of quantum transport in end-of-roadmap DG-MOSFETs using a fully self-consistent Wigner Monte Carlo approach. *IEEE Transactions on Nanotechnology* 5:737–744.

Rahman, A. and Lundstrom, M.S. 2002. A compact scattering model for the nanoscale double gate MOSFET. *IEEE Transactions on Electron Devices* 49:481–489.

Ravaioli, U., Winstead, B., Wordelman, C., and Kepkep, A. 2000. Monte-Carlo simulation for ultra-small MOS devices. *Superlattice Microstructures* 27:137–145.

Ren, Z., Venugopal, R., Datta, S., and Lundstrom, M.S. 2001. Examination of design and manufacturing issues in a 10 nm double gate MOSFET using non-equilibrium Green's function simulation. In *Proceedings of the International Electron Devices Meeting Technical Digest*, pp. 5.4.1–5.4.4. 2–5 December, Washington, DC, IEEE.

Ren, Z., Venugopal, R., Goasguen, S., Datta, S., and Lundstrom, M.S. 2003. nanoMOS2.5: A two-dimensional simulator for quantum transport in double-gate MOSFETs. *IEEE Transactions on Electron Devices* ED-50:1914–1925.

Roche, P. 1999. Etude du basculement induit par une particule ionisante dans une mémoire statique en technologie submicronique [Study of the soft error (SEU) induced by an ionizing particle in SRAM fabricated with deep submicron CMOS technologies]. PhD Thesis, Université de Montpellier II [in French].

Roche, P., Palau, J.-M., Belhaddad, K., Bruguier, G., Ecoffet, R., and Gasiot, J. 1998. SEU response of an entire SRAM cell simulated as one contiguous three dimensional device domain. *IEEE Transactions on Nuclear Science* 45:2534–2543.

Roche, P., Palau, J.-M., Tavernier, C., Bruguier, G., Ecoffet, R., and Gasiot, J. 1999. Determination of key parameters for SEU using full cell 3-D SRAM simulations. *IEEE Transactions on Nuclear Science* 46:1354–1362.

Rollins, J.G. and Choma Jr., J. 1988. Mixed-mode PISCES-SPICE coupled circuit and device solver. *Transactions on Computer-Aided Design of Integrated Circuits and Systems* 7:862–867.

Sai-Halasz, G.A. and Wordeman, M.R. 1980. Monte Carlo modeling of the transport of ionizing radiation created carriers in integrated circuits. *IEEE Electron Device Letters* EDL-1:211–213.

Sai-Halasz, G.A., Wordeman, M.R., and Dennard, R.H. 1982. Alpha-particle-induced soft error rate in VLSI circuits. *IEEE Transactions on Electron Devices* ED-29:725–731.

Selberrer, S. 1984. *Analysis and Simulation of Semiconductor Devices*. Wien, NY: Springer.

Shahidi, G.G., Antoniadis, D.A., and Smith, H.I. 1988. Electron velocity overshoot at room and liquid nitrogen temperatures in silicon inversion layers. *IEEE Electron Device Letters* 9:94–96.

Silvaco. 2004. *Silvaco Athena/Atlas User's Manual*. Santa Clara, CA: Silvaco International.

Song, Y., Vu, K.N., Cable, J.S., et al. 1988. Experimental and analytical investigation of single event, multiple bit upsets in poly-silicon load, 64 K×1 NMOS SRAMs. *IEEE Transactions on Nuclear Science* 35:1673–1677.

Spinelli, A.S., Benvenutti, A., and Pacelli, A. 1998. Self-consistent 2-D models for quantum effects in n-MOS transistors. *IEEE Transactions on Electron Devices* 45:1342–1349.

Stapor, W.J. and McDonald, P.T. 1988. Practical approach to ion track energy distribution. *Journal of Applied Physics* 64:4430–4434.

Stern, F. 1972. Self-consistent results for n-type Si inversion layers. *Physics Review B* 5:4891–4899.

Stettler, M., Alam, M., and Lundstrom, M. 1993. A critical examination of the assumptions underlying macroscopic transport equations for silicon devices. *IEEE Transactions on Electron Devices* 40:733–740.

Stratton, R. 1962. Diffusion of hot and cold electrons in semiconductor barriers. *Physical Review* 126:2002–2013.

Synopsys. 2014. Synopsys Sentaurus TCAD tools, manual. Available at: http://www.synopsys.com/tools/tcad/Pages/default.aspx.

Taur, Y. and Ning, T.H. 1998. *Fundamentals of Modern VLSI Devices*. Cambridge, UK: Cambridge University Press.

Thoma, R., Edmunds, A., Meinerzhagen, B., Peifer, H-J., and Engl, W. 1991. Hydrodynamic equations for semiconductor with nonparabolic band structure. *IEEE Transactions on Electron Devices* 38:1343–1353.

van Dort, M.J., Woerlee, P.H., and Walker, A.J. 1994. A simple model for quantization effects in heavily-doped silicon MOSFETs at inversion conditions. *Solid-State Electronics* 37:411–414.

Vasileska, D. and Goodnick, S.M. 2002. Computational electronics. *Materials Science Engineering Reports* 38:181–236.

Wettstein, A., Schenk, A., and Fichtner, W. 2002. Quantum device-simulation with Density-Gradient model. *IEEE Transactions on Electron Devices* 48:279–284.

Ziegler, J.F. and Puchner, H. 2004. *SER—History, Trends and Challenges*. San Jose, CA: Cypress Semiconductor.

9

Soft-Error Rate (SER) Monte Carlo Simulation Codes

9.1 General-Purpose Monte Carlo Radiation-Transport Codes

Monte Carlo (MC) methods are computational methods that use random numbers to model stochastic processes or to model deterministic processes that can be approximated by stochastic ones. The MC approach is the numerical method of choice for problems that model objects interacting with other objects or their environment based upon simple object–object or object–environment relationships (Bielajew 2001). Typical examples are transport of energetic particles with matter, which results from stochastic energy-loss processes, or transport of electrical carriers in semiconductors governed by random collisions (for this latter case, we already mentioned in Chapter 8 the direct solution of the Boltzmann transport equation by the MC method, which is well adapted for a rigorous solution of this problem).

In the case of energetic particles responsible for single-event effects in semiconductor devices, the interactions of these particles (neutrons, protons, muons, heavy ions, and alpha particles) with matter are generally well known and verified by experiment. MC simulation codes thus have solid foundations at microscopic level in this domain for predicting the trajectories and the energy deposition of particles through bulk materials or complex assemblies of materials, such as can be found in real architectures of devices and circuits. Among all the codes developed in this field, we would like to mention five major, well-proven, and very popular MC codes that provide general-purpose programs or packages for the simulation of the passage of particles through matter:

- Geant4 (GEometry ANd Tracking), a toolkit for the simulation of the passage of particles through matter (Agostinelli et al. 2003; Allison et al. 2006)
- FLUKA (FLUktuierende KAscade), a fully integrated particle-physics MC simulation package (Battistoni et al. 2007; Ferrari et al. 2005)
- MCNPX (Monte Carlo N-Particle eXtended), a software package for simulating nuclear processes (Pelowitz et al. 2007)
- MARS15, a set of MC programs that simulate the passage of particles through matter (Mokhov 2010)
- PHITS (Particle and Heavy Ion Transport code System), a general-purpose MC particle-transport code (Iwase et al. 2002; Niita et al. 2010; Sato et al. 2013)

Each of these codes has a long history of development and involves large communities of developers and users at international level (Mokhov 2011). Their application areas are also wide, including high-energy, nuclear, and accelerator physics, as well as studies in medicine, space science, and microelectronics. For this last application domain, several

TABLE 9.1

Main Characteristics of Five General-Purpose Radiation-Transport Simulation Codes

General	MCNPX	GEANT4	FLUKA	MARS	PHITS
Version	2.7.0	10.0	2011.2b	15	2.64
Laboratory affiliation	LANL	CERN, IN2P3, INFN, KEK, SLAC, TRIUMF, ESA	CERN, INFN	FNAL	JAEA, RIST GSI, Chalmers University
Language	Fortran 90/C	C++	Fortran 77	Fortran 95/C	Fortran 77
Cost	Free	Free	Free	Free	Free
Release format	Source and binary	Source and binary	Source and binary	Binary	Source and binary
Users	2500	~2000	~1000	220	220
Website	mcnpx.lanl. gov	cern.ch/geant4	www.fluka.org	www-ap.fnal. gov/MARS	phits.jaea.go.jp
Input format	Free	C++ main Fixed geometry	Fixed or free	Free	Free
Parallel execution	Yes	Yes	Yes	Yes	Yes

Source: Adapted after Mokhov, N.V., Simulation Tools and Computing Aspects, U.S. Particle Accelerator School Hampton, VA, 2011.

CERN, European Organization for Nuclear Research; ESA, European Space Agency; FNAL, Fermi National Accelerator Laboratory (United States); GSI, GSI Helmholtz Center for Heavy Ion Research (Germany); IN2P3, National Institute of Nuclear and Particle Physics (France); INFN, National Institute of Nuclear Physics (Italy); JAEA, Japan Atomic Energy Agency; KEK, High-Energy Accelerator Research Organization (Japan); LANL, Los Alamos National Lab; RIST, Research Organization for Information Science & Technology (Japan); SLAC, National Accelerator Laboratory (United States); TRIUMF, Canada's national laboratory for particle and nuclear physics.

more specialized programs for the simulation of single-event effects and the computation of the soft-error rate (SER) (see Section 9.2) are directly derived from these codes (e.g., as Geant4-compiled applications) or embed one of them as an internal simulation engine.

These general-purpose codes provide a complete set of capabilities for simulation of complex experiments or systems: 3D geometry, particle sources, tracking, detector response, run, event and track management, visualization, and user interface. The multidisciplinary nature of these toolkits requires that they supply an abundant set of physics processes to handle diverse interactions of particles with matter over a wide energy range. For many physics processes, a choice of different models is generally available. A comprehensive summary of the characteristics of these codes is beyond the scope of this chapter. We only give, in Table 9.1, a few main characteristics of these five codes. Table 9.1 also indicates the homepage address for the international collaborations developing these programs for further information, documentation, and code downloading.

9.2 Review of Recent Monte Carlo Codes Dedicated to the SER Issue

Full MC-based physical simulations of the SER solve the radiation problem in two main steps: (Step 1) the interaction of radiation with the device and the subsequent motion of charges and (Step 2) the resulting changes in nodal currents, voltages, or both within the device/circuit (Weller et al. 2009). Since this simulation chain is complex, due to its multiscale and multiphysics character, the same simulation engine cannot generally cover these two steps. For example, the interaction of radiation with the device and the subsequent motion of charges can be fully simulated using one of the general-purpose codes previously cited, but the resulting changes at device or circuit level require an electrical simulator or dedicated code. For computational cost reasons (central processing unit [CPU] time), the transport of the radiation-induced charges in Step 1 is also often processed by another program or a specially designed code based on a simplified transport model and optimized algorithm.

Several code developments have been reported in the literature in this domain. During the writing of this book, Reed et al. (2013) published an "Anthology of the Development of Radiation Transport Tools as Applied to Single Event Effects." This extended review paper contains contributions from 11 different groups, each developing or applying MC-based radiation-transport tools to simulate a variety of effects that result from energy transferred to a semiconductor material by a single particle event. The topics range from basic mechanisms for single-particle-induced failures to applied tasks such as developing websites to predict on-orbit single-event failure rates using MC radiation-transport tools (Reed et al. 2013). It seemed to us unnecessary to describe again all these codes, which are presented by the authors themselves in their paper. Table 9.2 lists the codes described in Reed et al.

TABLE 9.2

Radiation-Transport Simulation Codes Described in the Recent Anthology of the Development of Radiation-Transport Tools as Applied to Single-Event Effects

Program Name	Laboratory Affiliation	Acronym Definition/Short Description
CUPID	Clemson University	Clemson University Particle Interactions in Devices
NOVICE	EMPC	Radiation-transport/shielding code
SEMM/SEMM-2	IBM	Soft-Error Monte Carlo Model
CLUST-EVAP, PROPSET, PROTEST	NASA/Johnson Space Center	Monte Carlo proton reaction and transport codes
MRED	Vanderbilt University, NASA/GSFC	Monte Carlo Radiative Energy Deposition
MC-ORACLE	University of Montpellier-2	Monte Carlo predictive code for SET and SEUs
MUSCA SEP3	ONERA	MULti-SCAles Single Event Phenomena Predictive Platform/Monte Carlo SEE predictive tool
PHITS	JAEA, RIST, KEK and other institutes	Particle and Heavy Ion Transport code System/a general-purpose Monte Carlo particle-transport code
MCNP/MCNPX	Los Alamos National Laboratory	Monte Carlo N-Particle eXtended/a software package for simulating nuclear processes
GRAS	ESA	Geant4 for Radiation Analysis for Space
FLUKA	Fluka collaboration	General-purpose Monte Carlo energetic-particle reaction and transport code

Source: Reed, R.A., Weller, R.A., Akkerman, A., et al. *IEEE Trans. Nucl. Sci.* 60:1876–1911, 2013.
EMPC, Experimental and Mathematical Physics Consultants; ESA, European Space Agency; GSFC, Goddard Space Flight Center; JAEA, Japan Atomic Energy Agency; KEK, High-Energy Accelerator Research Organization (Japan); NASA, National Aeronautic and Space Administration; ONERA, National Institute of Aerospace Study and Research (France); RIST, Research Organization for Information Science & Technology (Japan).

(2013). Readers are invited to refer directly to this reference paper for further details about these codes.

As mentioned in Reed et al. (2013), there are three main families of MC tools used to compute single-event effects: (1) tools built from the *ground up*, using well-known radiation-transport physics (e.g., Clemson University Particle Interactions in Devices [CUPID]); (2) applications built using a radiation-transport toolkit designed by the high-energy physics (e.g., Geant4); and (3) programs developed from a precompiled MC tool (e.g., FLUKA). The breadth and depth of the application of each approach to solve single-event-effect problems vary dramatically, from almost nonexistent to a high level of detail (Reed et al. 2013). User access to the codes varies just as dramatically: some are widely available on the web to a large number of users, while others are exclusively used by the developers. In all cases, simulation time is of the order of a few minutes to a few days, depending on the desired statistics, the number of nodes used for the calculation, and the desired output (Reed et al. 2013).

In complement to this anthology, we provide in the following sections a succinct description of three additional radiation tools in the domain of SER simulation for which recent references are available in the literature: the Intel Radiation Tool (IRT), the Particle and Heavy Ion Transport Code System (PHITS)-Hyper Environment for Exploration of Semiconductor Simulation (HyENEXSS) code system, and the Tool suite for rAdiation Reliability Assessment (TIARA)-G4 program. This last code will also be described in more complete detail in Section 9.3.

9.2.1 Intel Radiation Tool (IRT)

IRT was presented in Foley et al. (2014). IRT is a Geant4-based simulation system capable of predicting upset rates for any radiation environment and cell design style. The implemented flow has a large degree of flexibility and extensibility stemming from its architecture, composed of a number of modules, each allowing different levels of precision based on the situation of interest. The overall IRT flow is shown in Figure 9.1 (Foley et al. 2014).

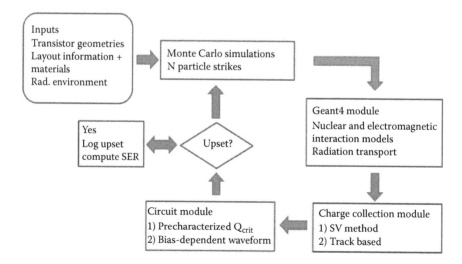

FIGURE 9.1
Simplified flow chart of the Intel Radiation Tool (IRT). (Reprinted from Foley, F., et al., IRT: A modeling system for single-event-upset analysis that captures charge sharing effects, in *Proceedings of the International Reliability Physics Symposium*, p. 5F1.1-5F1.9, 1–5 June, Waikolea, HI. © (2014) IEEE. With permission.)

Very schematically, a simulation starts with the creation of the 3D layout for the latch/circuit of interest, that is, the physical (geometric) layout that includes the circuit netlist and all sensitive latch nodes and spacings to accurately capture multitransistor, multibit, and possibly multicell effects. IRT only requires the material regions for devices, while the details of doping and so on are handled in the compact model that is the heart of IRT (see below). The radiation environment is also an input to the tool and consists of differential particle-flux distributions as a function of energy and particle type.

IRT then goes into a loop in which a single radiation event is created at each MC step.

The Geant4 library handles radiation transport and energy deposition in matter, as defined in the layout input structures. This includes nuclear and electromagnetic-interaction cross sections, which are specified for the entire periodic table. As the result, the charge profile is determined throughout the latch that is related to the charge collected at a circuit node through a compact model. IRT has two separate methods for relating the position dependence of the deposited charge in the charge-collection compact model: the sensitive volume method and the track-based method, a new physically motivated compact model that allows better predictive capabilities. Details of both models, including calibration, are given in Foley et al. (2014). Note that a tri-gate-specific charge-collection compact model with dependence on fin footprint, height, and doping levels below the wrap-around gate (fin) has been also developed within this work. Finally, the amount of collected charge is used to determine whether the current single event has caused an upset using either precomputed critical-charge values or full-circuit simulation.

IRT has been validated against measured high-energy proton and alpha-particle-induced SER data, collected on a wide range of memory devices manufactured in a 22 nm, high-κ, metal-gate tri-gate process. In particular, the authors have demonstrated that IRT is accurate in the case of a special class of reduced SER devices called reinforcing charge collection (RCC) devices, where accurate modeling of bias-dependent charge collection at more than one diffusion and node within the cell is crucial. Additional details about the IRT simulator and complete simulation results can be found in Foley et al. (2014).

9.2.2 PHITS-HyENEXSS Code System

Abe et al. have recently proposed a multiscale MC method for simulating soft errors by linking the particle transport code PHITS with the 3D technology computer-aided design (TCAD) simulator HyENEXSS. The first version of the resulting PHITS-HyENEXSS code system was presented in Abe et al. (2012a). Figure 9.2 illustrates the schematic flow of soft-error simulation using this code system. Schematically, PHITS is devoted to simulation of secondary-ion generation via nuclear interaction of an incident particle with constituent atoms in a device and the sequential charge deposition. HyENEXSS is used for simulation of the charge-collection process.

In more detail, PHITS outputs information about the secondary ions (i.e., species of ions, kinetic energy, generation position, and direction of motion) to *dump file* event by event. For a chosen single event, the initial charge distribution along each ion track is calculated by PHITS. Since it is time-consuming to perform a device simulation with HyENEXSS for all events, only the events primarily responsible for the soft errors are selected according to a filtering procedure, described later in Abe et al. (2012a). An interface tool between PHITS and HyENEXSS, called *takomesh*, has been developed to generate a mesh structure optimized for the event, in which some ion tracks extending to arbitrary directions are included. The Octree mesh method is used as a mesh-generation algorithm. The interface tool makes an input file for HyENEXSS that involves the linear energy transfer (LET)

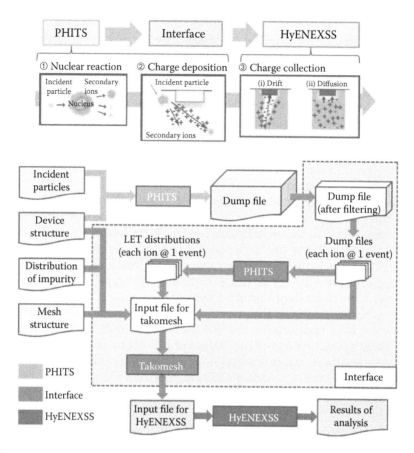

FIGURE 9.2

Simplified flow chart of SER analysis using PHITS-HyENEXSS code system. (Reprinted from Abe, S., et al., Neutron-induced soft error analysis in MOSFETs from a 65 nm to a 25 nm design rule using multi scale Monte Carlo simulation method, *Proceedings of the International Reliability Physics Symposium*, SE.3.1-SE.3.6, 15–19 April, Anaheim, CA. © (2012) IEEE. With permission.)

distribution of each secondary ion, device structure, spatial distribution of impurity, and mesh structure. Finally, the charge-collection process is simulated using the drift-diffusion method in HyENEXSS, and the transient-current response and collected charge can be derived for the selected event (Abe et al. 2012a).

The PHITS-HyENEXSS code system was applied in Abe et al. (2012b) to MC calculations of terrestrial-neutron-induced SERs for 65, 45, 32, and 25 nm technology N-channel metal-oxide-semiconductor field-effect transistors (NMOSFETs). The results show that the scaling trend of SERs per bit is still decreasing, similar to the other published predictions based on experimental or simulation results (see Abe et al. 2012b). From investigation of the impact of secondary-ion species on SER, it was found that He ions are a major cause of soft errors at the critical charge used in this work regardless of design rule. If the critical charge is further reduced, it is expected that H ions will impact soft errors considerably. The dependence of SER on incident neutron energy was also investigated. In consequence, it was found that about 80% of terrestrial-neutron-induced soft errors can be caused by neutrons up to 400 MeV regardless of design rule and critical charge. Thus, it is concluded that accurate estimation of production of secondary H and He ions in the neutron-energy range up to several hundreds of megaelectronvolts plays a key role in terrestrial-neutron-induced soft-error simulation for advanced microelectronic devices (Abe et al. 2012b).

9.2.3 TIARA-G4

TIARA-G4 has been developed in recent years jointly at Aix-Marseille University (IM2NP laboratory) and at STMicroelectronics (Central R&D, Crolles). TIARA-G4 is a general-purpose MC simulation code written in C++ and fully based on the Geant4 toolkit for modeling the interaction of Geant4 particles (including neutrons, protons, muons, alpha particles, and heavy ions) with various architectures of electronic circuits (Autran et al. 2012a). The primary ambition of TIARA is to embed in a unique simulation platform the state-of-the-art knowledge and methodology of SER evaluation. The initial version of TIARA was a stand-alone C++ native code dynamically linked with integrated circuit (IC) computer-aided design (CAD) flow through coupling with a Simulation Program with Integrated Circuit Emphasis (SPICE) solver (Uznanski 2011). The code has been developed such that the addition of new radiation environments, physical models, or circuit architecture should be quite simple. On the one hand, this first version was able to treat the transport and energy deposition of charged particles (heavy ions and alpha particles) without the need for a nuclear code such as Geant4; only Stopping and Range of Ions in Matter (SRIM) tables were used as input files to compute the transport of the particles in silicon and in a simplified back-end-of-line (BEOL) structure reduced to a single layer. On the other hand, for neutrons it used separate databases compiled using a specific Geant4 application to generate nuclear events in the simulation flow resulting from the interactions of incident neutrons with the circuit.

The new release of TIARA is called TIARA-G4, in reference to the fact that it has been totally rewritten in C++ using Geant4 classes and libraries and compiled as a full Geant4 application (Roche et al. 2014). Nuclear events are no longer provided from databases but are directly generated in the flow of the simulation code by Geant4. This now allows us to consider the entire complexity of the circuit in terms of materials, doping, and 3D geometry, using the Virtual Geometry Model (VGM) factory (http://ivana.home.cern.ch/ivana/VGM.html) and interface with both Geant4 for calculation and ROOT (http://root.cern.ch) for visualization. In other words, the main improvement of TIARA-G4 with respect to the first version of the code comes precisely from this transformation of the code in a Geant4 application, allowing the use of Geant4 classes for description of circuit geometry and materials (now including the true BEOL structure) and integration of particle transport and tracking directly in the simulation flow, without the need for external databases or additional files.

9.3 Detailed Description of the TIARA-G4 Code

This section describes in detail the content of the main modules of the TIARA-G4 code. Figure 9.3 shows the flow chart of the TIARA-G4 code and its different modules. In Section 9.4 we will illustrate the capabilities of TIARA-G4 for the SER evaluation of different static random-access memory (SRAM) complementary metal-oxide semiconductor (CMOS) bulk circuits (65 and 40 nm technologies) subjected to natural radiation at ground level.

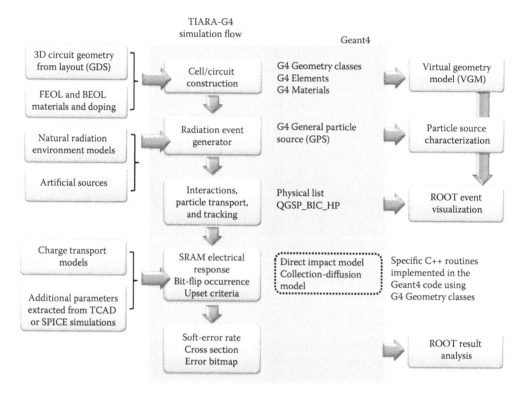

FIGURE 9.3

Schematics of the TIARA-G4 simulation flow showing the different code inputs and outputs and the links with Geant4 classes, libraries, models, or external modules and visualization tools. (Reprinted from Autran J.L., et al., Soft-error rate of advanced SRAM memories: Modeling and Monte Carlo simulation, In M. Andriychuk (ed.), *Numerical Simulation—From Theory to Industry*, Vienna: InTech, 2012. © 2012 Autran et al., licensee InTech.)

9.3.1 Circuit Architecture Construction Module

The first step of the TIARA-G4 simulation is to construct a model of the simulated circuit from Geant4 geometry classes and libraries of elements and materials. In the framework of Geant4, the circuit under simulation is considered as the *particle detector*. Structure creation in TIARA is based on 3D circuit-geometry information extracted from GDS formatted data classically used in the CAD flow of semiconductor-circuit manufacturing. To perform such an extraction from the GDS layout description, a separate tool has been developed (Uznanski 2011). It parses the GDS file, obtains coordinate points of CAD layers, and, using geometric computations, tracks the positions and dimensions of the transistor active areas, cell dimensions, p-type and n-type, and BEOL stack geometry. Based on this information and additional data concerning the depth of the wells, junctions, shallow trench isolation (STI) regions (obtained from TCAD or secondary-ion mass spectrometry [SIMS] measurements), and BEOL layer composition, TIARA creates a 3D structure of the elementary memory cell and, by repetition, of the complete portion of the simulated circuit. The real 3D geometry is simplified since it is essentially based on the juxtaposition of boxes of different dimensions, each box being associated with a given material (silicon, insulator, metal, etc.) or doped semiconductor (p-type, n-type).

Figure 9.4 (*left*) illustrates the geometry of a complete 65 nm SRAM architecture resulting from the circuit-construction process in TIARA-G4. Sensitive PMOS and NMOS drain

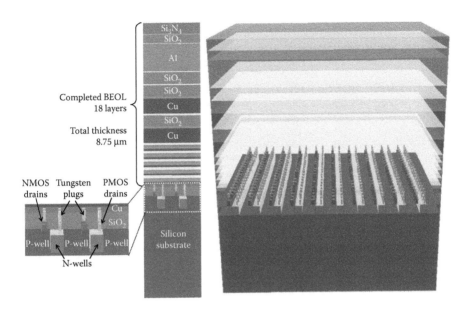

FIGURE 9.4
(Left and center) ROOT screenshots illustrating the geometry of the complete 65 nm SRAM architecture considered in TIARA-G4 simulation. *(Right)* 3D perspective view of a 10×20 SRAM cell array covered with the BEOL. (Reprinted from Autran J.L., et al., Soft-error rate of advanced SRAM memories: Modeling and Monte Carlo simulation, In M. Andriychuk (ed.), *Numerical Simulation—From Theory to Industry*, Vienna: InTech, 2012. © 2012 Autran et al., licensee InTech.)

regions are connected to the first metal layer (Cu) of the BEOL stack with tungsten plugs. The BEOL structure is composed of 18 uniform stacked layers with exact compositions and thicknesses. The 3D perspective view of a 10×20 SRAM cell array covered with the BEOL is shown in Figure 9.4 *(right)*. For better visibility, BEOL layers have been rendered semitransparent in this illustration.

9.3.2 Radiation-Event Generator

To numerically generate particles with spectral, spatial, and angular distributions mimicking all the characteristics of the natural background, as introduced and defined in Section 1.3.1, we use the G4 General Particle Source (Geant4-a), which is part of the Geant4 distribution. The module allows the user to define all the source parameters, in particular the energy of the emitted particles from a given energy distribution defined in a separate input file.

For the neutron flux, we consider the experimental atmospheric spectrum presented in Figure 1.11; for the other atmospheric particles, mainly muons and protons, we adopted the Quotid Atmospheric Radiation Model (QARM) or EXcel-based Program for calculating Atmospheric Cosmic-ray Spectrum (EXPACS) differential fluxes, also shown in Figure 1.9. Another important issue in MC simulation is the strong zenith angular dependence of atmospheric showers. To make our Geant4 GPS primary particle sources more realistic, we introduce in simulations the angular dependence of the primary flux intensity in the form $I(\theta) \sim \cos^n(\theta)$ (see Section 1.3.1), where θ is the zenith angle and n a parameter fixed to n = 3.5 for neutrons and n = 2 for muons.

For the simulation of α-particle emitters present in the IC materials, we directly generate in the code random positions and emission directions with uniform probability densities for each daughter element of the uranium and thorium decay chains (see Tables 3.3 and 3.4).

TABLE 9.3

Details of TIARA Simulation Results

	Monoenergetic or Broad-Spectrum Source								
	1 MeV	10 MeV	14 MeV	100 MeV	1 GeV	JEDEC	LANSCE	TRIUMF	ANITA
SBU	1837	3381	4290	3436	3202	3496	3271	3253	3133
MCU[2]	8	548	864	1876	1699	1125	967	1174	699
MCU[3]	0	87	225	459	479	251	212	235	134
MCU[4]	0	28	74	102	141	81	49	68	32
MCU[5]	0	2	24	48	37	20	22	35	14
MCU[6]	0	1	4	21	28	13	16	11	8
MCU[7]	0	0	1	12	8	6	3	10	1
MCU[8]	0	0	1	5	7	3	3	1	0
MCU[9]	0	1	1	3	7	2	1	0	1
MCU[10]	0	0	0	0	3	0	1	0	0
MCU[11]	0	0	0	0	2	0	1	0	0
MCU[12]	0	0	0	0	2	0	0	0	0
MCU[>12]	0	0	0	0	0	0	0	0	0
Total of cell flips	1853	5375	7157	9550	9205	7085	6307	6897	5195
% of MCU flips	0	37%	40%	64%	65%	51%	48%	53%	40%
Total of impacted NMOS (critical LET)	1348	2643	2726	1911	2203	1985	1867	1861	1774
Total of impacted PMOS (critical LET)	0	242	444	726	875	493	433	440	257
Total of NMOS flips (Imax-Tmax criterion)	400	2321	3707	6422	5680	4282	3695	4266	2920
Total of PMOS flips (Imax-Tmax criterion)	105	169	280	491	447	325	312	330	244
SER (FIT/Mbit)	70	203	270	366	353	266	238	260	162

Source: Data from Serre, S., Semikh, S., Uznanski, S, et al., *IEEE Trans. Nucl. Sci.* 59:714–722, 2012.

Note: Obtained on a 20×20 memory cell matrix (65 nm technology) for the different monoenergetic and broad-spectrum neutron-source databases (500×10⁶ incident neutrons). Results are given in number of single-bit upsets (SBUs) and multiple-cell upsets (MCUs) for the corresponding databases defined in Table 6.3.

ANITA, Atmospheric-like Neutrons from thIck Target, a Neutron Facility of the Svedberg Laboratory at Uppsala University; JEDEC, Joint Electron Device Engineering Council; LANSCE, Los Alamos Neutron Science Center; TRIUMF, Canada's national laboratory for particle and nuclear physics.

9.3.3 Interaction, Transport, and Tracking Module

Once an incident particle has been numerically generated with the radiation-event generator, the Geant4 simulation flow computes the interactions of this particle with the target (the simulated circuit) and transports step by step the particle and all the secondary particles eventually produced inside the world volume (the largest volume containing, with some margins, all other volumes contained in the circuit geometry). The transport of each particle occurs until the particle loses its kinetic energy to zero, disappears due to an interaction, or comes to the end of the world volume.

The G4ProcessManager class contains the list of processes that a particle can undertake. A physical process describes how particles interact with materials. The list of physical processes employed in our simulations is based on the physics list QGSP_BIC_HP (Geant4-b), one of the standard Geant4 lists covering the energy range of particles interacting in

```
Event #2359425

Incident particle: neutron

Energy (MeV): 5.664161e+01

Physical process: NeutronInelastic

Volume name: P-substrate

Reaction vextex positions (x,y,z): -5.656377e+00 -1.377065e+00 -5.347320e+00

-----------------------------------------------------------------------------------------------

Number of secondary particles produced: 5

Particle    Energy (MeV)    Px              Py              Pz

Alpha       3.197153e+01    6.097439e-01    -5.070998e-01   6.091488e-01

Neutron     1.697277e+00    -9.311596e-01   3.551987e-01    8.231417e-02

Gamma       9.051714e+00    3.281911e-01    -6.857286e-02   9.421191e-01

Gamma       1.377447e+00    2.331650e-01    8.485282e-01    4.750095e-01

Mg24[0.0]   2.554443e+00    -7.370831e-01   6.733902e-01    5.704476e-02
```

FIGURE 9.5

Example of a TIARA-G4 output in the case of particle interaction with the target (circuit). The present example describes a neutron-inelastic process (energy of the incident neutron of 56.64 MeV) with a silicon atom of the p-type substrate of the circuit described in Figure 9.4. This nuclear reaction produces five secondary particles at the reaction vertex position; for each produced particle, the particle energy and the three components of the normalized particle momentum (Px, Py, Pz) are indicated. (Reprinted from Autran J.L., et al., Soft-error rate of advanced SRAM memories: Modeling and Monte Carlo simulation, In M. Andriychuk (ed.), *Numerical Simulation—From Theory to Industry*, Vienna: InTech, 2012. © 2012 Autran et al., licensee InTech.)

low- to medium-energy ranges. This list uses binary-cascade, precompound, and various de-excitation models for hadrons, a standard electromagnetic package and a high-precision model for neutrons with kinetic energy below 20 MeV. This list is generally used for simulations in the fields of radiation protection, shielding, and medical applications.

Geant4 provides a way for the user to access the transportation process and to obtain the simulation results at the beginning and end of transportation, at the end of each step in transportation, and at the time when the particle is going into a given sensitive volume of the circuit. Figures 9.5 and 9.6 show two intermediate output results of TIARA-G4, describing, respectively, a particle interaction event (Figure 9.5: a nuclear inelastic event with a silicon atom of the p-type silicon substrate of the circuit; see Figure 9.4) and the tracking of two secondary particles impacting different sensitive volumes of the circuit (Figure 9.6). All these output data are saved as text files during the simulation and can be used later for event visualization or postprocessing. Finally, Figure 9.7 illustrates the visualization of an interaction event (here a negative muon capture by a silicon atom) using the data-analysis framework ROOT (http://root.cern.ch). Such a 3D perspective view is computed using a dedicated ROOT script that directly imports geometry and event data from a collection of files saved on the machine hard disk during simulation.

9.3.4 SRAM Electrical-Response Module

We detail in this section the model used to calculate the electrical response of the SRAM circuit subjected to irradiation. Starting a simulation sequence when a primary particle emitted

```
Event #5477703

EventStepData.size: 5

-----------------------------------------------------------

StepData object #0 ParticleName: Al28[0.0]

VolumeName: Psub #0

initPoint/um: (-2.234447e+00,-1.624657e+00,-8.960758e-01)

postPoint/um: (-2.239293e+00,-1.703238e+00,-8.400000e-01)

EnergyDep/MeV: 7.491555e-02

-----------------------------------------------------------

StepData object #1 ParticleName: Al28[0.0]

VolumeName: Pwell #8

initPoint/um: (-2.239293e+00,-1.703238e+00,-8.400000e-01)

postPoint/um: (-2.233404e+00,-2.258219e+00,-3.596903e-01)

EnergyDep/MeV: 4.304428e-01

-----------------------------------------------------------

StepData object #2 ParticleName: proton

VolumeName: Psub #0

initPoint/um: (-2.234447e+00,-1.624657e+00,-8.960758e-01)

postPoint/um: (-2.230874e+00,-1.572415e+00,-8.400000e-01)

EnergyDep/MeV: 2.421962e-04

-----------------------------------------------------------

StepData object #3 ParticleName: proton

VolumeName: Pwell #8

initPoint/um: (-2.230874e+00,-1.572415e+00,-8.400000e-01)

postPoint/um: (-2.196440e+00,-1.069293e+00,-3.000000e-01)

EnergyDep/MeV: 2.737365e-03

-----------------------------------------------------------

StepData object #4ParticleName: proton

VolumeName: Nmos #168

initPoint/um: (-2.181468e+00,-8.495000e-01,-6.405073e-02)

postPoint/um: (-2.177411e+00,-7.898294e-01,0.000000e+00)

EnergyDep/MeV: 2.871799e-04
```

FIGURE 9.6

Example of the tracking of two secondary particles (^{28}Al and proton) impacting different sensitive volumes (Psub, P-well, and NMOS) of the SRAM circuit. For each particle and each impacted sensitive volume, the (x, y, z) coordinates of the entry and exit points of the particle in this volume are indicated as well as the energy deposited by the particle in this same volume. (Reprinted from Autran J.L., et al., Soft-error rate of advanced SRAM memories: Modeling and Monte Carlo simulation, In M. Andriychuk (ed.), *Numerical Simulation—From Theory to Industry*, Vienna: InTech, 2012. © 2012 Autran et al., licensee InTech.)

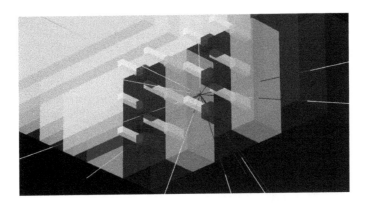

FIGURE 9.7
TIARA-G4 screenshot under ROOT visualization tool showing a part of the memory circuit (65 nm SRAM) subjected to negative-muon irradiation. The resulting interaction shown here is a muon capture by a silicon atom in the active circuit region (P-well) producing a shower of ten secondary particles. (Reprinted from Autran J.L., et al., Soft-error rate of advanced SRAM memories: Modeling and Monte Carlo simulation, In M. Andriychuk (ed.), *Numerical Simulation—From Theory to Industry*, Vienna: InTech, 2012. © 2012 Autran et al., licensee InTech.)

by the particle source enters into the world volume, we have already mentioned that Geant4 computes the interactions of this particle with the circuit and transports step by step the particle and all the secondary particles eventually produced until all these particles lose their kinetic energy to zero, disappear due to interaction, or come to the end of the world volume.

At the end of the sequence, TIARA-G4 examines the tracks of all the charged particles involved in this simulation step (including, eventually, the track of the incident primary particle, if it is charged) and determines the complete list of the different silicon volumes (drains, P-wells, N-wells, substrates, etc.) traversed by these particles. Two very general cases can be distinguished from this pure geometric analysis, as schematically shown in Figure 9.8:

1. *A single or several charged particles directly pass through a sensitive drain volume.* In this case, TIARA-G4 directly evaluates from Geant4 data the total energy deposited by these particles in the drain (ΔE), converts this value into a number of generated electron–hole pairs ($Q_{dep} = \Delta E/3.6$ eV for bulk silicon) and finally compares this value with the critical-charge value ($Q_{crit,P}$ for PMOS, $Q_{crit,N}$ for NMOS) of the simulated technology. If $Q_{dep} > Q_{crit}$, the memory cell is considered to be upset; in the contrary case, the electrical state of the cell is not changed (Munteanu and Autran 2008).

2. *A single or several charged particles impact one or several N-wells, P-wells, or the silicon substrate.* In this case, TIARA-G4 evaluates for each sensitive drain located in the impacted N-well(s) (for PMOS) or P-well(s) and substrate (for NMOS) the transient current I(t) resulting from the diffusion of carriers in excess in these regions and the collection of the charge by the sensitive nodes (see also Figure 9.9). Such calculations are performed using the *diffusion-collection* model detailed in the following. Until the I(t) characteristic is computed for all the considered sensitive drains, TIARA-G4 applies the I_{max}–t_{max} criterion, also described below, to determine whether or not the corresponding memory cell is upset.

In the diffusion-collection model (Hubert et al. 2006; Palau et al. 2001), the energy lost by a charged particle in silicon along its track is converted in a succession of elementary carrier

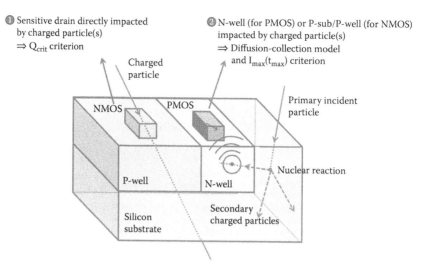

FIGURE 9.8
Schematics of the different cases envisaged in TIARA-G4 for the evaluation of cell upset in the simulated SRAM circuit. (Reprinted from Autran J.L., et al., Soft-error rate of advanced SRAM memories: Modeling and Monte Carlo simulation, In M. Andriychuk (ed.), *Numerical Simulation—From Theory to Industry*, Vienna: InTech, 2012. © 2012 Autran et al., licensee InTech.)

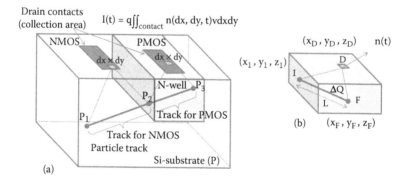

FIGURE 9.9
(a) Schematics of the diffusion-collection model used to compute the transient current I(t) resulting from the 3D spherical diffusion and then from the collection by a given reverse-biased drain of the charge in excess generated along a charged particle track. (b) Definition of the different points in Cartesian coordinates used to numerically evaluate n(t). (Reprinted from Autran J.L., et al., Soft-error rate of advanced SRAM memories: Modeling and Monte Carlo simulation, In M. Andriychuk (ed.), *Numerical Simulation—From Theory to Industry*, Vienna: InTech, 2012. © 2012 Autran et al., licensee InTech.)

densities δQ. The model then assumes that the behavior of these quasi-point charges is governed by a pure 3D spherical diffusion law:

$$\frac{\partial n}{\partial t} = D \cdot \Delta n \tag{9.1}$$

where:
  n = carrier density in excess generated in the silicon
  D = ambipolar diffusion coefficient

The temporal and spatial concentration n(r,t) resulting from the diffusion of a quasi-punctual charge δQ in the silicon at the distance r from this charge is thus described by the following equation:

$$n(r,t) = \frac{\delta Q}{\sqrt[3]{4\pi Dt}} \times \exp\left(-\frac{r^2}{4Dt} - \frac{t}{\tau}\right) \tag{9.2}$$

where:
  τ = carrier lifetime
  r = distance from the element δQ
  t = time

In the present implementation of the diffusion-collection model, δQ is directly evaluated from Geant4 data, considering the energy lost by the particle in a given geometry volume. Figure 9.9a illustrates the general case of a given volume impacted by a particle. The particle enters the volume at point I and exits at point F. Because drain and well volumes have reduced dimensions (typically expressed in tenths of a micron), the electrical charge δQ deposited per elementary length dl can be approximated by $Q_{dep} \times dl/L$, where L is the length of the segment IF. Considering the Cartesian coordinates of the geometric points I, F, and D defined in Figure 9.9, the total collected charge at point D due to the contribution of the complete segment IF can be analytically evaluated from the following expression:

$$n(t) = \frac{\delta Q/q}{8\pi DtL} \times \exp\left(-\frac{l_0^2}{4Dt} - \frac{t}{\tau}\right) \times \exp\left(-\frac{K^2}{L^2 4Dt}\right) \times \left\{ erf\left[\frac{1}{2\sqrt{Dt}}\left(L + \frac{K}{L}\right)\right] - erf\left[\frac{K}{2L\sqrt{Dt}}\right] \right\} \tag{9.3}$$

where quantities L, l_0, and K are defined from the Cartesian coordinates of points I, F, and D:

$$\begin{cases} L = \sqrt{(x_F - x_I)^2 + (y_F - y_I)^2 + (z_F - z_I)^2} \\ l_0^2 = (x_D - x_I)^2 + (y_D - y_I)^2 + (z_D - z_I)^2 \\ K = (x_I - x_D)(x_F - x_I) + (y_I - y_D)(y_F - y_I) + (z_I - z_D)(z_F - z_I) \end{cases} \tag{9.4}$$

The total electrical charge from the particle track collected at the level of a sensitive drain electrode is obtained by integrating Equation 9.3 on the total drain surface, as illustrated in Figure 9.9. Then, the charge is converted into a current by multiplying n(t) by the elementary charge and by the average collection velocity via space-charge region of the reverse-biased drain:

$$I(t) = q \iint_{Contact} n(t) \cdot v \cdot dxdy \tag{9.5}$$

Once the I(t) characteristic has been computed for all drains of the sensitive transistors in the SRAM cell matrix, TIARA-G4 applies the I_{max}–t_{max} criterion (Correas et al. 2007; Uznanski et al. 2010) to determine the cell upsets. This criterion is separately obtained from TIARA-G4 simulation and requires the combination of TCAD and SPICE analysis. The calculated I_{max}–t_{max} characteristic delimitates two current-time domains, as illustrated in Figure 9.10. If

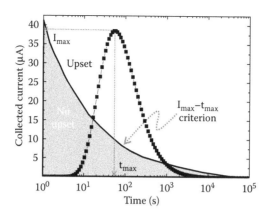

FIGURE 9.10
Example of a transient-current characteristic superimposed on the I_{max}–t_{max} upset criterion. The transient current has been calculated from the diffusion-collection model for a ^{24}Mg particle (10 MeV) that perpendicularly impacts a P-well at the distance of 0.25 μm from the NMOS drain contact. (Reprinted from Autran J.L., et al., Soft-error rate of advanced SRAM memories: Modeling and Monte Carlo simulation, In M. Andriychuk (ed.), *Numerical Simulation—From Theory to Industry*, Vienna: InTech, 2012. © 2012 Autran et al., licensee InTech.)

the transient-current peak is located above this curve, an upset occurs; in the contrary case, the extracted transient current from the sensitive node is not able to sufficiently disturb the electrical state of the bi-stable flip-flop and consequently to upset the memory point.

9.3.5 Soft-Error Rate Calculation Module

At the end of the simulation flow, the last module of the TIARA-G4 code evaluates the SER of the SRAM circuit from the following expression (JESD89A 2006):

$$\text{SER}\left[\text{FIT/Mbit}\right] = \frac{1024 \times 1024}{\text{CellNum}} \times \frac{\text{IntFlux}\left[\text{cm}^{-2}\right]}{\text{NumPart}} \times L_X\left[\text{cm}\right] \times L_Y\left[\text{cm}\right] \times 10^9 \times \text{TotalUpsetNum}$$

(9.6)

where:

TotalUpsetNum	= total number of cell upsets obtained during the simulation
CellNum	= number of memory cells of the simulated circuit
L_X and L_Y	= circuit dimensions (in cm)
NumPart	= total number of primary particles incident on the simulated circuit
IntFlux	= integral flux (cm^{-2}) of the particle source used to generate the incident particles. For example, considering the atmospheric neutron spectrum of Figure 1.11, IntFLux = 7.6 neutrons cm^{-2} for Part 1, 16 neutrons cm^{-2} for Part 2, and 20 neutrons cm^{-2} for Part 3 of the spectrum.

9.4 Experimental versus Simulation Results: Discussion

In this last section, the capabilities of the TIARA-G4 code are illustrated though a few dedicated studies on the simulation of 65 or 40 nm CMOS bulk SRAM circuits subjected

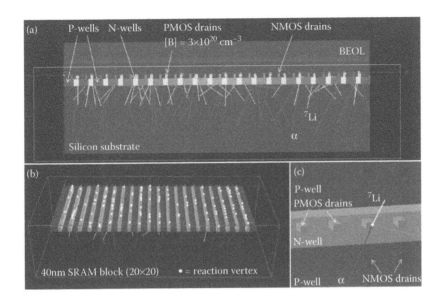

FIGURE 9.11
(a,b) TIARA-G4 screenshots under ROOT illustrating the results of 2×10^9 thermal neutrons incident on the 40 nm single-port SRAM (20×20 memory cell block). White points correspond to the reaction vertex (only localized in the PMOS drain volumes). For clarity, BEOL and substrate have been removed from the perspective view in (b). (c) Detail of a ^{10}B fission reaction occurring in the volume of a PMOS drain (note that the α-particle and the lithium nucleus are emitted in opposite directions to conserve momentum). (Reprinted from Autran, J.L., et al., Soft-error rate induced by thermal and low energy neutrons in 40 nm SRAMs, *IEEE Trans. Nucl. Sci.* 59:2658–2665. © (2012) IEEE. With permission.)

to different sources of atmospheric particles. These four independent examples concern: (i) the impact of thermal and low-energy neutrons on a 40 nm SRAM circuit; (ii) comparison between TIARA and TIARA-G4 to evaluate the impact of the BEOL materials on the SER; (iii) the SER estimation of a 65 nm SRAM under high-energy atmospheric neutrons; and (iv) the effects of low-energy muons on the same 65 nm circuit.

9.4.1 Impact of Thermal and Low-Energy Neutrons on a 40 nm SRAM Circuit

As previously described in Section 5.1.4, the interaction of thermal neutrons with the ^{10}B isotope of boron has been identified as a major source of soft errors in electronic circuits. Although this isotope has been quasi-definitively eliminated from the BEOL materials of modern technologies, recent work has demonstrated a substantial SER sensitivity with low neutron energies for many current SRAM circuits, as explained in Section 5.1.4. Using TIARA-G4 code, we explored the question of thermal and low-energy-neutron-induced soft errors in state-of-the-art 40 nm SRAMs (Autran et al. 2012b). Such a study can only be conducted with a code taking into account the real geometry at silicon level, including silicon doping with natural boron in p-type regions containing 19.9% ^{10}B. We thus constructed a 40 nm SRAM matrix with exact doping levels at the level of P-wells ($[B] = 10^{16}$ cm^{-3}) and PMOS drains ($[B] = 3 \times 10^{20}$ cm^{-3}).

Figure 9.11 illustrates the simulation results obtained on a 20×20 SRAM matrix. This circuit was irradiated with thermal and low-energy neutrons generated by the Geant4 GPS source, considering Part 1 of the reference atmospheric spectrum shown in Figure 1.11. In order to obtain a sufficient event statistic (interaction events are relatively rare), we

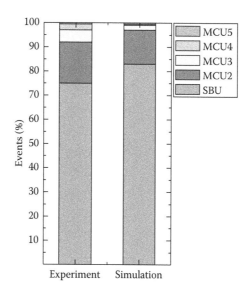

FIGURE 9.12

Event-multiplicity distributions obtained for the 40 nm SRAM subjected to thermal neutrons and deduced from both experiment and numerical simulation, respectively, conducted at the LLB facility and obtained with the new release of the TIARA/Geant4 code. (Reprinted from Autran, J.L., et al., Soft-error rate induced by thermal and low energy neutrons in 40 nm SRAMs, *IEEE Trans. Nucl. Sci.* 59:2658–2665. © (2012) IEEE. With permission.)

pushed the number of incident particles up to 2×10^9 thermal neutrons. A total of 116 single-bit upsets (SBU) and 24 multiple-cell upsets (MCU) were detected: they resulted exclusively from ^{10}B fission events localized in the drain volumes of the PMOS transistors (see Figure 9.11b; the vertexes of the reactions are indicated by the white dots).

Figure 9.12 shows that both the SER value and the event multiplicity distributions are in excellent agreement with experimental data performed at the Laboratoire Léon Brillouin (LLB) facility, located at CEA Saclay, Gif-sur-Yvette, France (http://www-llb.cea.fr/). The experiment was conducted on the G3-2 beam line under a thermal neutron flux reduced to 7.88×10^8 n cm^{-2} s^{-1} (beam surface area of 25×50 mm^2, neutron energies in the range 1.8–10 meV). For the purposes of the study, we considered a 7 Mbit 40 nm SRAM array with a layout cell area of 0.374 μm^2. We obtained an experimental thermal-neutron-induced SER of 4 FIT/Mbit for the SRAM, 75% of events consisting of SBU, 17% MCU with a multiplicity of 2%, and 8% with multiplicities ranging from 3 to 5. All these MCU events correspond to physically adjacent bitcells in the memory plan. For comparison, results obtained with TIARA-G4 give an SER equal to 4.5 FIT/Mbit with 83% of SBU, 14% of MCU with a multiplicity of 2%, and 4% of events with multiplicities ranging from 3 to 5.

Figure 9.13 illustrates the convergence of the soft-error rate during the TIARA-G4 simulation as a function of the number of incident primary neutrons. The code asymptotically converges toward a unique SER value, demonstrating the invariance of the extracted SER when the statistics become satisfactory, typically above 1.5×10^9 primary neutrons in this case.

Finally, Figure 9.14 shows a synthesis of both experimental and simulation results obtained for the soft-error rate (expressed in bit flips) of the 40 nm single-port SRAM. Simul. Part 1, Simul. Part 2, and Simul. Part 3 correspond to the SER extracted from TIARA-G4 simulations considering the parts labeled 1, 2, and 3, respectively, of the

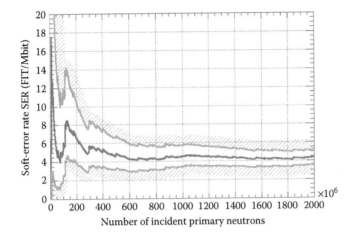

FIGURE 9.13
Convergence of the soft-error rate as a function of the number of incident primary neutrons obtained from TIARA-G4 simulation. The upper and lower limits of the SER confidence interval for 90% based on the χ^2 distribution are also plotted. (Reprinted from Autran J.L., et al., Soft-error rate of advanced SRAM memories: Modeling and Monte Carlo simulation, In M. Andriychuk (ed.), *Numerical Simulation—From Theory to Industry*, Vienna: InTech, 2012. © 2012 Autran et al., licensee InTech.)

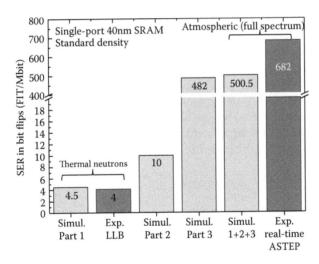

FIGURE 9.14
Synthesis of both experimental and simulation results obtained for the soft-error rate (expressed in bit flips) of the 40 nm single-port SRAM subjected to atmospheric neutrons. (Reprinted from Autran, J.L., et al., Soft-error rate induced by thermal and low energy neutrons in 40 nm SRAMs, *IEEE Trans. Nucl. Sci.* 59:2658–2665. © (2012) IEEE. With permission.)

reference atmospheric spectrum (Figure 1.11) as the primary source of particles. Note that the contributions of parts 1 and 2 of the neutron spectrum in the SRAM SER are very small with respect to the high-energy Part 3, and only represent 3% of the total SER value. Simul. 1+2+3 is equal to the sum of these three SER simulated values and corresponds to the SER estimation for the full atmospheric spectrum, estimated around 500 FIT/Mbit. The Exp. real-time ASTEP value (682 FIT/Mbit) corresponds to the neutron-SER extracted from a real-time experiment (conducted on the ASTEP platform at an altitude of 2252 m; see Chapter 7) and corrected for the contribution of internal chip radioactivity (α-particle emission). The comparison of these results evidences a ~30% discrepancy between simulation and experimental results. Such an underestimation of the total SER by simulation in such ultrascaled technology can be easily explained by the fact that the bipolar amplification mechanism has not yet been included in the SRAM electrical-response module of the simulation code. In the next validation example (Section 9.4.2) considering a less integrated technology (65 nm), the impact of bipolar amplification will be significantly reduced and its impact on SER value quasi-negligible.

9.4.2 Comparison between TIARA and TIARA-G4: Impact of the BEOL on the SER

In this second example, we simulated the complete 65 nm SRAM architecture previously defined in Figure 9.4. The objective was to compare neutron-induced SER results obtained by TIARA-G4 with simulation data previously obtained from the initial code TIARA (Uznanski 2011) (see the introduction in Section 9.2.3). Another objective was to perform simulation with and without taking into account the complete BEOL stack in order to evidence the impact of this BEOL on the neutron-SER.

Figure 9.15 shows the comparison of the neutron-induced SER computed with the different versions of TIARA for a 20×20 65 nm SRAM array. The atmospheric neutron source considered for these simulations corresponds to Part 3 of the reference neutron spectrum of Figure 1.11 (high-energy neutrons below 1 MeV). Very close values are obtained with TIARA and TIARA-G4 without taking into account the complete BEOL structure (a single SiO₂ layer is used as a simplified BEOL stack in this case): 266 and 251 FIT/Mbit, respectively, evidencing the equivalence of the two approaches in terms of global SER values. Recall that the initial version of TIARA computes neutron–silicon (n–Si) interactions from precalculated databases using Geant4, while TIARA-G4 is a full Geant4 application. Taking into account the complete BEOL structure (defined in Figure 9.4) in the new code TIARA-G4 results in a significant increase of the SER (337 FIT/Mbit). This +30% variation of the SER can be attributed to additional secondary particles produced by the

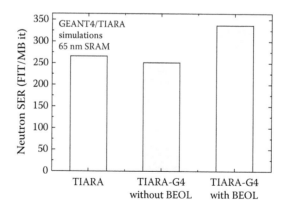

FIGURE 9.15

Comparison of simulation results obtained with the initial (TIARA) and new (TIARA-G4) versions of the code for the evaluation of the neutron-induced SER in the 65 nm SRAM architecture. Data are also plotted for TIARA-G4 with and without taking into account the real BEOL structure. (Reprinted from Autran J.L., et al., Soft-error rate of advanced SRAM memories: Modeling and Monte Carlo simulation, In M. Andriychuk (ed.), *Numerical Simulation—From Theory to Industry*, Vienna: InTech, 2012. © 2012 Autran et al., licensee InTech.)

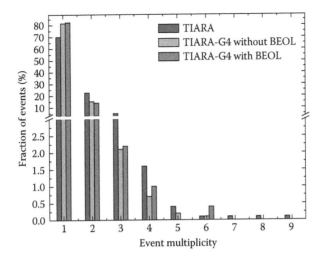

FIGURE 9.16
Event-multiplicity distributions obtained with the initial (TIARA) and new (TIARA-G4) versions of the code for the evaluation of the neutron-induced SER in the 65 nm SRAM architecture. Data are also plotted for TIARA-G4 with and without taking into account the real BEOL structure. (Reprinted from Autran J.L., et al., Soft-error rate of advanced SRAM memories: Modeling and Monte Carlo simulation, In M. Andriychuk (ed.), *Numerical Simulation—From Theory to Industry*, Vienna: InTech, 2012. © 2012 Autran et al., licensee InTech.)

interactions of incoming neutrons with the different BEOL materials (mainly SiO_2, Cu, and Al), these secondary particles being able to deposit electrical charges in the active silicon regions.

A more detailed analysis, shown in Figure 9.16, evidences slight differences in the distributions of the events as a function of the event size or event multiplicity (which corresponds to the number of simultaneous cell flips induced by a single primary-particle interaction with the circuit). TIARA-G4 is found to generate more single-bit upsets and, conversely, fewer multiple-cell upsets than the initial TIARA code. The presence of the complete BEOL structure above silicon is also found to induce more single-bit upsets and high-multiplicity events than the simplified BEOL structure. This can be attributed to the production of new nuclei and recoil nuclei up to the atomic number $Z = 74$, corresponding to tungsten, which is present in the BEOL at the level of the plugs for the interconnection of drains with the first metal layer. Detailed analysis of these newly produced nuclei shows that the majority of those inducing an upset are the result of neutron–copper interactions in the first metal layers close to the active silicon.

9.4.3 SER Estimation of a 65 nm SRAM under High-Energy Atmospheric Neutrons

In a direct continuation of the study described in Section 6.5, we interfaced TIARA-G4 with the different nuclear-event databases describing n–Si interactions for different synthetic (LANSCE, TRIUMF, ANITA) and monoenergetic sources of neutrons (see Section 6.5). TIARA-G4 was used to evaluate, for each atmospheric or monoenergetic neutron source, the SER related to the 65 nm SRAM circuit previously constructed in TIARA-G4 (see Figure 9.4).

From extensive simulations successively considering the eight databases as the input file (plus an additional monoenergetic database for the particular case of 14 MeV neutrons), we obtained the results compiled in Table 9.3 for a 20×20 memory-cell matrix (after irradiation of 5×10^8 incident neutrons). This table gives the total number of SBU and MCU (the number in brackets corresponds to the MCU multiplicity), the total number of cell flips, the number of NMOS and PMOS drains directly impacted by a particle (the critical LET is considered as the upset criterion), the number of upsets triggered by the I_{max}–T_{max} criterion for both NMOS and PMOS, and, finally, the value of the corresponding SER in FIT/Mbit normalized for the New York City (NYC) location (JEDS89A 2006).

For monoenergetic sources, the total number of flips increases with neutron energy below ~50 MeV, and then saturates above ~100 MeV (additional simulations at 50 and 150 MeV, not shown here, have been performed). In the same time, the percentage of MCU increases monotonously with energy, ranging from 0% for 1 MeV to 65% for 1 GeV. From these data, we can make the assumption that a correlation exists between the percentage and the highest nonnull-multiplicity order of the MCU events (Table 9.3), on the one hand, and the shower distribution and multiplicity for the different neutron sources investigated in Section 6.5 (Figure 6.12), on the other hand. Indeed, the creation of several fragments is clearly favorable to the occurrence of MCUs with a significant probability, as evidenced in the case of the present 65 nm SRAM technology with multiplicities ranging from 2 to 12.

For broad-spectrum neutron sources, close results are obtained for JEDEC, LANSCE, and TRIUMF sources in terms of SBU, MCU, and SER values. It must be noted that a typical 10% agreement is very acceptable within uncertainty margins for such complex SER estimations (due to several potential sources of errors resulting from simplifications and approximations in the modeling/simulation approach). Concerning ANITA, the obtained SER is clearly lower than the others, evidencing a reduced number of MCUs (only 40% against ~50% for the other sources), logically due to a lower high-energy-neutron flux (above 100 MeV) reducing more than a decade the number of high-multiplicity (>6) showers (Figure 6.12) and thus reducing the MCU event distribution (clear deficit of MCU for all multiplicities; see Table 9.3).

To conclude on Table 9.3, note that the additional values obtained for a monoenergetic source of 14 MeV very reasonably fit the JEDEC ones, globally, in terms of total flips and SER values. However, analysis of simulation data in terms of SBU and MCU numbers and respective contributions to the SER shows that 14 MeV results exhibit a ~20% excess of SBU and a global ~10% deficit of MCU. This result, obtained for our 65 nm SRAM characterized by a critical charge of 1.8 fC, agrees well with the observations reported by Normand and Dominik (2010) and the conclusions of a recent work by Clemens et al. (2011). Indeed, in Normand and Dominik (2010) a systematic comparison was made of the SEU cross sections measured using a monoenergetic 14 MeV neutron source and the LANSCE facility. It was observed that, for multiple SRAMs from various technologies with feature sizes $\geq$ 130 nm, the SEU cross section measured using 14 MeV neutrons was within a factor of two of that measured using the LANSCE wide-neutron source. Our result here shows similar good agreement between global SER values obtained from 14 MeV and JEDEC or LANSCE simulations. In Clemens et al. (2011) and using calibrated MC calculations, the results showed that, for 65 nm SRAMs with a critical charge < 27 fC, a monoenergetic 14 MeV neutron beam can be used to estimate the SEU response to the terrestrial-neutron environment within a factor of two, in agreement with Normand and Dominik (2010) and data shown in Table 9.3. However, a 14 MeV neutron beam was found inadequate to estimate the MCU response to the terrestrial neutron environment due to an underestimation of the MCU cross section, except for devices characterized by a very low critical charge (<1.2 fC). Our data show the same trend (MCU underestimation) even if it is difficult to conclude for MCU multiplicity $\geq$ 3 due to an insufficient statistic (the simulation should be increased to ten times the number of incident neutrons in a future work).

Finally, Figure 6.14, previously introduced in Chapter 6, shows a direct comparison between the experimental SER value (259 FIT/Mbit) obtained for the 65 nm SRAM on ASTEP (checkerboard pattern, room temperature, nominal $V_{dd}=1.2$ V) and the different estimated values (for both the four monoenergetic values and the four atmospheric-like sources) resulting from TIARA-G4 simulation. It must be noted that the SER value obtained from real-time testing at altitude is corrected for α-SER contribution and is given for NYC. Figure 6.14 shows that JEDEC, LANSCE, and TRIUMF simulations agree well with the experimental value. As noted previously, and in the particular case of the present 65 nm SRAM technology, the SER value deduced from 14 MeV simulations also fits the experimental SER very well. For ANITA, the obtained SER value is clearly lower than the previous ones by a factor of 0.6.

9.4.4 Effects of Low-Energy Muons on a 65 nm SRAM Circuit

This last example illustrates the capability of TIARA-G4 to simulate the impact of low-energy muons on SRAM memories (Serre et al. 2012). For the computation of muon interactions and transport, the physical list used in TIARA-G4 invokes the following classes and models. First, the G4MuIonization class provides the continuous energy loss due to muon ionization and simulates the discrete component of the ionization, that is, delta rays (δ-electrons) produced by muons. Inside this class, the following models are used:

- The G4BraggIonModel for E < 0.2 MeV. Muon energy losses are derived from the tabulated proton energy loss using scaling relation for the stopping power of heavy particles (ICRU 1993), which is a function only of the particle velocity.
- The G4BetheBlochModel for 0.2 MeV < E < 1 GeV. The Bethe–Bloch formula with shell and density corrections is applied.

To handle muon decay and nuclear muon capture, we invoke, respectively, the G4MuonDecayChannel and G4MuonMinusCaptureAtRest processes included in any physics lists of Geant4, such as QGSP_BIC. In the algorithm of the G4MuonMinusCaptureAtRest process, the $\mu^- + p \rightarrow n + \nu_\mu$ reaction is used with energy conservation. More precisely, the muon is captured by a bound proton with a Fermi momentum of =250 MeV/c. The off-shell mass of such a proton is about 45 MeV less than a free proton mass. The muon mass is added to the off-shell mass of the proton, and the resulting compound system becomes only about 60 MeV heavier than a free neutron. This compound system decays into a free neutron and a neutrino. The neutrino takes the majority of the energy (about 55 MeV), and the rest of the muon mass (about 50 MeV) is transferred to the nucleus. Nuclear de-excitation is performed by the Geant4 Pre-compound model.

For this study, we considered low-energy (<1 MeV) negative and positive muons susceptible to directly depositing charge by ionization or to being captured (negative muons) after they stop in silicon. The Geant4 general particle source was then used to generate monoenergetic muons incident on the 65 nm SRAM architecture (with complete BEOL) previously defined in Figure 9.4.

Figure 9.17 illustrates different possible scenarios of negative and positive-muon interactions with the structure. Figure 9.17a shows a negative-muon decay in the top layers of the BEOL structure; this cannot lead to an upset, since the muon disintegrates into light particles unable to deposit any significant charge in silicon. Figure 9.17b shows a similar

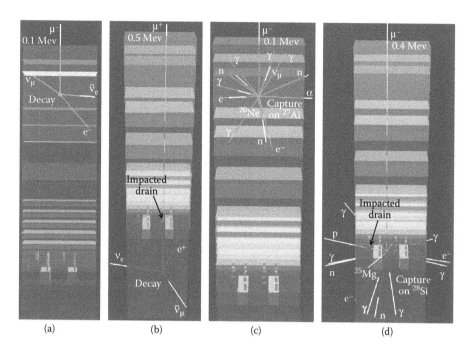

FIGURE 9.17

(a–d) Visualization of four events illustrating the interactions of low-energy negative and positive muons with a 65 nm SRAM structure. From left to right: μ⁻ decay in the BEOL (Al layer), μ⁺ upsetting a drain by direct charge deposition through the structure followed by muon decay in the substrate, μ⁻ capture on an aluminum atom in the BEOL, μ⁻ capture on a silicon atom in the active circuit region (P-well) leading to a drain upset via a direct impact by a secondary particle (a proton in this case). (Reprinted from Autran J.L., et al., Soft-error rate of advanced SRAM memories: Modeling and Monte Carlo simulation, In M. Andriychuk (ed.), *Numerical Simulation—From Theory to Industry*, Vienna: InTech, 2012. © 2012 Autran et al., licensee InTech.)

event, but occurring in the silicon substrate. In this case, the incoming positive muon traverses the complete BEOL structure and, statistically, can cross a sensitive drain. If the charge deposited in the impacted drain is higher than the critical charge for this transistor type and for this technology, the corresponding memory cell is upset. Figures 9.17c and 9.17d show two-negative-muon capture events occurring in the BEOL and in silicon, respectively. These events produce large secondary particle showers, containing one or more charged particles capable of reaching the active silicon region and inducing an upset or even an MCU. Of course, the probability of inducing an upset is maximum when the muon-capture-induced shower is produced in the immediate vicinity of the sensitive drain layer, as illustrated in Figure 9.17d. This case corresponds to a reduced energy interval for the incoming muons insofar as the penetration depth of the muons in the structure, and then the capture location, primarily depends on the muons' kinetic energy.

In order to illustrate this effect, we plotted in Figure 9.18 the distribution inside the SRAM structure of the vertex positions related to the negative-muon-capture reactions for three different values of the incident muon's kinetic energy: 0.1, 0.3, and 0.5 MeV. We clearly evidence in this figure a dependency of the capture position (depth) on the muon's kinetic energy. As a result, the soft-error occurrence, and consequently the soft-error rate induced by negative-muon irradiation, should present a maximum precisely when muon

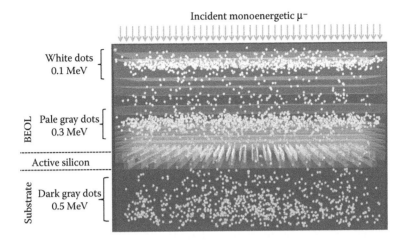

FIGURE 9.18
3D distribution inside the SRAM circuit of the vertex positions related to the negative-muon-capture reactions for three different values of the incident muon's kinetic energy: 0.1 MeV (white dots), 0.3 MeV (pale gray dots), and 0.5 MeV (dark gray dots). (Reprinted from Autran J.L., et al., Soft-error rate of advanced SRAM memories: Modeling and Monte Carlo simulation, In M. Andriychuk (ed.), *Numerical Simulation—From Theory to Industry*, Vienna: InTech, 2012. © 2012 Autran et al., licensee InTech.)

captures occur at the depth of the layer containing sensitive drains (i.e., the active silicon region). This point will be analyzed and discussed in Section 10.4.2, together with a comparison with other experimental and simulation results.

References

Abe, S., Watanabe, Y., Shibano, N., et al. 2012a. Multi-scale Monte Carlo simulation of soft errors using PHITS-HyENEXSS code system. *IEEE Transactions on Nuclear Science* 59:965–970.

Abe, S., Watanabe, Y., Shibano, N., et al. 2012b. Neutron-induced soft error analysis in MOSFETs from a 65 nm to a 25 nm design rule using multi-scale Monte Carlo simulation method. In *Proceedings of the International Reliability Physics Symposium*, SE.3.1–SE.3.6. 15–19 April, Anaheim, CA, IEEE.

Agostinelli, S., Allison, J., Amako, K., et al. 2003. Geant4—A simulation toolkit. *Nuclear Instruments and Methods in Physics Research Section A: Accelerators, Spectrometers, Detectors and Associated Equipment* 506:250–303. See also http://geant4.cern.ch.

Allison, J., Amako, K., Apostolakis, J., et al. 2006. Geant4 developments and applications. *IEEE Transactions on Nuclear Science* 53:270–278.

Autran, J.L., Semikh, S., Munteanu, D., Serre, S., Gasiot, G., and Roche, P. 2012a. Soft-error rate of advanced SRAM memories: Modeling and Monte Carlo simulation. In M. Andriychuk (ed.), *Numerical Simulation—From Theory to Industry*. Vienna: InTech.

Autran, J.L., Serre, S., Semikh, S., Munteanu, D., Gasiot, G., and Roche, P. 2012b. Soft-error rate induced by thermal and low energy neutrons in 40 nm SRAMs. *IEEE Transactions on Nuclear Science* 59:2658–2665.

Battistoni, G., Muraro, S., Sala, P.R., et al. 2007. The FLUKA code: Description and benchmarking. *AIP Conference Proceedings of the Hadronic Shower Simulation Workshop 2006* 896:31–49, 6–8 September, Chicago. IL.

Bielajew, A.F., 2001. Fundamentals of the Monte Carlo method for neutral and charged particle transport, University of Michigan. Available at: http://www-personal.umich.edu/~bielajew/MCBook/book.pdf

Clemens, M.A., Sierawski, B.D., Warren, K.M., et al. 2011. The effects of neutron energy and high-Z materials on single event upsets and multiple cell upsets. *IEEE Transactions on Nuclear Science* 58:2591–2598.

Correas, V., Saigné, F., Sagnes, B., et al. 2007. Innovative simulations of heavy ion cross sections in 130 nm CMOS SRAM. *IEEE Transactions on Nuclear Science* 54:2413–2418.

Ferrari A., Sala, P.R, Fasso, A., and Ranft, J. 2005. FLUKA: A multi-particle transport code. CERN 2005-10, INFN/TC_05/11, SLAC-R-773.

Foley, F., Seifert, N., Velamala, J.B., Bennett, W.G., and Gupta, S. 2014. IRT: A modeling system for single event upset analysis that captures charge sharing effects. *Proceedings of the International Reliability Physics Symposium*, p. 5F1.1–5F1.9, 1–5 June, Waikolea, HI.

Geant4-a. Geant4 general particle source. Available at: http://geant4.web.cern.ch/geant4/UserDocumentation/UsersGuides/ForApplicationDeveloper/html/ch02s07.html.

Geant4-b. Geant physics lists. Available at: http://geant4.web.cern.ch/geant4/support/proc_mod_catalog/physics_lists/referencePL.shtml.

Hubert, G., Bougerol, A., Miller, F., et al. 2006. Prediction of transient induced by neutron/proton in CMOS combinational logic cells. *Proceedings of the IEEE International On-Line Testing Symposium (IOLTS'06)*. IEEE.

ICRU 1993. Stopping powers and ranges for protons and alpha particles. ICRU Report 49.

Iwase, H., Niita K., and Nakamura, T. 2002. Development of general-purpose particle and heavy ion transport Monte Carlo code. *Journal of Nuclear Science and Technology* 39(11):1142–1151.

JEDS89A 2006. Measurement and reporting of alpha particle and terrestrial cosmic ray-induced soft errors in semiconductor devices, JEDEC standard JESD89A, October 2006. Available at: http://www.jedec.org/download/search/jesd89a.pdf.

Mokhov, N.V. 2010. Recent Mars15 developments: Nuclide inventory, DPA and gas production. Fermilab-Conf-10-518-APC. Available at: http://www-ap.fnal.gov/users/mokhov/papers/2010/Conf-10-518-APC.pdf.

Mokhov, N.V. 2011. Simulation tools and computing aspects. U.S. Particle Accelerator School Hampton, VA, 2011. Available at: http://uspas.fnal.gov/materials/11ODU/L5_SimulationTools.pdf.

Munteanu, D. and Autran, J.L. 2008. Modeling of digital devices and ICs submitted to transient irradiations. *IEEE Transactions on Nuclear Science* 55:1854–1878.

Niita, K., Matsuda, N., Iwamoto, et al. 2010. PHITS: Particle and heavy ion transport code system, Version 2.2. JAEA-Data/Code 2010-022 2010. Available at: http://phits.jaea.go.jp/index.html.

Normand, E. and Dominik, L. 2010. Cross comparison guide for results of neutron SEE testing of microelectronics applicable to avionics. *Proceedings of the IEEE Radiation Effects Data Workshop*, p. 50. 20–23 July, Denver, CO, IEEE.

Palau, J.M., Hubert, G., Coulie, K., Sagnes, B., Calvet, M.C., and Fourtine, S. 2001. Device simulation study of the SEU sensitivity of SRAMs to internal ion tracks generated by nuclear reactions. *IEEE Transactions on Nuclear Science* 48:225–231.

Pelowitz, D.B. (ed.) 2007. MCNPX user's manual version 2.6.0. LA-CP-07-1473, Los Alamos National Laboratory.

Reed, R.A., Weller, R.A., Akkerman, A., et al. 2013. Anthology of the development of radiation transport tools as applied to single event effects. *IEEE Transactions on Nuclear Science* 60:1876–1911.

Roche, P., Gasiot, G., Autran, et al. 2014. Application of the TIARA radiation transport tool to single event effects simulation, *IEEE Transactions on Nuclear Science* 61:1498–1500.

Sato, T., Niita, K., Matsuda, N., et al. 2013. Particle and heavy ion transport code system PHITS, Version 2.52. *Journal of Nuclear Science and Technology* 50(9):913–923.

Serre, S., Semikh, S., Uznanski, S., et al. 2012. Geant4 analysis of n-Si nuclear reactions from different sources of neutrons and its implication on soft-error rate. *IEEE Transactions on Nuclear Science* 59:714–722.

Uznanski, S. 2011. Monte-Carlo simulation and contribution to understanding of single event upset (SEU) mechanisms in CMOS technologies down to 20 nm technological node. PhD Thesis, Aix-Marseille University, Marseille, France.

Uznanski, S., Gasiot, G., Roche, P., Autran, J.L., and Tavernier, C. 2010. Single event upset and multiple cell upset modeling in commercial bulk 65 nm CMOS SRAMs and flip-flops. *IEEE Transactions on Nuclear Science* 57:1876–1883.

Weller, R.A., Schrimpf, R.D., Reed, R.A., et al. 2009. Monte Carlo simulation of single event effects. European Workshop on Radiation Effects on Components and Systems. Short course text, 14–18 September, Bruges, Belgium.

Section IV

Soft Errors in Emerging Devices and Circuits

10

Scaling Effects and Their Implications for Soft Errors

10.1 Introduction

The continuous scaling of CMOS is driving information-processing technology into a broadening spectrum of new applications. Many of these applications are enabled by performance gains or increased complexity realized by scaling (ITRS 2012). In its early editions, the International Technology Roadmap for Semiconductors (ITRS) emphasized *miniaturization* and its associated benefits in terms of performance, the traditional parameters in Moore's Law (Arden et al. 2010). This trend for increased performance will continue, while performance can always be traded against power, depending on the individual application, sustained by the incorporation of new materials into devices and the application of new transistor concepts. This direction for further progress is labeled *More Moore*.

Concerning this More Moore trend, the ITRS predicts a drastic reduction of both CMOS supply voltage and physical gate length (Hung 2010). Also, according to the ITRS roadmap, in the next 10 years the number of transistors per integrated system-on-chip (SoC) will be multiplied by 12. Soft errors in these SoCs will thus grow accordingly if a steady soft-error rate (SER)/bit is assumed. Soft errors are and, unfortunately, will continue to be a key reliability topic for most mass-customer applications (often responsible for the highest failure rate of all the reliability mechanisms). In addition, dimensional scaling, core-voltage reduction, and increasing frequency of circuit operation are three important facets of the famous Moore's Law that have direct consequences for the evolution of SER with integration. These parameters do not necessarily all influence the SER in the same way, and some are in competition. These aspects will be described and discussed in detail in the following.

10.2 Feature-Size Scaling

10.2.1 Geometric Scaling

The reduction of device feature sizes, combined with the increase of circuit integration (i.e., the number of transistors per unit area), has important implications for soft errors, as discussed in Massengill et al. (2012):

- Reduction of the per-bit cross section presented to an incident ionizing particle
- Reduction of the energy-deposition volume traversed by the particle
- Increase of the particle's region of influence in the circuit plan

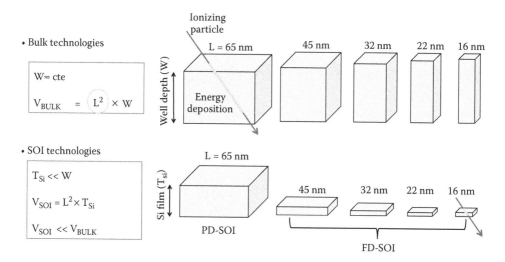

FIGURE 10.1
Schematic illustration of charge-collection volumes in both bulk and SOI technologies. (Adapted from Sierawski, B.D., Low-energy proton single-event upsets in SRAMs. In *3rd Annual NEPP Electronic Technology Workshop [ETW], NASA Electronic Parts and Packaging [NEPP],* 2012.) Planar cross section and active silicon thickness are smaller in SOI compared with bulk. The resulting charge-collection volume presents a projected target for ionizing-particle energy deposition that diminishes with scaling. (Reprinted from Roche, P. et al., Technology downscaling worsening radiation effects in bulk: SOI to the rescue, In *International Electron Devices Meeting Technical Digest,* 31.1.1–31.1.4. © (2013) IEEE. With permission.)

It should be noted that, for bulk technologies, the scaling of planar dimensions has not been accompanied by the same scaling of vertical dimensions in front-end-of-line (FEOL) processing, such as wells or epitaxial depths. Consequently, the efficiency of energy transfer from an incident ionizing-particle track to circuit nodes has scaled at a rate closer to feature size squared rather than feature size cubed (Massengill et al. 2012). This important point is schematically illustrated in Figure 10.1 for current bulk and SOI technology nodes.

The recent migration from bulk to SOI technologies, characterized by thin or ultrathin top silicon layers that are totally isolated, is an important geometric factor that limits single-event deposition volumes; from a purely volumetric scaling point of view, the SER should decline if no opposing phenomenon is involved in the soft-error susceptibility of these new technologies.

Figure 10.2 illustrates the increase of the particle's region of influence in a circuit, which is one of the most spectacular effects observed in recent technologies. Because the impact of a single event is not *point like* but has a certain radial extension (resulting in a radial charge distribution, illustrated and discussed in the next section), this spot of influence can intersect a more or less important portion of the circuit, depending on the technology node considered. In Figure 10.2, the spot of influence of an alpha particle intersects up to six memory cells in a 45 nm SRAM, whereas its impact was previously limited to a single cell for the 130 nm node. Such a purely geometric effect is responsible in part (and in part only) for multiple-cell upsets (MCU), observed in all recent technologies, typically below the 130–90 nm technological nodes.

10.2.2 Ion-Track Spatial Structure versus Device Dimensions

With increasing integration, the domain of influence of a single event does not only affect a single device (or cell) but can now have an influence at circuit level, since the characteristic

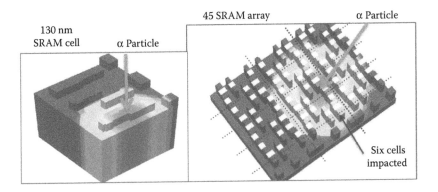

FIGURE 10.2
Illustration of the domain of influence of a single alpha particle striking a SRAM in both 130 and 45 nm technologies. Only a single cell is impacted in 130 nm, whereas a cluster of six adjacent cells are impacted in 45 nm. (Reprinted from Roche, P. et al., Technology downscaling worsening radiation effects in bulk: SOI to the rescue, In *International Electron Devices Meeting Technical Digest*, 31.1.1–31.1.4. © (2013) IEEE. With permission.)

radial dimension of a ion-track structure is of the same order of magnitude or larger as the device (or cell) feature size. Figure 10.3 shows the comparison between two charge distributions induced in silicon by energetic heavy ions with the feature sizes of 0.25 μm and 50 nm SOI transistors. For the shortest device in particular, it is clear that a nonnegligible part of the deposited charge is located outside the device. In other words, the deposited charge cannot be represented using a simple radial function, for example, a simple cylindrical or Gaussian charge-generation function with a uniform charge distribution and a constant LET along the ion path.

However, a real ion-track structure has a more complex radial profile than a simple Gaussian function, as illustrated in Figure 10.4 (Raine 2011). In addition, the track structure

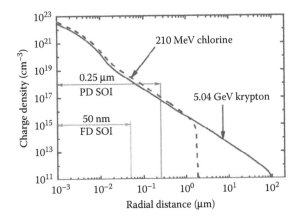

FIGURE 10.3
Comparison of the radial charge distributions induced by 210 MeV chlorine and 5.04 GeV krypton ions in silicon with the feature size of 0.25 μm and 50 nm SOI transistors. (Courtesy of Raine, M., PhD Thesis, Université Paris-Sud 11, 2011.)

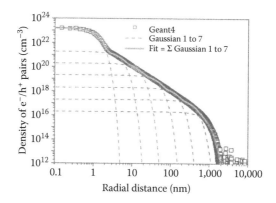

FIGURE 10.4

Reconstruction of a radial charge distribution (induced by the passage of an energetic ion in silicon) obtained using Geant4 with a sum of seven Gaussian functions. (Courtesy of Raine, M., PhD Thesis, Université Paris-Sud 11, 2011.)

varies in both space and time when the particle passes through the matter. As previously explained in Chapter 5, immediately after the particle strike, the core of the track is characterized by the production of highly energetic primary electrons (called δ rays). They generate a very large density of electron–hole pairs in a very short time and a very small volume around the ion trajectory, referred to as the ion track. These carriers are collected by both drift and diffusion mechanisms, and also recombine by different mechanisms of direct recombination (radiative, Auger) in the very dense core track, which strongly reduces the peak carrier concentration. All these mechanisms modify the track distribution in time and space. As the particle travels through matter, it loses energy, the δ rays become less energetic, and the electron–hole pairs are generated closer to the ion path. Then, the incident particle generates a characteristic cone-shaped charge plasma in the device (Dodd 2005).

The real ion-track structure has been calculated using various Monte Carlo methods (Hamm et al. 1979; Martin et al. 1987; Oldiges et al. 2000), including Geant4 code (Raine 2011). These simulations highlighted important differences between the track structure of low-energy and high-energy particles, in particular when the LET is the same (for details, see Dodd 2005; Dodd et al. 1998).

To take into account these differences in device or circuit simulations, different practical approaches can be envisaged. The first is to consider analytical models for ion-track structure. Several models have been proposed in the literature and implemented in simulation codes. One of the most interesting models is the *non-uniform power law* track model, based on the Katz theory (Kobetich and Katz 1968) and developed by Stapor and McDonald (1988). Based on their theory, an analytical model was proposed by Waligorski et al. (1986) for ion tracks in water and later adapted to silicon by Fageeha et al. (1994).

In this model, the ion track has a radial distribution of excess carriers expressed by a power-law distribution and allows the charge density to vary along the track (i.e., the LET is not constant along the track) (Dussault et al. 1993). Other analytical models propose a constant-radius nonuniform track or a Gaussian-distribution nonuniform track. But, as mentioned in (Raine et al. 2011b), all these models led to a radial distribution of energy deposition per unit volume proportional to the inverse square of the radial distance to the ion path. This dependence was then shown to be inaccurate, particularly for the track core region.

Rodbell et al. (2011) implemented into the IBM simulation code SEEM-2 an ion-track model following a $1/r^2$ law (at large radii) and assuming a maximum radius of 1000 nm. They studied the impact of such a radial ion distribution on the magnitude of the SEU cross sections for SOI latches in 32 and 45 nm technologies. This work shows that Monte Carlo modeling with a realistic track structure is necessary to correctly reproduce experimental irradiation data, the classical line-charge approximation (i.e., with no radius dependence) leading to a clear underestimation of the SEU cross sections (Rodbell et al. 2011).

To consider more realistic track structures, an interesting approach has been proposed by Raine et al. (2011a): it consists in building a database of ion-track structures obtained from Geant4 simulations and fitting a given track with multiple Gaussians. This technique is illustrated in Figure 10.4 (Raine 2011). The resulting fitting coefficients are then used as input data in the Synopsys Sentaurus device simulator. TCAD simulation at device level can then be performed to evaluate the influence of the exact ion-track structure on transistor electrical operation.

The same authors have also recently investigated the influence of the radial dimension of the ion tracks at memory-cell level using a Monte Carlo simulator (Raine et al. 2011b). In this work, two ion-track descriptions have been compared: the *punctual* approach, which considers an ion track as a series of punctual deposited charges (only taking into account the evolution of the LET with depth, with no radial dimension) and the new *radial* approach, which proposes to use the realistic track structures obtained with Geant4. In this case, the distribution of deposited charge is discretized in both directions. The authors obtained cartographies of ion impacts leading to an SEU for a 10 MeV per nucleon krypton-ion incident in four SOI SRAM technology nodes, using either the punctual or the radial approach. Visually, the per-bit cross section is larger for the radial approach as compared with the value obtained with the punctual description of the ion track. From this study, three major trends have been highlighted: (1) the ratio between the radial and the punctual SEU cross sections increases with decreasing energy per nucleon (as long as the track is wide enough); (2) for a given technology node and energy per nucleon, the ratio increases with ion mass and LET; (3) for a given ion and energy, the ratio increases with technology integration (Raine et al. 2011b).

10.2.3 Carrier Channeling in Wells and Electrical Related Effects

The reduction of circuit feature sizes has resulted in the emergence of regions where carriers in excess, as generated by a single event, can be confined, or more exactly channeled, in a certain region of the circuit at FEOL level. This particularly concerns the narrow wells of a given semiconductor type implanted in a semiconductor region of opposed type (e.g., an N-well in a p-type substrate or vice versa).

The energy deposition along the particle track, and the resulting generation of carriers in excess in such a well structure, creates a charge injection, thus transiently perturbing the electrostatic potential distribution in the well. Moreover, the propagation of this charge in excess along the well extension can have a sufficient magnitude to trigger parasitic bipolar injection from source to drain regions of transistors implanted in the same well. This effect has been observed and simulated by Giot et al. in SRAM memories, in particular for technology nodes below 90 nm (Giot 2009; Giot et al. 2008). Figure 10.5 illustrates this phenomenon in the case of an SRAM circuit with and without a technological option called *triple well* (TW) or *deep N-well* (DNW). This layer corresponds to either an N+ or a P+ buried

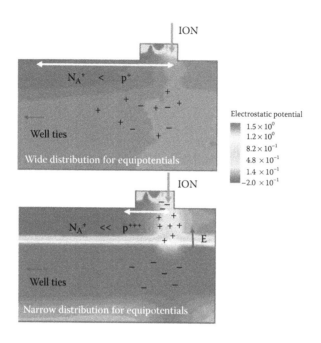

FIGURE 10.5

Electrostatic potential distribution around the ion track for a 65 nm SRAM memory cell away from well ties without triple well (*top*) and with triple well (*bottom*). Without the TW, holes are deposited and diffuse in the P-well directly toward well taps. With the TW, high hole confinement is created (p+++) by both the narrow P-well transverse dimensions and the P-well/TW junction, which quickly carries holes from the TW into the P-well. The positive charge gradient created between deposited holes (p+++) and well doping creates a narrow distribution for the electrostatic potential. N_A^+ is the doping concentration in the P-well. (Courtesy of D. Giot.)

layer in a P- or N-doped substrate, respectively. On the one hand, it has been used for years to decrease the single-event latchup (SEL) sensitivity, since the base resistance of the PNP parasitic bipolar is strongly reduced. TW accordingly makes it more difficult to trigger the latchup thyristor on. On the other hand, the presence of TW electrically isolates the p-type wells related to the NMOS transistors of the SRAM cells, enhancing carrier channeling in such P-wells in the case of single events. The narrow distribution for electrostatic potentials with the TW is explained by the intense hole-current densities. This high current is created near the ion strike by the P-well triple-well junction, which rapidly carries holes into the P-well. This hole current creates a high positive-charge gradient (high concentration of deposited holes p+++ near fixed charge in concentration in the P-well) in a small volume (narrower P-well transverse section with TW). This hole confinement within active areas (transistors) and TW induces high electric fields, which explains the high variations of electrostatic potential observed around the ion track (Giot et al. 2008). Just a few tens of femtoseconds after the ion strike and in the presence of TW, the electrostatic potential in the P-well around the struck bitcell is so high that the source is still injecting electrons into the P-well. The logic state of the drain allows it to recover a reverse-biased junction rapidly and to collect a high number of electrons from the source. Without the TW, both the potential and current magnitudes have decreased. Thus, there is no longer any injection of electrons from the source into the P-well (Giot 2009).

Another example related to a more recent technology (40 nm SRAM) is illustrated in Figure 10.6. This figure shows the physical bitmaps of different MCUs detected during a real-time experiment conducted in altitude on the ASTEP platform (see Chapter 7). These

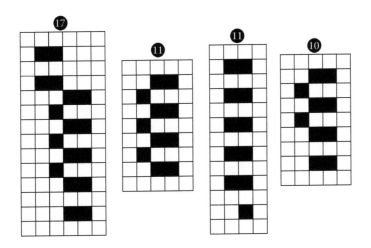

FIGURE 10.6
Physical bitmaps of the different MCUs detected for 40 nm single-port SRAM during a real-time experiment conducted at altitude on the ASTEP platform. Circled numbers indicate event multiplicity. (Reprinted from Autran, J.L. et al., Real-time soft-error testing of 40 nm SRAMs, In *Proceedings of the IEEE International Reliability Physics Symposium*, 3C.5.1–3C.5.9. © (2012) IEEE. With permission.)

large MCU events (≥10) correspond to high-density SRAM with a 25% reduced cell area with respect to the standard 40 nm node. This demonstrates the increase of MCU sensitivity when increasing the technological integration and, thus, reducing the memory-cell area. The topological shape of the MCUs detected can be explained by the combination of the SRAM layout (alternative structure of vertical P-wells and N-wells) with the checkerboard pattern used to fill the memory plan (Autran et al. 2012b). This is the reason why one can observe, for example, numerous horizontal pairs of adjacent cells (impact on the sensitive NMOS drain of two adjacent cells, these two drains being located in the same vertical P-well) vertically aligned and a systematic alternation of sensitive and nonsensitive horizontal rows (effect of the physical checkerboard). For all events characterized by a large multiplicity (≥10), it is clear that MCUs occur preferentially in columns, due to the mechanisms of charge diffusion and channeling that propagate the perturbation (and consequently trigger bipolar amplification) into the well directly impacted by the ionizing particle at the origin of the observed MCU.

To conclude, Roche et al. (2013) compiled the percentage of MCUs experimentally determined from accelerated tests for different generations of SRAMs manufactured in bulk CMOS with and without the deep N-well technological option. This compilation of results is shown in Figure 10.7. This latter clearly evidences the negative impact of the deep N-well on the MCU rate for all generations of investigated circuits, confirming that, in bulk CMOS, the resulting excess carriers can be confined and channeled far away from the particle strike. SER can even double with deep N-well in 65 nm SRAMs. The MCU rate increases faster than for single-cell upsets with scaling because of concurrent critical-charge reduction, higher probability of charge deposition in cells nearby, and enhanced carrier-channeling effect in the case of deep N-well architectures protecting bulk SRAMs against SEL. This trend would look bleak if one were limited to planar bulk devices for future technologies. However, a trade-off to obtain both zero SEL and low SER consists in using DNW combined with a high density of well contacts, a low-resistivity substrate, or both. In the case of SOI technologies, fortunately, devices are totally isolated from the substrate by the buried oxide and this effect does not exist.

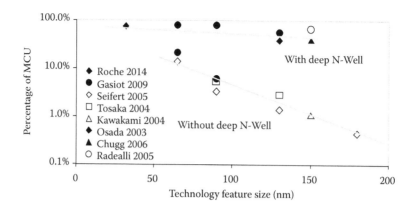

FIGURE 10.7
Scaling of multiple-cell upsets (MCUs) for different generations of SRAMs in bulk CMOS with and without "Deep N-well." (See Roche, P. et al., Technology downscaling worsening radiation effects in bulk: SOI to the rescue, In *International Electron Devices Meeting Technical Digest*, 31.1.1–31.1.4, 2013 for the references included in this graph.) (Reprinted from Roche, P. et al., Technology downscaling worsening radiation effects in bulk: SOI to the rescue, In *International Electron Devices Meeting Technical Digest*, 31.1.1–31.1.4. © (2013) IEEE. With permission.)

10.2.4 Variability and SEE

Variability is an important challenge for current and future CMOS technologies, which increases as device dimensions are scaled down. Two main sources of variability can be distinguished: global and local sources. As mentioned in Arnaud et al. (2011), global variability sources correspond to all the process steps inducing a spread at transistor level, such as threshold voltage and saturation current. Usually the main variability sources are geometric, such as physical gate length, offset spacers, and spacer widths. They are systematic factors modulating transistor parameters inside a die, across a wafer, and across lots. On the other hand, local variability sources are related to random effects. In Si/SiO_2-based bulk technologies, well-known sources are random dopant fluctuation and line-edge roughness. These two mechanisms degrade the matching factor of the transistors, which is fundamental for both SRAM functionality at low voltage and analog block performance as current mirrors. The recent introduction of high-κ dielectric/metal gate stacks brings a new local variability, that is, the work-function modulation.

Both global and local variability sources can have a direct impact on the SER of circuits via the variations of device geometry or electrical characteristics, resulting in variations of the critical charge, for memories in particular. Two recent works illustrate such variability effects in SRAMs. In the first example (Jahinuzzaman et al. 2009), the impact of process-induced variability on the SRAM critical (Q_{crit}) charge was studied from an analytical formulation of Q_{crit} derived for nanometric SRAM (see Section 10.3). Figure 10.8a shows the effects of threshold voltage (V_{TH}) variations of the driver and load transistors on Q_{crit}. The variations of Q_{crit} with slow and fast process corners ($\pm 6\%$ V_{TH} for PMOS, $\pm 3\%$ V_{TH} for NMOS) are also reported in Figure 10.8b. Calculated values of Q_{crit} are found to be in excellent agreement with SPICE simulation.

Such critical-charge variations inevitably induce SER variations, as illustrated in the second example, recently published by Gasiot et al. (2012). In this work, the variability encountered in a commercial 90 nm CMOS process was characterized by studying die-to-die variation on a large population of dies from a single wafer.

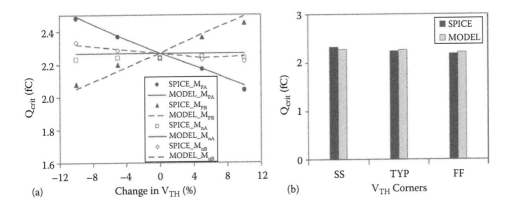

FIGURE 10.8
Critical-charge (Q_{crit}) variations as a function of threshold-voltage (V_{TH}) variations and V_{TH} corners as deduced from an analytical model for soft-error critical charge of nanometric SRAMs. (Reprinted from Jahinuzzaman, S.M. et al., An analytical model for soft error critical charge of nanometric SRAMs, *IEEE Trans. Very Large Scale Integr. (VLSI) Syst.* 17, 1187–1195. © (2009) IEEE. With permission.)

The dies were characterized with alpha particles and atmospheric neutrons, and SER spread was investigated as a function of their original position on the wafer. Moreover, experimental SER spread was compared with circuit simulations based on the manufacturer's process-design kit, which by default takes into account parameter spread due to process variability. SER distributions were derived from the critical-charge statistical distribution, computed using an analytical model calibrated with experimental data. Figure 10.9 shows the comparison between experimental and simulated distributions for neutron- and alpha-SER related to 60 instances of 90 nm SRAMs. The good agreement between the two sets of experimental and simulated data demonstrates the potential of the method to derive SER variability from the critical-charge statistical distribution determined from process-variation estimators.

10.3 Critical Charge

As introduced in Section 5.7, the concept of critical charge (Q_{crit}) is a first-order metric that has been introduced to quantify the susceptibility of a static memory to being upset from one logical state to the other. It corresponds to the minimum amount of electrical charge that can flip the data bit stored in a memory cell. Its relationship with the circuit SER is exponential, as illustrated by the analytical model developed by Hazucha and Svensson (2000):

$$SER = K \times A \times F \times \exp\left(-\frac{Q_{crit}}{Q_S}\right) \quad (10.1)$$

where:
  K   = scaling factor
  F   = particle flux ($cm^{-2} \times s^{-1}$)

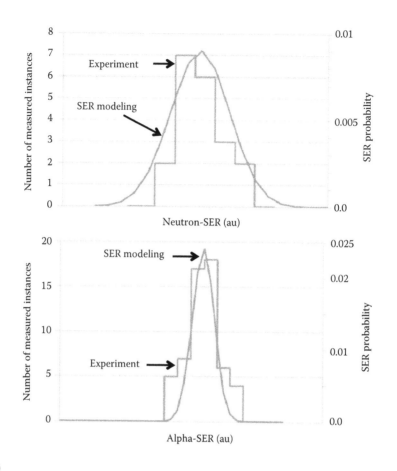

FIGURE 10.9
Experiments: experimental neutron- and alpha-SER of 60 SRAM instances measured on the same wafer and sorted by intervals of 20 FIT/Mb for 90 nm SRAMs. Modeling: probability densities of both neutron- and alpha-SER obtained from critical-charge distributions using an analytical model calibrated with experimental data. (Adapted from Gasiot, G. et al., Process variability effect on soft error rate by characterization of large number of samples. *IEEE Trans. Nucl. Sci.* 59, 2914–2919. © (2012) IEEE.)

A = area of the circuit sensitive to particle strikes (cm²)
Q_{crit} = critical charge
Q_S = charge-collection efficiency of the device (same units as Q_{crit})

Two key parameters for SER are the critical charge (Q_{crit}) of the SRAM cell and the charge-collection efficiency (Q_S) of the circuit. A lower Q_{crit} means a priori more soft errors. Q_S and Q_{crit} are determined by the process technology (Hazucha and Svensson 2000), whereas Q_{crit} also depends on characteristics of the circuit, particularly the supply voltage and the effective capacitance of the drain nodes. Q_{crit} and Q_S are essentially independent, but both decrease with decreasing feature size. Equation 10.1 highlights that changes in the ratio $-Q_{crit}/Q_S$ will have a very large impact on the resulting SER. The SER is also proportional to the area of the sensitive region of the device, and therefore it decreases proportionally to the square of the device size.

Different analytical/semianalytical models can be used for estimating the value of Q_{crit} in complement to full SPICE or TCAD approaches. We summarize in the following the brief review by Jahinuzzaman et al. (2009). To a first order, Q_{crit} is simply modeled as a sum of capacitance and conduction components:

$$Q_{crit} = C_N V_{DD} + I_{DP} T_F \qquad (10.2)$$

where:
C_N = equivalent capacitance of the struck node
V_{DD} = supply voltage
I_{DP} = maximum current of the on-state PMOS transistor
T_F = cell flipping time

While both capacitance and conductance components do indeed contribute to this critical charge, the first term is generally overestimated because the flipping threshold of an inverter is less than V_{DD} ($V_{DD}/2$ for perfectly matched NMOS and PMOS). In addition, the conductance term only considers the peak value of the current, which is not realistic. A more correct way of estimating the critical charge has been proposed by Xu et al. (2004):

$$Q_{crit} = \int_0^{V_{trip}} C_N dV + \eta I_P T_{pulse} \qquad (10.3)$$

where:
V_{trip} = static tripping point of the SRAM cell
η = correction factor
I_P = driven current of the on-state PMOS transistor
T_{pulse} = duration of the particle-induced current pulse I_{pulse}

Equation 10.3 provides a better estimation of the capacitance component of Q_{crit}, particularly the effect of junction capacitance and the addition of the backend metal-insulator-metal (MIM) capacitor. However, this model fails to incorporate the dynamics of voltage transient at the struck node, the quantitative description of I_{pulse}, and the contributions of the different transistors that constitute the cell. As a result, the accuracy of Equation 10.3 in estimating Q_{crit} is limited.

Improved analytical techniques with reduced discrepancies ($\leq 10\%$ and below) with respect to full SPICE simulations have been proposed in recent years by Zhang et al. (2006) and more recently by Jahinuzzaman et al. (2009). This last model takes into account the dynamic behavior of the cell and demonstrates a simple technique to decouple the non-linearly coupled storage nodes. Decoupling of storage nodes enables solving associated current equations to determine the critical charge for a classically used double-exponential pulse current. The critical-charge model thus developed consists of both NMOS and PMOS transistor parameters. Critical-charge values calculated by the model have been found to be in good agreement with SPICE simulations for a commercial 90 nm CMOS process, with a maximum discrepancy of less than 5%.

Previous models, TCAD, or mixed-mode or full SPICE simulations can be used to study the scaling of Q_{crit} as a function of transistor/circuit feature sizes. In recent work, Massengill et al. (2012) provides a compilation of Q_{crit} values for 6T SRAMs as a function of feature size (Figure 10.10). As remarked by the authors, the early 1980s prediction by Petersen of Q_{crit} scaling with feature size squared (Petersen et al. 1982) has remained

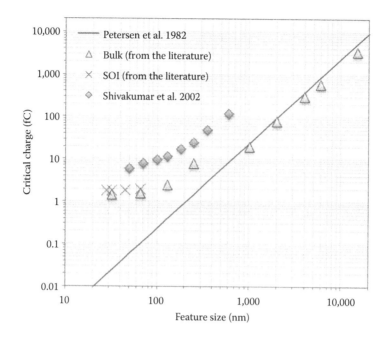

FIGURE 10.10
Critical-charge scaling as a function of feature size. Data from Shivakumar et al. and analytical predictions from the scaling model of Petersen et al. are compared with experimental values published in literature for both bulk and SOI technologies. (Data from Shivakumar P., Kistler, M., Keckler, S.W., Burger, D., and Alvisi, L., Proceedings of the International Conference on Dependable Systems and Networks, pp. 389–398. © (2002) IEEE; Petersen, E.L., Shapiro, P., Adams Jr, J.H., and Burke, E.A., IEEE Trans. Nucl. Sci., 29, 2055–2063. © (1982) IEEE.)

remarkably accurate, even across decades of generational changes in substrates, lithography, and devices. Figure 10.10 shows that, for recent technologies, Q_{crit} values are now below a femtocoulomb, a value well below the amount of charge deposited by a singly charged ionizing particle in silicon.

10.4 Increasing Sensitivity to Background Radiation

10.4.1 Low-Energy Protons

The impact of low-energy protons on modern electronics is an important issue for both space and terrestrial applications. We mentioned this reliability concern in Section 6.3.3. Low-energy protons are typically generated during scattering of high-energy protons or neutrons, and are primarily a concern in space-radiation environments (Seifert et al. 2011). While shielding materials can easily absorb low-energy protons, scattering of high-energy protons can yield significant fluxes of low-energy protons, impacting electronics. Moreover, because secondary low-energy protons can be generated in nuclear spallation reactions of high-energy neutrons with silicon and other materials present in modern semiconductor devices, low-energy-proton-induced SERs also need to be accounted for in terrestrial radiation environments (Rodbell et al. 2007).

In recent years, several authors have reported on evidence of direct ionization from low-energy protons in SRAMs and latches (Heidel et al. 2008, 2009; Insoo et al. 2003; Puchner et al. 2011; Rodbell et al. 2007; Sierawski et al. 2009). As recalled by Seifert et al. (2011), low-energy-proton energies in this context refer to proton energies that are at or below the lowest threshold energy for nuclear reactions of protons with silicon (Si), that is, below the MeV (Insoo et al. 2003). The effect has been reported in both 65 nm bulk and 65 and 45 nm silicon-on-insulator (SOI) devices (Heidel et al. 2008, 2009; Insoo et al. 2003; Rodbell et al. 2007; Sierawski et al. 2009). These studies showed that critical charges are well below 1 fC for the examined devices and linear energy transfer (LET) values are sufficient to cause upsets via direct ionization. In Sierawski et al. (2009), the authors investigated by different techniques (low-energy accelerated tests, TCAD, and Monte Carlo simulation) this mechanism of direct ionization for a 65 nm CMOS SRAM. Their TCAD simulations clearly establish that the electronic stopping of protons, which have a peak LET near 0.5 MeV cm^{-2} mg^{-1}, may induce upsets in the memory if the peak occurs near the sensitive device regions.

Figure 10.11 shows experimental proton test results measured on bulk CMOS SRAMs manufactured in 40 nm and 28 nm technologies by two different semiconductor manufacturers. Low-energy-proton measurements were performed at the TRIUMF test facility for the two circuits (see Section 6.3.3). The SEU results are very different for these two SRAMs as compared with SRAMs fabricated in previous technology generations. Specifically, no upset threshold is observed as the proton energy is decreased down to a few megaelectronvolts, and a sharp rise in the upset cross section is observed below a few megaelectronvolts. The increase below 10 MeV is clearly attributed to upsets caused by direct ionization from the low-energy protons.

10.4.2 Atmospheric Muons

Until recently, the effect of muons on electronics has been the subject of only a very small number of investigations. We can cite the pioneering work in the 1980s of Ziegler and Lanford (1979) and a few experimental characterization studies of memories (SRAM, DRAM) using artificial muon beams (Dicello et al. 1983; Duzellier et al. 2001; Gelderloos et al. 1997).

Recently, Sierawski et al. (2010, 2011) conducted the first major work on the subject, combining measurements and numerical simulations on the effect of low-energy (<3 MeV) and atmospheric positive muons on advanced technologies. They demonstrated and quantified the effects of muon direct ionization for different bulk SRAMs of different technology nodes (65, 55, 45, and 40 nm). The data presented in Figure 10.12 show the probability of

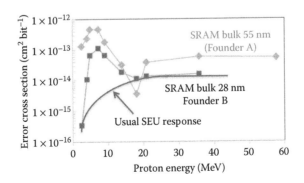

FIGURE 10.11
Experimental proton-induced SEU data from 1 to 500 MeV related to a 65 nm SOI SRAM. (Reprinted from Roche, P. et al., Technology downscaling worsening radiation effects in bulk: SOI to the rescue, In *International Electron Devices Meeting Technical Digest*, 31.1.1–31.1.4. © (2013) IEEE. With permission.)

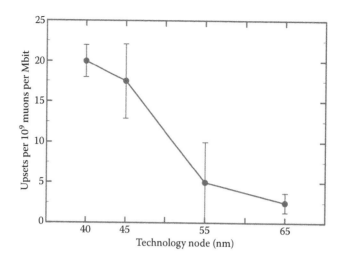

FIGURE 10.12

Experimental muon-induced single-event-upset probability for different SRAMs under test operated at nominal supply voltage and irradiated using a 21.6 MeV c⁻¹ positive muon beam. (Reprinted from Sierawski, B.D. et al., Effects of scaling on muon-induced soft errors, In *Proceedings of the IEEE International Reliability Physics Symposium*, 3C.3.1–3C.3.6, 10–14 April, Monterey, CA. © (2011) IEEE. With permission.)

upset at nominal bias for four different SRAMs subjected to μ+ irradiation using the M20B surface muon beam at TRIUMF.

As stated by Sierawski et al. (2011), these data show a clear increase in the SEU susceptibility and significance of energy deposition by muons for scaled technologies. To a first order, the reduction of the device area results in a decrease in the number of particles passing through the cell and capable of producing an upset. This scaling would reduce the probability of a bit being upset by the beam. The increase in upset probability is therefore attributed to differences in the geometry of the charge collection, an increase in the fluence of energy deposition events exceeding the critical charge, or both. Further, in these experiments, the incident muons have a distribution of kinetic energies and therefore a distribution of stopping powers. As the technology node decreases, the charge required to upset a single memory cell decreases. The effect of this trend is an increase in the fraction of the distribution that is able to induce an upset. For the 40 nm SRAM, a larger portion of the 21.6 MeV c⁻¹ beam exceeds the stopping-power threshold as compared with the 65 nm SRAM. While the probability of upset increases for future devices, it cannot be ascertained from these data whether the trend is increasing linearly or superlinearly.

In complement to these experimental results, the authors performed a series of simulations with the Vanderbilt Monte Carlo Radiative Energy Deposition (MRED) code (Weller et al. 2009). They estimated sea-level error rates for 32, 22, and 16 nm representative sensitive volumes. Muon-induced SER was found to increase for lower values of generated charge, as one would expect a more sensitive device to have a higher error rate. Below the 32 nm technology node, the critical charge is sufficiently decreased that this threshold now permits a significant rate of errors to occur. The results also indicate the potential for large variations in the error rate for 16 nm devices. The dramatic increase in the error rate for devices with a threshold below 0.2 fC suggests that even minor design differences may have a large impact on the reliability of the memory (Sierawski et al. 2011).

Serre et al. (2012) studied the complementary effect of low-energy (<1 MeV) negative muons on SRAM memories and demonstrated the importance of the negative-muon capture mechanism as an additional mechanism of charge deposition for negative muons that can be stopped in silicon. Some of these results have been already presented and discussed in the previous chapter (see Section 9.4.4). As a reminder, Figure 9.17 shows different possible scenarios of negative and positive-muon interactions with a 65 nm SRAM structure as simulated using the TIARA-G4 simulation code. This figure qualitatively illustrates that the probability of inducing an upset is maximal when the muon-capture-induced shower is produced in the immediate vicinity of the sensitive drain layer. This case corresponds to a reduced energy interval for the incoming muons, insofar as the penetration depth of the muons in the structure and the capture location primarily depend on the muon's kinetic energy (see also Figure 9.18).

In order to quantify this effect, Serre et al. calculated the muon-induced SER as a function of the incident muon's kinetic energy. Results are shown in Figure 10.13 (Autran et al. 2012a). As expected, the occurrence of soft errors, and consequently the SER induced by negative-muon irradiation, presents a maximum precisely where muon captures occur at the depth of the layer containing sensitive drains (i.e., the active silicon region). This figure also gives the percentage of cell upsets induced by muon-capture reactions or directly by muon impacts on sensitive drains (i.e., direct charge deposition in drain volumes). When the kinetic energy of primary particles increases, the fraction of upsets induced by muon capture rapidly decreases as soon as captures occur deeper in silicon, below the active layer. In this case, upsets become mainly induced by direct charge deposition from incident muons.

Comparison of this simulation result with experimental data previously obtained by Sierawski et al. (2011) is very interesting, since it shows similarities in the dependence of the soft-error occurrence on the muon's incident kinetic energy.

Figure 1 (top) of Sierawski et al. (2011) shows a maximum occurrence of upsets for the tested 65 nm SRAM near 700 keV for low-energy (<3 MeV) positive muons. At this incident energy,

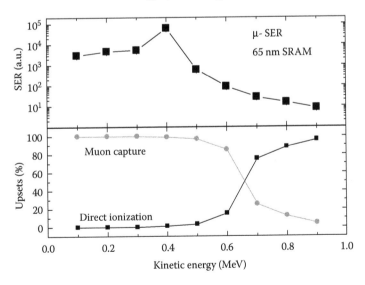

FIGURE 10.13
Estimated monoenergetic negative-muon-induced soft-error rate versus muon kinetic energy for the 65 nm SRAM. The percentage of cell upsets induced by the secondary particles produced by muon capture reactions or directly by muon impacts on sensitive drains (i.e., direct charge deposition in drain volumes) are also plotted. (Reprinted from Autran, J.L. et al., Soft-error rate of advanced SRAM memories: Modeling and Monte Carlo simulation, In M. Andriychuk (ed.), *Numerical Simulation—From Theory to Industry*, Vienna: InTech, 2012. © 2012 Autran et al., licensee InTech.)

the range of the positive muon beam through the SRAM BEOL is such that a large portion of the muons traversing the active silicon region verify the Bragg peak condition (Sierawski et al. 2011): the deposited charge is then maximal and sufficient to exceed the critical charge of the SRAM under test, leading to a maximum upset rate at this energy. Our simulation result (obtained for a similar but different 65 nm SRAM circuit), shown in Figure 10.13, is in good agreement with these experimental data, the percentage of upsets induced by negative muons with energies above ~0.7 MeV being dominated by direct charge deposition. Below this energy, our result evidences an additional mechanism of charge deposition for negative muons resulting from the capture of the incident muons that can be stopped in silicon in the active silicon region. Such an additional channel for muon-induced charge deposition is found to be maximal at 0.4 MeV for the SRAM architecture considered here.

In conclusion, these recent investigations clearly demonstrate the importance of low-energy atmospheric muons as a new radiation constraint at ground level for the most advanced CMOS technologies. Further analyses will be necessary in the future to investigate in particular the exact proportion of positive and negative atmospheric muons that can significantly deposit charge in silicon, with respect to not only the circuit architecture but also the local environment (shielding) of the circuit, capable of profoundly impacting the distribution of such low-energy atmospheric muons below ~1 MeV. The current lack of both experimental and theoretical knowledge related to atmospheric muon distributions below 1 MeV therefore represents a limitation to accurately estimating the impact of muons on electronics at ground level.

10.4.3 Low-Alpha-Material Issue

Alpha particles emitted from radioactive materials or radioactive impurities are a nonnegligible part of the natural background radiation at terrestrial level. With the downscaling of CMOS technologies, the sensitivity of ICs to ultratraces of alpha-particle-emitter contamination is a crucial question because of the constant reduction of the supply voltage and node capacitance, as previously discussed. As also examined in Section 3.7, there will clearly be a need for lower-alpha-emitting materials used at the different levels of the fabrication process for upcoming and future technologies. This includes not only packaging materials, which generally represent the prime source of alpha contamination, but also materials and alloys employed at both FEOL and back-end-of-line (BEOL) levels. Due to their even greater proximity to circuit sensitive nodes, FEOL materials with low levels of alpha contamination can therefore represent a nonnegligible source of soft errors. As an illustration of the severity of alpha-particle contamination at silicon level, Figure 10.14 shows the results of real-time cave-SER testing of 65 nm SRAMs conducted at the underground laboratory of Modane (LSM, France; see Chapter 7) over more than 3 years (Martinie et al. 2012). At the depth of the LSM, cosmic rays are totally screened out, and one can consider that soft errors are quasi-exclusively due to the internal chip radioactivity, that is, the alpha-particle emitters present in chip materials (ultralow-emissivity packages were used). The experiment involved 3226 Mbits of stand-alone single-port memories assembled in low-alpha PBGA packages. A total of 90 bit flips were detected in approximately 24,000 h of cave measurements. These results have been perfectly reproduced using a Monte Carlo SER simulation code and assuming 0.1 ppb of ^{238}U contamination in the silicon bulk of the circuits, demonstrating that a low-alpha material (in this case the silicon wafers) can therefore be responsible for detectable soft errors (Martinie et al. 2012).

In addition, and as noted by Clark (2012), the trend toward flip-chip and 3D IC architecture, in particular, has increased the need for reliable low-alpha packaging materials. In

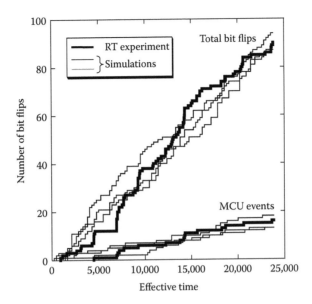

FIGURE 10.14
Cumulative number of bit flips versus test duration for underground real-time SER testing of 3.2 Gbits of 65 nm single-port SRAM (*bold line*). Several error distributions (*thin lines*) obtained by Monte Carlo simulations for 0.1 ppb of ^{238}U contamination in silicon are also shown. (Reprinted from Roche, P. et al., Technology downscaling worsening radiation effects in bulk: SOI to the rescue, In *International Electron Devices Meeting Technical Digest*, 31.1.1–31.1.4. © (2013) IEEE. With permission.)

these designs, packaging features such as wafer-level solder bumps and copper-pillar solder caps are located close to the transistors of the device. This increases the transistor vulnerability to alpha emissions from these features and can lead to higher SERs. Figure 10.15 shows the acceptable alpha-emission limits for packaging materials as a function of the technological node. These projections illustrate the transition from standard low-alpha (LA) to ultralow-alpha (ULA) and even to super-ultralow-alpha (SULA) purities that characterizes the recent evolution of microelectronics as it enters the nanometer area. An effort to synthesize (chemical aspects) and characterize (metrology tools) such ULA/SULA materials certainly represents a challenge for the microelectronics industry as a whole.

10.5 Trends and Summary for Ultrascaled Technologies

To conclude this chapter, we summarize in the following the impact of the main factors previously discussed on the SER of ICs manufactured in bulk versus (SOI) technologies (Roche et al. 2013).

For bulk technologies, Figure 10.16 illustrates the evolution of the measured single-bit SRAM SER as a function of current technology node. The SER per bit has peaked at the 130 nm technology node and since then has been decreasing. This current reduction tendency can be explained by the dramatic reduction of charge-collection volumes due to scaling and compensating for the opposite influence of lower critical charges. At chip level, since the cells are more densely packed with scaling, the charge from a single particle is more easily shared among several cells. This reduces the amount of charge effectively collected by an individual cell. The

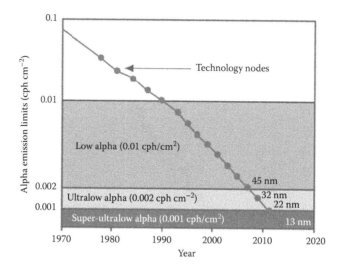

FIGURE 10.15
Acceptable alpha-emission limits for packaging materials as a function of the technological node. (Reprinted with permission from Clark, B., The distribution and transport of alpha activity in tin. White paper. Honeywell International Inc., 2012. Available at: http://www.honeywell-radlo.com, 2012. © 2012 Honeywell International Inc.)

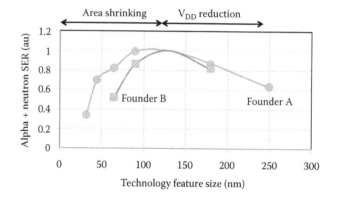

FIGURE 10.16
Scaling of neutron- and alpha-SER as a function of feature size for bulk technologies down to 32 nm technology node. (Reprinted from Roche, P. et al., Technology downscaling worsening radiation effects in bulk: SOI to the rescue, In *International Electron Devices Meeting Technical Digest*, 31.1.1–31.1.4. © (2013) IEEE. With permission.)

net effect is an improved per-bit SER of scaled technologies, as observed in DRAMs, dynamic logic, and SRAMs, but not for latches and combinatorial logic (Roche et al. 2013). Finally, at circuit or system-on-chip levels, SER keeps increasing with the growing amount of memories and latches, with a particularly strong SER impact on latches. The decrease of supply voltages for higher power efficiency (e.g., for cloud computing) is an aggravating factor and will constitute a great reliability challenge.

For SOI technologies, the sensitive volume traversed by an ionizing particle is even further reduced by the isolation barriers (Figure 10.1), resulting in a stronger intrinsic SER resilience in SOI compared with bulk. This total isolation of the device also removes the parasitic thyristor (see also Chapter 12) at the origin of the SEL. SOI technologies are thus immune to SEL by construction, and this immunity is also verified for hybrid bulk devices in FDSOI 28 nm

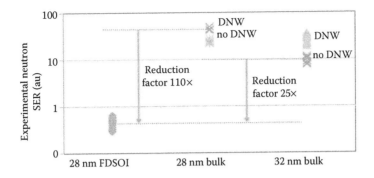

FIGURE 10.17
High-energy-neutron-SER test results (TRIUMF test facility, Vancouver) for SRAMs in ultrathin box and body (UTBB) FDSOI 28 nm, bulk 32 and 28 nm, with and without N+ Deep N-Well (DNW). (Reprinted from Roche, P. et al., Technology downscaling worsening radiation effects in bulk: SOI to the rescue, In *International Electron Devices Meeting Technical Digest*, 31.1.1–31.1.4. © (2013) IEEE. With permission.)

(Roche et al. 2013). However, a stronger parasitic bipolar in SOI can degrade SER in spite of small critical charges and small sensitive volumes. Connecting the partially depleted (PD)-SOI internal body to either source or ground greatly improves SER, but induces area penalty. In fully depleted (FD)-SOI 28 nm, the bipolar gain is very low (<3 in the worst case) and full benefit can be obtained from the ultrasmall sensitive volume, thus greatly minimizing SER. Figure 10.17 shows that up to a 110× reduction factor in the high-energy-neutron SER has been experimentally measured in FDSOI 28 nm compared with bulk 28 nm, with the lowest failure-in-time (FIT) rate ever observed for SRAMs (<10 FIT/Mbit) (Roche et al. 2013).

To conclude, Table 10.1 summarizes the expected SER, SEL, and total-ionizing-dose (TID) performances between CMOS bulk, FinFET, and SOI technologies. This table has been

TABLE 10.1

Expected SER, SEL, and TID Performances between CMOS Bulk, FinFET, and SOI Technologies

With Respect to Bulk CMOS	PD-SOI	Bulk FinFET	SOI-FinFET Experimental	UTBB-FDSOI
Critical charge (fC) Minimal charge to upset	0.1	<0.1	<0.1	<0.1
Sensitive volume Charge deposition and collection	Small	Very small	Very small	Ultrasmall
Parasitic bipolar Charge amplification	Significant w/o body ties <20	Ultralow substrate tied to body	Low ~2–8	Ultralow <3
Alpha/neutron-SER	÷5 to ÷20	÷2.5 to ÷3.5	÷10	÷100
Muon-SER			New SER risk (1000×)	
Thermal-neutron-SER			New SER risk (2×)	
Low-energy-proton-SER			New SER risk (to be evaluated)	
Single-event latchup (SEL) Ion-induced latchup	Immune by construction	No data yet in literature	Immune by construction	Immune including hybrid devices
Total ionizing dose (TID) Gamma and x-rays	Megarads with body ties/taps	Hundreds of kilorads with large fins	Megarads with narrow fins	Hundreds of kilorads

Source: Adapted from Roche, P. et al., Technology downscaling worsening radiation effects in bulk: SOI to the rescue, In *International Electron Devices Meeting Technical Digest*, 31.1.1–31.1.4, 2013.

compiled from references cited in Roche et al. (2013) and offers a global overview of the performances, drawbacks, and main challenges for the upcoming technologies. Certain issues for ultrascaled SOI devices or FinFET architectures will be investigated in depth and discussed in Chapter 12.

References

Arden, W., Brillouët, M., Cogez, P., et al. 2010. Towards a "More-than-Moore" roadmap. Report from the CATRENE Scientific Committee. Available at: http://public.itrs.net.

Arnaud, F., Pinzelli, L., Gallon, C., Rafik, M., Mora, P., and Bœuf, F. 2011. Challenges and opportunity in performance, variability and reliability in sub-45 nm CMOS technologies. *Microelectronics Reliability* 51:1508–1514.

Autran, J.L., Semikh, S., Munteanu, D., Serre, S., Gasiot, G., and Roche, P. 2012a. Soft-error rate of advanced SRAM memories: Modeling and Monte Carlo simulation. In M. Andriychuk (ed.), *Numerical Simulation—From Theory to Industry*. Vienna: InTech.

Autran, J.L., Serre, S., Munteanu, D., et al. 2012b. Real-time soft-error testing of 40 nm SRAMs. In *Proceedings of the IEEE International Reliability Physics Symposium*, pp. 3C.5.1–3C.5.9. 15–19 April, Anaheim, CA, IEEE.

Clark, B. 2012. The distribution and transport of alpha activity in tin. White paper. Honeywell International, 2012. Available at: http://www.honeywell-radlo.com.

Dicello, J.F., McCabe, C.W., Doss, J.D., and Paciotti, M. 1983. The relative efficiency of soft-error induction in 4K static RAMs by muons and pions. *IEEE Transactions on Nuclear Science* NS-30:4613–4615.

Dodd, P.E. 2005. Physics-based simulation of single-event effects. *IEEE Transactions on Device and Materials Reliability* 5:343–357.

Dodd, P.E., Musseau, O., Shaneyfelt, M.R., et al. 1998. Impact of ion energy on single-event upset. *IEEE Transactions on Nuclear Science* 45:2483–2491.

Dussault, H., Howard Jr, J.W., Block, R.C., Pinto, M.R., Stapor, W.J., and Knudson, A.R. 1993. Numerical simulation of heavy ion charge generation and collection dynamics. *IEEE Transactions on Nuclear Science* 40:1926–1934.

Duzellier, S., Falguère, D., Tverskoy, M., Ivanov, E., Dufayel, R., and Calvet, M.C. 2001. SEU induced by pions in memories from different generations. *IEEE Transactions on Nuclear Science* 48:1960–1965.

Fageeha, O., Howard, J., and Block, R.C. 1994. Distribution of radial energy deposition around the track of energetic particles in silicon. *Journal of Applied Physics* 75:2317–2321.

Gasiot, G., Castelnovo, A., Glorieux, M., Abouzeid, F., Clerc, S., and Roche, P. 2012. Process variability effect on soft error rate by characterization of large number of samples. *IEEE Transactions on Nuclear Science* 59:2914–2919.

Gelderloos, C.J., Peterson, R.J., Nelson, M.E., and Ziegler, J.F. 1997. Pion-induced soft upsets in 16 Mbit DRAM chips. *IEEE Transactions on Nuclear Science* 44:2237–2242.

Giot, D. 2009. Numerical simulation study of the sensitivity to the terrestrial radiation environment of SRAM memories in 90 nm to 65 nm technologies. PhD Thesis, Aix-Marseille University (in French).

Giot, D., Roche, P., Gasiot, G., Autran, J.L., and Harboe-Sørensen, R. 2008. Heavy ion testing and 3-D simulations of multiple cell upset in 65 nm standard SRAMs. *IEEE Transactions on Nuclear Science* 55:2048–2054.

Hamm, R.N., Turner, J.E., Wright, H.A., and Ritchie, R.H. 1979. Heavy ion track structure in silicon. *IEEE Transactions on Nuclear Science* 26:4892–4895.

Hazucha, P. and Svensson, C. 2000. Impact of CMOS technology scaling on the atmospheric neutron soft error rate. *IEEE Transactions on Nuclear Science* 47:2586–2594.

Heidel, D.F., Marshall, P.W., LaBel, K.A., et al. 2008. Low energy proton single-event-upset test results on 65 nm SOI SRAM. *IEEE Transactions on Nuclear Science* 55:3394–3400.

Heidel, D.F., Marshall, P.W., Pellish, J.A., et al. 2009. Single-event upsets and multiple-bit upsets on a 45 nm SOI SRAM. *IEEE Transactions on Nuclear Science* 56:3499–3504.

Hung, Y.C. 2010. CMOS nonlinear signal processing circuits. In P.K. Chu (ed.), *Advances in Solid State Circuit Technologies*, Vienna: InTech.

Insoo, J., Xapsos, M.A., Messenger, S.R., et al. 2003. Proton nonionizing energy loss (NIEL) for device applications. *IEEE Transactions on Nuclear Science* 50:1924–1928.

ITRS (International Technology Roadmap for Semiconductors). 2012. International Technology Roadmap for Semiconductors. Available at: http://public.itrs.net.

Jahinuzzaman, S.M., Sharifkhani, M., and Sachdev, M. 2009. An analytical model for soft error critical charge of nanometric SRAMs. *IEEE Transactions on Very Large Scale Integration (VLSI) Systems* 17:1187–1195.

Kobetich, E.J. and Katz, R. 1968. Energy deposition by electron beams and δ rays. *Physical Review* 170:391–396.

Martin, R.C., Ghoniem, N.M., Song, Y., and Cable, J.S. 1987. The size effect of ion charge tracks on single event multiple-bit upset. *IEEE Transactions on Nuclear Science* 34:1305–1309.

Martinie, S., Autran, J.L., Sauze, S., et al. 2012. Underground experiment and modeling of alpha emitters induced soft-error rate in CMOS 65 nm SRAM. *IEEE Transactions on Nuclear Science* 59:1048–1053.

Massengill, L.W., Bhuva, B.L., Holman, W.T., Alles, M.L., and Loveless, T.D. 2012. Technology scaling and soft error reliability. In *Proceedings of the IEEE International Reliability Physics Symposium*, pp. 3C.1.1–3C.1.7. 15–19 April, Anaheim, CA, IEEE.

Oldiges, P., Dennard, R., Heidel, D., Klaasen, B., Assaderaghi, R., and Ieong, M. 2000. Theoretical determination of the temporal and spatial structure of α-particle induced electron-hole pair generation in silicon. *IEEE Transactions on Nuclear Science* 47:2575–2579.

Petersen, E.L., Shapiro, P., Adams Jr, J.H., and Burke, E.A. 1982. Calculation of cosmic-ray induced soft upsets and scaling in VLSI devices. *IEEE Transactions on Nuclear Science* 29:2055–2063.

Puchner, H., Tausch, J., and Koga, R. 2011. Proton-induced single event upsets in 90 nm technology high performance SRAM memories. In *Proceedings of the IEEE Radiation Effects Data Workshop*, pp. 1–3. 25–29 July, Las Vegas, NV, IEEE.

Raine, M. 2011. Etude de l'effet de l'énergie des ions lourds sur la sensibilité des composants électroniques [Study of the effect of heavy ion energy on the sensitivity of electronic components]. PhD Thesis, Université Paris-Sud 11 (in French).

Raine, M., Gaillardin, M., Paillet, P., Duhamel, O., Girard, S., and Bournel, A. 2011a. Experimental evidence of large dispersion of deposited energy in thin active layer devices. *IEEE Transactions on Nuclear Science* 58:2664–2672.

Raine, M., Hubert, G., Gaillardin, M., et al. 2011b. Impact of the radial ionization profile on SEE prediction for SOI transistors and SRAMs beyond the 32-nm technological node. *IEEE Transactions on Nuclear Science* 58:840–847.

Roche, P., Autran, J.L., Gasiot, G., and Munteanu, D. 2013. Technology downscaling worsening radiation effects in bulk: SOI to the rescue. In *Proceedings of the International Electron Devices Meeting Technical Digest*, pp. 31.1.1–31.1.4, 9–11 December, Washington, DC.

Rodbell, K.P., Heidel, D.F., Pellish, J.A., et al. 2011. 32 and 45 nm radiation-hardened-by-design (RHBD) SOI latches. *IEEE Transactions on Nuclear Science* 58:2702–2710.

Rodbell, K., Heidel, D., Tang, H., Gordon, M., Oldiges, P., and Murray, C. 2007. Low-energy proton-induced single-event-upsets in 65 nm node, silicon-on-insulator, latches and memory cells. *IEEE Transactions on Nuclear Science* 54:2474–2479.

Seifert, N., Gill, B., Pellish, J.A., Marshall, P.W., and LaBel, K.A. 2011. The susceptibility of 45 and 32 nm bulk CMOS latches to low-energy protons. *IEEE Transactions on Nuclear Science* 58:2711–2718.

Serre, S., Semikh, S., Autran, J.L., Munteanu, D., Gasiot, G., and Roche, P. 2012. Effects of low energy muons on electronics: Physical insights and Geant4 simulation. In *Proceedings of the European Conference on Radiation and its Effects on Components and Circuits (RADECS)*, 24–28 September, Biarritz, France. Available at http://www.im2np.fr/news/articles/RADECS2012_Muons_Proceedings.pdf.

Shivakumar, P., Kistler, M., Keckler, S.W., Burger, D., and Alvisi, L. 2002. Modeling the effect of technology trends on the soft error rate of combinational logic. In *Proceedings of the International Conference on Dependable Systems and Networks*, pp. 389–398, 23–26 June, Washington, DC, IEEE.

Sierawski, B.D. 2012. Low-energy proton single event upsets in SRAMs. In *3rd Annual NEPP Electronic Technology Workshop (ETW), NASA Electronic Parts and Packaging (NEPP)*, 11–13 June, Greenbelt, MD.

Sierawski, B.D., Mendenhall, M.H., Reed, R.A., et al. 2010. Muon-induced single event upsets in deep-submicron technology. *IEEE Transactions on Nuclear Science* 57:3273–3278.

Sierawski, B.D., Pellish, J.A., Reed, R.A., et al. 2009. Impact of low-energy proton induced upsets on test methods and rate predictions. *IEEE Transactions on Nuclear Science* 56:3085–3092.

Sierawski, B.D., Reed, R.A., Mendenhall, M.H., et al. 2011. Effects of scaling on muon-induced soft errors. In *Proceedings of the IEEE International Reliability Physics Symposium*, pp. 3C.3.1–3C.3.6. 10–14 April, Monterey, CA, IEEE.

Stapor, W.J. and McDonald, P.T. 1988. Practical approach to ion track energy distribution. *Journal of Applied Physics* 64:4430–4434.

Waligorski, M.P.R., Hamm, R.N., and Katz, R. 1986. The radial distribution of dose around the path of a heavy ion in liquid water. *Nuclear Tracks and Radiation Measurements* 11:309–319.

Weller, R.A., Schrimpf, R.D., Reed, R.A., et al. 2009. Monte Carlo simulation of single event effects. In *European Conference on Radiation and Its Effects on Components and Circuits (RADECS)*, Short-Course Text, 14–18 September, Bruges, Belgium.

Xu, Y.Z., Puchner, H., Chatila, A., et al. 2004. Process impact on SRAM alpha-particle SEU performance. In *Proceedings of the IEEE International Reliability Physics Symposium*, pp. 294–299. 25–29 April, Phoenix, AZ, IEEE.

Zhang, B., Arapostathis, A., Nassif, S., and Orshansky, M. 2006. Analytical modeling of SRAM dynamic stability. In *Proceedings of the IEEE/ACM International Conference on Computer-Aided Design*, pp. 315–322. 5–9 November, San Jose, CA, IEEE.

Ziegler, J.F. and Lanford, W.A. 1979. Effect of cosmic rays on computer memories. *Science* 206:776–788.

11

Natural Radiation in Nonvolatile Memories: A Case Study

11.1 Introduction

Among all the integrated circuits used in the many application areas for which a high reliability level is required (medical, space, automotive, networking, and nuclear), nonvolatile memories (NVMs) are known for their relative robustness to single events, even if the different components of a flash memory circuit (on the one hand the memory cell array, on the other hand the peripheral control circuitry) exhibit distinct levels of radiation sensitivity. In addition, the specific question of their sensitivity to the terrestrial radiation environment has been little studied until now. Cellere et al. (2008, 2009) and Gerardin et al. (2010, 2012) were the first to clearly state, using accelerated tests, that atmospheric-neutron-induced soft-error occurrence is possible in flash memories, although with extremely low probability at ground level.

In this context, and complementary to previous works based on accelerated tests, Just et al. (2013) proposed a new type of experiment, based on a long-term exposure of fully processed wafers (with functional chips) to natural radiation background, in order to experimentally quantify the impact of atmospheric neutrons on flash memory circuits. The key point of this approach is to use a high density of devices at wafer level instead of a collection of stand-alone packaged circuits; this is possible because floating-gate (FG) memories do not require any circuitry or power supply except during the writing and read operations. One can thus envisage exposing wafers to natural radiation and then periodically reading wafers on which a large number of memory arrays have been initially programmed and characterized. In the same study, Just et al. (2013) also developed a numerical simulation code capable of computing the SER of FG flash memories. Our simulation platform, named TIARA-G4 and described in (Autran et al. 2012a), has been adapted to flash memory architectures (TIARA-G4 NVM release for *nonvolatile memories*) by modifying the device/circuit 3D geometries and by implementing a model for charge loss from the FGs induced by ionizing particles (Autran et al. 2014). This chapter presents in detail the modeling and simulation approach as well as the code validation by comparison of numerical results with experimental data reported in Just et al. (2013).

11.2 Flash Memory Architectures and Electrical Operation

This section provides a brief introduction to the architecture and operating principles of FG flash memories at both device and circuit levels. Flash memory is an electronic nonvolatile storage device that can be electrically erased and reprogrammed; it offers fast read access times, as fast as dynamic RAM, although not as fast as static RAM or ROM. Flash memories are used in a wide variety of electronic devices for general storage, configuration data storage, or data transfer. Modern flash memories store logical information in an array of memory cells built from FG transistors. In traditional single-level cell devices, each cell stores only one bit of information. A newer type of flash memory, known as a multilevel cell device, can store more than one bit per cell by choosing between multiple levels of electrical charge to apply to the FGs of its cells.

As illustrated in Figure 11.1, the memory cell consists of a single n-channel transistor with a control gate (CG) and an electrically isolated polysilicon FG. The two gates are separated by an oxide–nitride–oxide dielectric stack (ONO), often called *inter-polyoxide*. Data can be stored in the cell by adding or removing electrons in the FG, which induces changes of the threshold voltage of the cell transistor. Charge injection into the FG through the tunnel oxide (TO) is governed by the electrical signals applied to the CG due to the electrostatic coupling existing between the two gates. Indeed, the electrostatic potential of the FG (V_{FG}; see Figure 11.2a) is directly determined by the potential of the CG and the amount of electrical charge stored in the FG. These operations require high-voltage signals produced on chip using special DC-to-DC converters (charge pumps) that use capacitors as energy storage elements to create higher voltages from the circuit external supply voltage. Two threshold voltage levels (V_T^0 and V_T^1; see Figure 11.2b) are considered to store one bit of information in the cell. The difference between the two levels, ΔV_T, is directly linked to the variation of the charge amount in the FG and to the coupling capacitance between the CG and FG electrodes. A reference voltage value V_T^{REF}, intermediate between V_T^0 and V_T^1, is considered as a demarcation level between the two logical states "0" and "1" (Figure 11.2b).

To form dense circuits with storage capacities up to several millions or billions of bits, elementary memory cells are arranged in a matrix, that is, in rows and columns. In addition, cells are generally grouped to form a hierarchical organization of increasing size: groups, blocks, pages, and so on. Lines, called *wordlines*, are connected to the CGs and columns, called *bitlines*, are connected to the drain terminals. Around the matrix memory is the peripheral control circuitry, composed of additional circuits for decoding cell addresses, generating high-voltage signals (charge pumps), reading cells (sense amplifiers), and managing circuit information. There are two main types of circuit architectures at the memory-plan level, called NOR and NAND gate flash memories, each corresponding to a certain configuration of associating several cells. The construction and operation of NOR and NAND flash memories are briefly described in Sections 11.2.1 and 11.2.2.

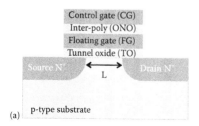
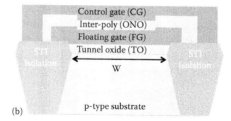

FIGURE 11.1

Schematic cross sections of a floating-gate transistor, acting as the elementary memory element in a flash memory circuit, along its length (a) and its width (b). (Reprinted from Autran, J.L. et al., Computational modeling and Monte Carlo simulation of soft errors in flash memories, in J. Awrejcewicz (ed.), *Computational and Numerical Simulations*, InTech, Vienna, 2014. © 2014 Autran et al., licensee InTech.)

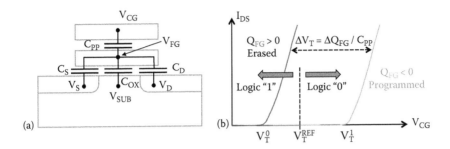

FIGURE 11.2
(a) Equivalent capacitance network of the floating-gate transistor with four terminals and definition of the main voltage and capacitance values. (b) Electrical characteristics $I_{DS}(V_{CG})$ of the floating-gate transistor with two different values of the floating-gate charge corresponding to erased and programmed states. (Reprinted from Autran, J.L. et al., Computational modeling and Monte Carlo simulation of soft errors in flash memories, in J. Awrejcewicz (ed.), *Computational and Numerical Simulations*, InTech, Vienna, 2014. © 2014 Autran et al., licensee InTech.)

11.2.1 NOR Architecture

The organization of the NOR gate flash, shown in Figure 11.3a, is as follows. Several cells are connected to a bitline: each cell has the source terminal connected directly to ground, and the drain terminal is connected directly to a bitline. The drain contacts of individual transistors connected to the bitline are shared between two adjacent cells. This setting of elementary devices is called *NOR flash* because it operates as a NOR gate. The default state of a single-level NOR flash cell is logically equivalent to a binary "1" value: when a suitable voltage is applied to the CG, the current flows through the channel and the bitline voltage is pulled down. The programming operation of a NOR flash cell (i.e., setting to a binary "0" value) is done by injection of hot carriers from the channel. The high current required by this mechanism limits the parallelism of the operation (not all cells can be programmed at the same time) (Gerardin et al. 2013). The programming procedure is as follows. A voltage increase (typically > 5 V) is applied to the CG, which turns on the channel, and the electrons can flow from source to drain (for an n-channel MOS transistor). The source-drain current is sufficiently high that a certain number of electrons of high energy are able to pass through the insulating layer on the FG by a hot-electron-injection mechanism. The erasing operation of a NOR flash cell (resetting it to the "1" state) is performed by the Fowler–Nordheim (FN) mechanism. For this purpose, a high voltage of the opposite polarity is applied between the CG and the source terminal, pulling out the electrons from the FG by FN tunneling. This organization of the NOR flash allows fast random access (~100 ns). The programming operation is carried out at block level and is much slower (~5 µs). The erasing operation is also carried out at block level and is even slower, typically 200 ms (Gerardin et al. 2013). Taking these characteristics into account, the NOR flash is used principally as a read-only memory, mainly for code storage, for which random access time is important, but where programming/erasing operations are rarely carried out (Gerardin et al. 2013). In the NOR architecture, the manufacturer guarantees that all individual bits are functional and meet retention and endurance specifications, as explained in Gerardin et al. (2013); no implementation of error correction code (ECC) is needed from the user side. In some cases (e.g., multilevel architecture), an internal ECC, totally transparent to the user, may be present. NOR devices typically have separate buses for addresses and data (Gerardin et al. 2013).

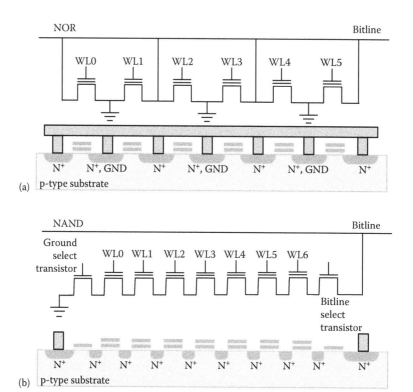

FIGURE 11.3
(a,b) Schematic representations of NOR and NAND architectures. (Reprinted from Autran, J.L. et al., Computational modeling and Monte Carlo simulation of soft errors in flash memories, in J. Awrejcewicz (ed.), *Computational and Numerical Simulations*, InTech, Vienna, 2014. © 2014 Autran et al., licensee InTech.)

11.2.2 NAND Architecture

The organization of the NAND gate flash is shown in Figure 11.3b. In this configuration, several groups of FG transistors are connected in series. These groups are then connected via some additional transistors to a NOR-style bitline array in the same way as single transistors are linked in the NOR flash. Due to this arrangement, the bitline is pulled low only if all wordlines are pulled high (above the threshold voltage of the transistors). This organization of elementary transistors is called NAND flash because transistors are connected in a way that is similar to a NAND gate. Compared with NOR flash, replacing single transistors with serial-linked groups adds an extra level of addressing. As explained in Gerardin et al. (2013), the series arrangement and the great level of parallelism that is achieved with this organization thanks to the low program/erase currents give rise to poor random access time but very good serial access. Programming of the NAND flash is performed by FN tunneling at the page level (which is typically a few kBytes) and is carried out in about 0.2 ms (Gerardin et al. 2013). The erasing operation is performed at the block level (typically a few MBytes) and takes about 2 ms (Gerardin et al. 2013). Block erasure is also carried out by FN tunneling, but by using opposite polarity. Thanks to these characteristics, flash NAND is better adapted to data storage, in which the problems of latency are minor and the random access time is not very important. In this configuration, the use of external ECC is mandatory (which increases the latency), because the manufacturer does not guarantee each single bit and the commercial devices may contain a few defective blocks (Gerardin et al. 2013).

11.3 Radiation Effects in Floating-Gate Memories

FG memories are sensitive to ionizing radiation, both total ionizing dose (TID) and single-event effects (SEEs). Very schematically, ionizing radiation induces charge loss in the FG and charge trapping in the different dielectric layers of the transistor stack; it can also generate interface states. The induced current transients and such parasitic charges and defects cause degradation of circuit functionality and loss of logical information stored in the FG array, in addition to possible global circuit performance degradation. Detailed results regarding both TID and SEEs in flash memories are available in recent papers and review presentations (Cellere et al. 2008, 2009; Gerardin et al. 2010, 2012, 2013; Gerardin and Paccagnella 2010; Paccagnella et al. 2009).

In this chapter, we will exclusively focus on soft errors induced by atmospheric neutrons in the FG array of flash memories. SEEs in FG memories are due to highly energetic particles that directly (heavy ions) or indirectly (neutrons) induce charge loss from the FG. Other effects may be possible, such as single-event functional interruptions (SEFI) or destructive events such as single-event gate rupture (SEGR) at the level of the FG array or in the peripheral circuitry. Note that SEEs only affect FG cells impacted by at least one particle, whereas TID uniformly impacts the programmed FG cells.

As explained in Chapter 5, neutrons are not ionizing (they do not directly create e^-/h^+ pairs in matter) and, specifically due to their neutral character, they can penetrate deeply into the chip's atomic structure. Only the resulting products of the neutron–silicon (n–Si) (or other atoms of the circuit, O, W, Al, etc.) collisions are ionizing and, in consequence, only the impact of such secondary products on the FG can result in charge loss. This is the reason why, in the following, we will focus on the underlying physical mechanisms of charge loss induced by ionizing particles, but the link with atmospheric neutrons remains evident.

Figure 11.4 illustrates the effects of ionizing-particle irradiation on the threshold voltage distributions of a large array of FG devices. Before irradiation, threshold voltages of

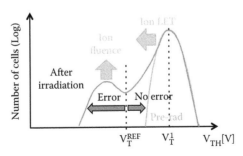

FIGURE 11.4
Schematic illustration of the effects of highly ionizing particle irradiation on the threshold voltage distributions of a floating-gate array. (Reprinted from Autran, J.L. et al., Computational modeling and Monte Carlo simulation of soft errors in flash memories, in J. Awrejcewicz (ed.), *Computational and Numerical Simulations*, InTech, Vienna, 2014. © 2014 Autran et al., licensee InTech.)

individual cells are distributed following a typical Gaussian distribution, sharply centered on the programmed value V_T^1. A secondary peak and a tail appear after irradiation: these structures correspond to all cells that have been hit by incident ionizing particles. The position (V_T shift) of the peak with respect to the initial distribution gives the average threshold voltage shift: it is directly linked to the ion linear energy transfer (LET) and to the electric field in the TO. The height of the peak is related to the irradiation fluence. Finally, the tail is related to all memory cells for which the V_T values are intermediate between the secondary peak and the initial distribution. Of course, all cells initially programmed at V_T^1 and having their post-irradiation V_T value below V_T^{REF} have been upset.

In their IRPS 2008 paper (Butt and Alam 2008), Butt and Alam reviewed several models of charge loss due to a radiation-particle strike. Different physical mechanisms have been proposed in the literature for modeling the charge loss from FGs after single radiation-particle strikes. The authors summarized these earlier models and underlined their strengths and limitations: such classical models include trap-assisted tunneling (TAT), the conductive pipe model, generation–recombination–transport in oxide, and electron emission. The most important limitation of these approaches is their failure to quantitatively predict the charge loss on the basis of a set of physics-based equations without any fitting parameter or phenomenological assumption. The authors have proposed a new model, called the transient carrier flux (TCF) model, which quantitatively explains the observed charge loss in FG memories irradiated with heavy ions. Figure 11.5 illustrates all these different physical mechanisms and models, which are briefly detailed in the following list.

1. *Generation/Recombination/Transport in Oxide*: The ion strike produces hot holes in the TO or inter–poly dielectric, and a certain fraction of these hot holes are not recombined in the prompt recombination phase. This model considers that the holes that are not recombined (Ma and Dressendorfer 1989) may drift into the FG. This may be possible, since the negative electron charge stored on the FG itself produces an electrical field across the oxide, which attracts holes. Holes that drift into the FG are recombined with electrons stored in the FG and cause a reduction in its negative charge. At the same time, the electrons produced by the ionizing particle are quickly transported to the silicon bulk or to the CG due to their high mobility. However, this model lacks sufficient experimental validation, because it does not agree quantitatively with data loss measured in FG flash memory cells (Cellere et al. 2006). Indeed, the number of holes that survive the prompt recombination after a heavy-ion strike in a 10 nm TO is fewer than 100, while the data show that the charge loss is a few thousand electrons (Butt and Alam 2008).

2. *Electron Emission*: This phenomenon was originally proposed by Snyder et al. as one of the main mechanisms of charge loss in FG electrically erasable programmable read-only memory (EEPROM) cells under gamma-ray irradiation (Snyder et al. 1989). The charge loss is explained by the fact that electrons stored in the FG can gain energy from ionizing radiation and can be emitted over the oxide barrier in the CG or in the silicon substrate. This mechanism is also called photoemission. The emission over the oxide has been empirically modeled. However, this mechanism has not been physically modeled or extended to heavy ions or other particles (Snyder et al. 1989). Moreover, in this model, the photoemission is limited only to electrons stored in the FG, which has no physical justification (Snyder et al. 1989). In fact, an ionizing-particle strike can generate a number of electrons much larger than the net number of electrons stored in the FG. Some of the electrons

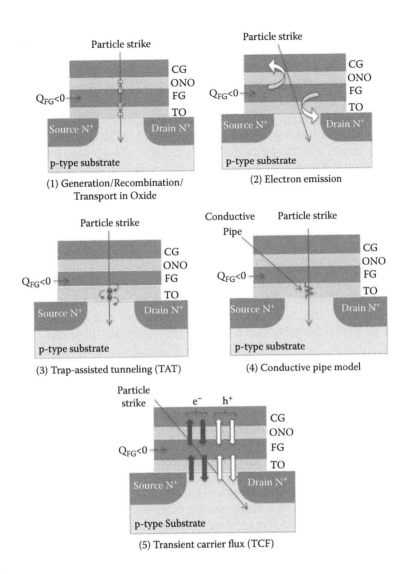

FIGURE 11.5
Schematic illustration of different models of charge loss due to a radiation particle strike. (Reprinted from Autran, J.L. et al., Computational modeling and Monte Carlo simulation of soft errors in flash memories, in J. Awrejcewicz (ed.), *Computational and Numerical Simulations*, InTech, Vienna, 2014. © 2014 Autran et al., licensee InTech.)

generated by the particle strike may have enough energy to be emitted over the oxide barriers (Snyder et al. 1989).

3. *Trap-Assisted Tunneling (TAT)*: This mechanism is one of the most important causes of oxide wearout due to electrical stress of program/erased cycles of a FG cell. When an ionizing particle strikes the cell, defects are created in the TO. These defects may provide a percolation path for the electrons, which can thus pass by a tunneling effect through the TO; therefore, this mechanism is called trap-assisted tunneling (TAT). It has been shown that TAT is responsible for retention problems in at least a certain percentage of irradiated devices (Snyder et al. 1989). However, a very long time is required to discharge a FG cell by the TAT mechanism (a few

hours to a few weeks) (Cellere et al. 2006). Therefore, the TAT mechanism cannot be responsible for SEU in the FG cell, taking into account that SEU data are taken immediately after cell irradiation and then do not change with time (Snyder et al. 1989). TAT may nevertheless result in hard errors that cause retention problems in the FG cell.

4. *Conductive Pipe Model*: This model has been proposed by Cellere et al. to explain charge loss due to heavy-ion strikes (Cellere et al. 2004; Ma and Dressendorfer 1989). This model assumes that the dense plasma of e–h pairs generated by the ion strike creates a temporary, very thin (~10 nm) conductive path in the TO for a short (subpicosecond) time after the strike. This is accompanied by a local lowering of the oxide energy barrier, which allows the electrons stored in the FG to pass through this conducting pipe. This phenomenological model reproduces well the experimental data of charge loss. However, there is a lack of physical explanation of the mechanisms governing both the resistance of the conductive path and the lowering of the oxide barrier (Butt and Alam 2008).

5. *Transient-Carrier-Flux (TCF) model*: This model was proposed by Butt and Alam (2008) to explain the charge loss due to an SEU in FG memory cells. In this model, it is assumed that the dominant physical mechanism that causes the FG charge loss due to a particle strike is the net flux of hot carriers flowing within a short time (~1 ps) over the oxide barrier at the FG/oxide interfaces. After a particle strike, a dense cluster of hot electron–hole pairs is generated, with carriers having broad energy distributions that return to thermal equilibrium in ~1 ps (Butt and Alam 2008). The tail of the high-energy distribution induces a transient carrier flow in and out of the FG over the tunnel and inter-polyoxides. In the case of a zero electric field in the oxide, the incoming and outgoing carrier flow balance each other at both oxide/FG interfaces, and therefore the net flux is zero. On the contrary, in the programmed state, the electron negative charge stored in the FG induces a relatively high electric field in the oxide. Due to this electric field, the electron flux leaving the FG is greater than the electron flux entering the FG. In addition, the incoming holes flux is greater than the holes flux exiting the FG. The net flux therefore causes a reduction of the number of electrons stored in the FG. A small imbalance between the incoming and outgoing fluxes may be sufficient to disturb the state of the memory cell, for which the tolerance of charge loss can be 100 electrons or fewer (Butt and Alam 2008). Butt and Alam validated their model by numerical simulation using a high-energy particle-physics-based toolkit—Geant4—for generation and initial energy distributions in the high-energy range (from approximately 10 eV to kiloelectronvolts). The hydrodynamic model, coupled with MC simulations, was used for carrier relaxation in the low-energy (<10 eV) range, in order to accurately take into account the energy relaxation due to phonon scattering and impact ionization (Butt and Alam 2008). The transient fluxes of hot carriers flowing in and out of the FG over the barrier oxides are calculated by self-consistently solving a system of equations, including the transmission probability through the oxides and the Poisson equation, until the carriers relax and reach thermal equilibrium. These fluxes are then used to obtain the charge loss in flash memory cells due to alpha particles and cosmic neutron strikes. Butt and Alam finally demonstrated that the TCF model is in very good agreement with experimental data from (Cellere et al. 2006), as will be shown in Section 11.4.2.

11.4 Modeling and Simulation of Nonvolatile Memories Using TIARA-G4 Platform

In this section, we describe in detail our modeling and numerical simulation approaches to computing the SER related to the FG array of a flash memory circuit (Just et al. 2013).

11.4.1 Description of TIARA-G4 NVM Platform

The Tool Suite for Radiation Reliability Assessment (TIARA) platform has been extensively detailed in Sections 9.2.3 and 9.3, and illustrated in Section 9.4. Up to now, TIARA-G4 has been used to simulate the interaction of various particles (including high-energy and thermal neutrons, protons, muons, alpha particles, and heavy ions) with different types of SRAM and Flip-Flop architectures (Autran et al. 2012a). In Just et al. (2013), the code has been partially rewritten and extended to the case of nonvolatile memories with FG devices as the sensitive memory elements. This new version of the code has been termed TIARA-G4 NVM. Figure 11.6 shows a schematic of the new TIARA-G4 NVM simulation flow structured into several independent modules and integrating new dedicated modules/subroutines to FG NVM devices. In particular, we wrote a new cell/circuit construction model to reproduce the flash-chip geometry (FG array) with high fidelity. A second

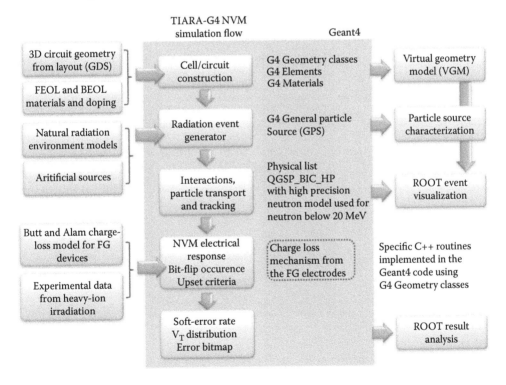

FIGURE 11.6
Schematics of the TIARA-G4 NVM simulation flow showing the different code inputs and outputs and the links with Geant4 classes, libraries, models, or external modules and visualization tools. (Reprinted from Autran, J.L. et al., Computational modeling and Monte Carlo simulation of soft errors in flash memories, in J. Awrejcewicz (ed.), *Computational and Numerical Simulations*, InTech, Vienna, 2014. © 2014 Autran et al., licensee InTech.)

dedicated module implementing a physical model for radiation-induced charge loss from the FG has been also developed, as detailed in Section 11.4.2.

To test the capability of the code to consider a real geometry, we based our developments on an NOR FG flash-memory architecture designed and fabricated by STMicroelectronics using a 90 nm CMOS process. This process is based on a borophosphosilicate glass (BPSG)-free back end of line (BEOL) that eliminates the major source of ^{10}B in the circuits and drastically reduces the possible interaction between ^{10}B and low-thermal-energy neutrons (Baumann and Smith 2001). Figure 11.7a shows a transmission electron microscopic (TEM) cross section of the FG devices along the transistor channel and Figure 11.7b shows a portion of the cell array layout at metal1/metal2 level. The elementary memory cell has an area of 0.18 μm². In TIARA-G4 NVM, the different transistor domains have been modeled as simple axis-aligned box volumes (Geant4 elements) of different materials (silicon, silicon dioxide, ONO, and BEOL stack), as illustrated in Figure 11.7c for a portion of the memory array. Figure 11.7d also shows a larger view of the array with different particle tracks interacting with certain FG stacks.

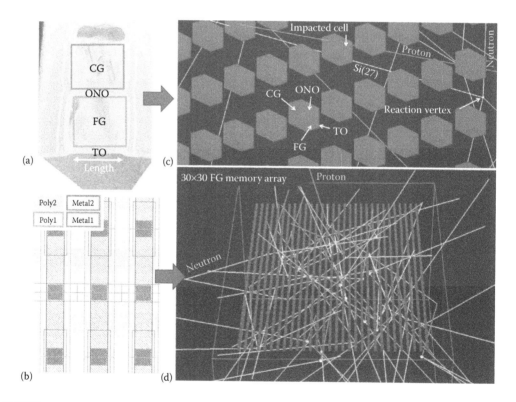

FIGURE 11.7

90 nm NOR floating-gate flash memory architecture considered in this work. (a) TEM cross section of the floating-gate transistor geometry along the transistor channel and (b) layout of the cell array at metal1/metal2 level. (c) and (d): ROOT (http://root.cern.ch) screenshots of a TIARA-G4 simulation showing detailed (c) and global (d) views of the memory array and different particle tracks resulting from interaction of atmospheric neutrons with circuit materials. For a better view at FG-cell level, BEOL materials (six metal levels), silicon substrate, and intracell silicon and dielectrics are not shown. (Reprinted from Autran, J.L. et al., Computational modeling and Monte Carlo simulation of soft errors in flash memories, in J. Awrejcewicz (ed.), *Computational and Numerical Simulations*, InTech, Vienna, 2014. © 2014 Autran et al., licensee InTech.)

11.4.2 Physical Model Considered

In complement to geometric aspects, we also implemented in TIARA-G4 NVM a new module describing the charge loss from FGs after single radiation-particle strikes. From the review of the different available models in literature presented in Section 11.3, our initial choice was to adopt the full physical model of the TCF proposed by Butt and Alam (2008). The original approach of these authors is based on complex simulations, in particular for the computation of carrier relaxation in the low-energy (<10 eV) range, using coupled hydrodynamic and MC simulations in order to correctly account for energy relaxation due to phonon scattering and impact ionization. This requires outsourcing from the main code the calculation of the charge loss from FG as a function of the incident particle properties.

An example of Butt and Alam's simulations is illustrated in Figure 11.8. Simulated curves reproduce experimental data very well without any fitting parameter (data extracted from figure 12 of Butt and Alam 2008). From a practical point of view, and in the absence of a relatively simple computational solution to implement Butt and Alam's model, we adopted a pragmatic approach, assuming that an ionizing particle of LET striking the FG produces a number of electron loss (NEL) given by the following analytical function:

$$NEL = A \times LET^2 + B \times LET \tag{11.1}$$

Figure 11.8 shows that Equation 11.1 is able to reproduce data very well. Of course, fitting coefficients A and B must be carefully evaluated for each device considered for the simulation from experimental measurements (heavy-ion irradiation) or complementary numerical simulation using Butt and Alam's complete computational procedure. The next release of TIARA-G4 NVM will integrate such an external dedicated module to confer on the code the capability to simulate a wide variety of NVM devices.

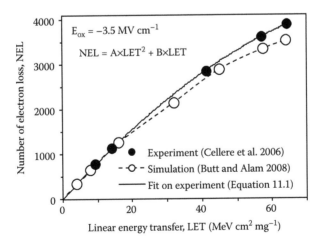

FIGURE 11.8
Number of electron loss (NEL) as a function of the particle LET for device T3 of Cellere et al. (*Journal of Applied Physics*, 99, 074101, 2006) under an oxide electric field of 3.5 MV cm^{-1}. Simulation results from Butt and Alam (*Proceedings of the IEEE International Reliability Physics Symposium*, pp. 547–555, 27 April–1 May, Phoenix, AZ. © (2008) IEEE.) are also reported. The full line corresponds to the fitting function (Equation 11.1) on experimental data. (Reprinted from Autran, J.L. et al., Computational modeling and Monte Carlo simulation of soft errors in flash memories, in J. Awrejcewicz (ed.), *Computational and Numerical Simulations*, InTech, Vienna, 2014. © 2014 Autran et al., licensee InTech.)

In the particular case of the present study, and by chance, Figure 11.8 is based on data from Cellere et al., who worked precisely on STMicroelectronics FG arrays. It has been found that device T3 in Cellere et al. (2006) is technologically very close to our circuit, with the same thicknesses for the different layers composing both the front-end-of-line (FEOL) and BEOL stacks. In order to consider the values given by Equation 11.1 for our memory devices, we introduced a scaling factor coefficient to take into account the difference in the dimensions of the FG polysilicon electrodes between the devices considered in Figure 11.8 and the present memory-cell architecture (a simple ratio of the volumes). Without any other calibration, we use Equation 11.1 to directly derive the threshold voltage shift, ΔV_T, resulting from a single particle strike in the FG domain using

$$\Delta V_T = \frac{q \times NEL}{C_{pp}} \tag{11.2}$$

where C_{pp} is the coupling capacitance between the FG and CG electrodes (see Figure 11.2a).

11.4.3 Simulation Results

Using TIARA-G4 NVM, we performed extensive MC simulations on large arrays of memory cells (up to 10^5 cells), considering the JEDEC atmospheric neutron source for high-energy incident neutrons above 1 MeV (see Figure 1.11) (JEDS89A 2006). Other simulations have also been performed using random generation of alpha particles inside the silicon material to mimic the presence of ^{238}U contamination at ppb level (we considered in this case the eight alpha-particle emitters of the ^{238}U decay chain) (Gedion et al. 2011).

Figure 11.7c and d show two simulation screenshots for a reduced matrix of 30×30 cells (to give a clearer view). A larger simulation view is shown in Figure 11.9 for several hundred

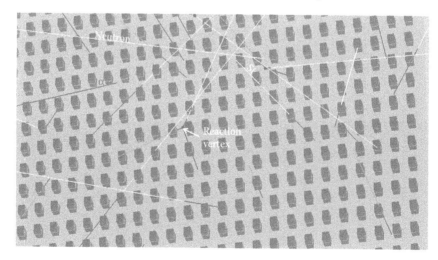

FIGURE 11.9
ROOT screenshots of a TIARA-G4 simulation showing several hundred memory cells and different particle tracks resulting from interaction of atmospheric neutrons with circuit materials. BEOL materials, intracell silicon, and dielectrics are not shown; silicon substrate corresponds to the gray background. (Reprinted from Autran, J.L. et al., Computational modeling and Monte Carlo simulation of soft errors in flash memories, in J. Awrejcewicz (ed.), *Computational and Numerical Simulations*, InTech, Vienna, 2014. © 2014 Autran et al., licensee InTech.)

cells. The screenshots illustrate the interaction of atmospheric neutrons with the circuit materials and the way in which the neutron-induced secondary particles can impact the memory cells (direct strikes on the FG electrodes). A large number of the events are induced by secondary particles generated in the proximity of the FEOL/BEOL interface, predominantly protons and silicon recoil nuclei. The BEOL stack is found to contribute marginally (<2%) to the total SER in spite of the presence of several layers and vias of high-density materials (W, Cu, Ta).

Figure 11.10 shows the simulated V_T distributions for 10^5 cells before and after irradiation. The initial distribution corresponds to a Gaussian distribution with the mean and standard deviation values calibrated on experimental data (see Section 11.5, Figure 11.13). The final distribution is the result of 10^9 incident neutrons (JEDEC spectrum) on the cell matrix, which corresponds to 50×10^6 h (i.e., more than 5700 years!) under natural atmospheric radiation at New York City, the reference location defined by a high-energy neutron flux of 20 n cm^{-2} h^{-1} (neutron energies above 1 MeV). One can observe the emergence of a typical neutron-induced tail on a large domain of V_T values below 7 V. This tail indicates that the V_T value has sufficiently decreased for a certain number of cells to appear outside the Gaussian distribution. Among them, some cells have shifted below the sense value fixed at 5.7 V: their state has thus changed from a logical point of view ($0 \rightarrow 1$ transition) and their number must be taken into account in the evaluation of the neutron-SER. For the other cells of the distribution tail, ΔV_T values are not sufficient to decrease their V_T below 5.7 V but are large enough to shift the cells outside the initial curve.

In complement to Figure 11.10, Figure 11.11 shows the threshold-voltage-shift distribution for all the cells of the simulated array. The peak at $\Delta V_T = 0$ V indicates that the great majority of the cells have not been impacted during the simulation run. For $\Delta V_T > 0$ V, the distribution decreases when ΔV_T increases. This directly reflects (cf. Equations 11.1 and 11.2) the LET distribution of the secondary particles (i.e., neutron by-products) striking the FGs: the lightest particles (protons, alphas) with low LET values (typically below 1.5 MeV cm^{-2} mg^{-1}) induce a large number of events characterized by a small or moderate ΔV_T shift

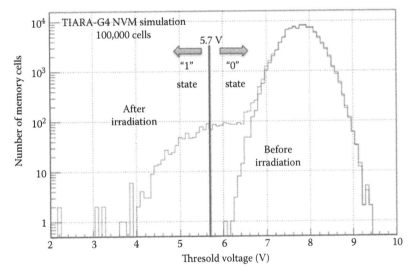

FIGURE 11.10
Distributions of V_T values computed by TIARA-G4 NVM for a population of 100,000 memory cells before and after irradiation with atmospheric neutrons. (Reprinted from Just, G. et al., Soft errors induced by natural radiation at ground level in floating gate flash memories, in *Proceedings of the IEEE International Reliability Physics Symposium*, 3D.4.1–3D.4.8, 14–18 April, Monterey, CA. © (2013) IEEE. With permission.)

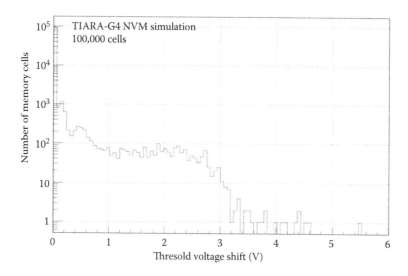

FIGURE 11.11

Distributions of ΔV_T values extracted from data of Figure 11.10. (Reprinted from Autran, J.L. et al., Computational modeling and Monte Carlo simulation of soft errors in flash memories, in J. Awrejcewicz (ed.), *Computational and Numerical Simulations,* InTech, Vienna, 2014. © 2014 Autran et al., licensee InTech.)

(<1 V); in contrast, particles with the highest LET values, much less numerous, induce the largest ΔV_T (>3 V). From the number of cells verifying $V_T < 5.7$ V after irradiation, the neutron-SER at sea level has been numerically evaluated as 7.7 (in arbitrary units, taking into account the common arbitrary scaling factor for confidentiality reasons). This value is expressed for the reference location (New York City).

For the alpha SER, a value of 0.12 (arbitrary units [a.u.]) has been obtained, considering a concentration of 0.2 ppb of ^{238}U uniformly distributed in the volume of circuit materials at both FEOL and BEOL levels. This concentration was directly deduced from experimental emissivity measurements (see Section 11.5).

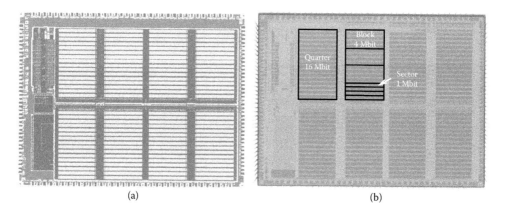

(a) (b)

FIGURE 11.12

Layout (a) and die (b) of the ANNA test chip (area 9.230×7.044 mm^2) fabricated by STMicroelectronics in CMOS 90 nm technology. The memory array is segmented into 32 blocks of 4 Mbits or 128 sectors of 1 Mbits (total capacity of 128 Mbits per chip). (Reprinted with permission from Just, G. et al., Soft errors induced by natural radiation at ground level in floating gate flash memories, in *Proceedings of the IEEE International Reliability Physics Symposium,* 3D.4.1–3D.4.8, 14–18 April, Monterey, CA. © (2013) IEEE.)

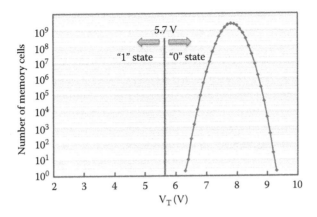

FIGURE 11.13
Initial distribution of the measured threshold voltage V_T values for all the programmed FG memory cells (all "0" pattern) related to a series of five wafers. (Reprinted from Autran, J.L. et al., Computational modeling and Monte Carlo simulation of soft errors in flash memories, in J. Awrejcewicz (ed.), *Computational and Numerical Simulations*, InTech, Vienna, 2014. © 2014 Autran et al., licensee InTech.)

11.5 Experimental Characterization

In parallel to this work of modeling and numerical simulation, previously described, we launched an experimental verification procedure to estimate the circuit SER from direct measurements. For this, we considered a large collection of NOR FG flash-memory circuits fabricated by STMicroelectronics using a 90 nm CMOS process. Circuits were directly operated and characterized at wafer level.

The test chip, named macrocell ANNA and shown in Figure 11.12, is a 128 Mbit array of memory cells organized in 1 Mbit sectors, 4 Mbit blocks and 16 Mbit quarters without ECC. Several tens of macrocells are available per test wafer (200 mm wafers); more than 50 Gbits (~20 wafers) were used and fully characterized for the present experiment.

The test began with an initial wafer-level characterization at ST-Rousset (near Marseille) of all the circuits using a high-performance tester (Verigy® V93000 platform). The test platform uses high-precision voltage sources and parameter analyzers calibrated before each measurement campaign: the accuracy of V_T extraction is guaranteed to be less than 10 mV. Memory arrays have been written (all "0" pattern) and then read several times, allowing the compilation of a reference threshold voltage (V_T) mapping for all the test chips, cell per cell and wafer per wafer. The corresponding numerical data are stored on a hard disk bay. During this initial characterization, all the wafers were also submitted to a 24 h bake at 250°C followed by a new V_T characterization in order to identify (and thus to eliminate) all the test chips exhibiting electrical instabilities or abnormal FG charge loss. Figure 11.13 shows a typical V_T distribution, sharply centered around 7.8 V for a population of memory cells corresponding to all functional test chips for a series of five wafers (same technological lot). The reproducibility (i.e., repeatability) of such an electrical characterization has been attested by the fact that repeated measurements on the same wafer show exactly the same V_T distribution within measurement margins (<10 mV), cell per cell.

In addition to this initial electrical characterization, we also performed alpha-emissivity measurements at wafer level using a XIA UltraLo-1800 alpha-particle counter (described

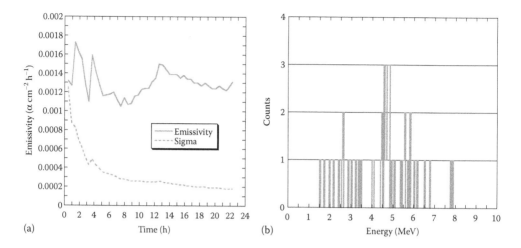

FIGURE 11.14

Alpha-particle-emissivity characterization of 90 nm flash-memory wafers using a XIA UltraLo-1800 alpha-particle counter (described in Section 4.4). (a) Emissivity and measurement error sigma as a function of measurement duration. (b) Energy distribution of the detected alpha particles emitted from the surface of the wafers (fully processed wafers). (Reprinted from Autran, J.L. et al., Computational modeling and Monte Carlo simulation of soft errors in flash memories, in J. Awrejcewicz (ed.), *Computational and Numerical Simulations*, InTech, Vienna, 2014. © 2014 Autran et al., licensee InTech.)

in Section 4.4). Figure 11.14 shows the results of this characterization in terms of emissivity and measurement error (Figure 11.14a) and energy distribution (Figure 11.14b) of the emitted alpha particles from the fully processed wafers. An emissivity level of 0.0013 α cm^{-2} h^{-1} was measured, which corresponds to a concentration of 0.2 ppb ^{238}U uniformly distributed in the volume of circuit materials at both FEOL and BEOL levels. This correspondence was estimated using a recently developed reverse-alpha-particle-emissivity analytical model (Martinie et al. 2011).

After the initial characterization, approximately one half of the total number of wafers were stored in Rousset and the second half were delivered to an altitude test site by express mail and exposed to natural radiation. Figure 11.15 shows the flowchart of this test method, which illustrates the sequencing of the different characterization and wafer-transportation steps. Two different radiation environments were thus considered: the first at sea level in Rousset for reference and the second at altitude on the ASTEP platform (www.astep.eu). The two sites are characterized by a relative atmospheric neutron flux of 1.04 and 6.02, respectively, with respect to New York City (Lei 2012; Autran et al. 2012b). After an exposure period of several months, the wafers stored at ASTEP (see Figure 11.16) were delivered to ST-Rousset for complete electrical characterization. Those remaining in Rousset were also measured at the same time. The complete characterization loop (Rousset → ASTEP → Rousset) was repeated three times for the present study.

Figure 11.17 shows the results for the two series of wafers exposed in Rousset and on the ASTEP platform. Three reading operations were performed on the wafers stored in Rousset, after 5, 12, and 18 months of exposure, respectively. Similarly, two reading operations were performed on the ASTEP wafers, after 5 and 12 months of natural irradiation at altitude.

For the wafers exposed at sea level, only a single memory cell on a total of more than several tens of gigabits was detected, with a V_T value changing at $t_0 + 12$ months and becoming inferior to the reference value ($V_T^{REF} = 5.7$ V) delimiting the "0" and "1" logical states. For this memory cell, the threshold voltage shifted from 8.0 to 4.5 V. Likewise, two and three

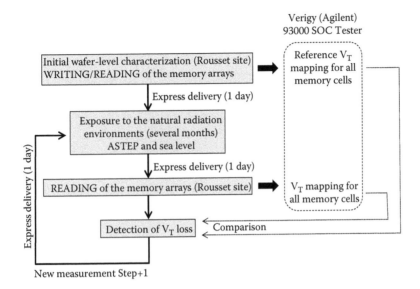

FIGURE 11.15
Flowchart of the multisite characterization technique developed to evaluate the soft-error rate of flash memories written and read at wafer level using a Verigy® V93000 platform. (Reprinted from Just, G. et al., Soft errors induced by natural radiation at ground level in floating gate flash memories, in *Proceedings of the IEEE International Reliability Physics Symposium*, 3D.4.1–3D.4.8, 14–18 April, Monterey, CA. © (2013) IEEE. With permission.)

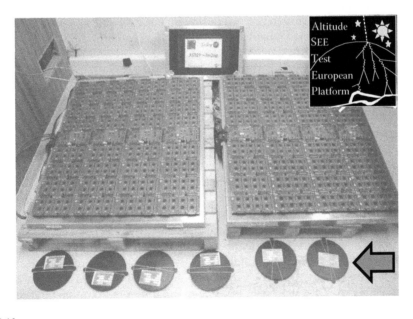

FIGURE 11.16
Global view of one of the ASTEP experimental rooms showing, in the foreground, six wafers of flash memories stored on the ground during their exposure to natural radiation on the ASTEP platform and, in the background, a real-time test setup based on 40 nm SRAM circuits presented in Section 7.4.2. (Reprinted from Just, G. et al., Soft errors induced by natural radiation at ground level in floating gate flash memories, in *Proceedings of the IEEE International Reliability Physics Symposium*, 3D.4.1–3D.4.8, 14–18 April, Monterey, CA. © (2013) IEEE. With permission.)

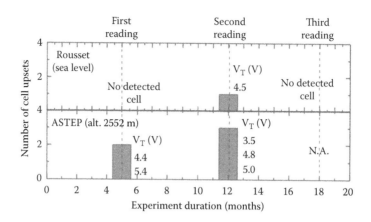

FIGURE 11.17
Number of memory cells with shifted V_T below the reference value (5.7 V) delimiting the "0" and "1" logical states and detected during the first and the second wafer readings. (Reprinted from Just, G. et al., Soft errors induced by natural radiation at ground level in floating gate flash memories, in *Proceedings of the IEEE International Reliability Physics Symposium*, 3D.4.1–3D.4.8, 14–18 April, Monterey, CA. © (2013) IEEE. With permission.)

shifted-V_T cells were detected on the ASTEP wafers after 5 months and 1 year of exposure, respectively. Measured V_T values for these flipped cells are also reported in Figure 11.17.

A detailed analysis of the analogic V_T bitmaps (not shown) for all these impacted cells showed that these latter correspond to isolated cells (i.e., adjacent cells were not impacted) randomly distributed in the FG array and on the exposed wafers. A more detailed investigation of the complete V_T distributions shows that several other cells were potentially impacted during their exposure to natural radiation. Figure 11.18 shows such a distribution for the whole cell population exposed at ASTEP. At $t_0 + 12$ months, two groups of impacted cells can be distinguished: a first group of three cells, labeled A, which corresponds to the $0 \rightarrow 1$ flipped cells reported in Figure 11.17 (bottom graph, at $t_0 + 12$), and a second group of six cells, labeled B, for which the V_T have shifted but not enough to cross the limit of 5.7 V delimiting the two binary states "0" and "1."

From the data shown in Figure 11.17, obtained at two different locations, the global SER and its two components can be determined, as suggested in Puchner (2011) (see also Section 7.2.2). The two components are the n-SER, taking into account the contribution of atmospheric neutrons to the SER, and the so-called α-i-SER, accounting for all the internal failure mechanisms in the chips, including the possible contribution of alpha-particle emitters. Indeed, several physical intrinsic mechanisms can be invoked to explain the long-term charge loss generally observed in FG devices, in particular different leakage mechanisms through the TO or through the ONO inter–poly dielectric based on various possible trap/defect-assisted tunneling mechanisms (Kim and Choi 2004). These latter are not inevitably related to radiation effects, but can also be linked to material properties or induced by the technological process or by an electrical stress. This is the reason why the second contribution to the SER is here called α-i-SER and not only α-SER. We thus have a system with two equations and two unknown quantities:

$$\alpha - i - SER + AF_{Rousset} \times n - SER = SER_{Rousset} \tag{11.3}$$

$$\alpha - i - SER + AF_{ASTEP} \times n - SER = SER_{ASTEP} \tag{11.4}$$

where $AF_{Rousset} = 1.04$ and $AF_{ASTEP} = 6.02$ are the neutron flux AFs, as previously reported.

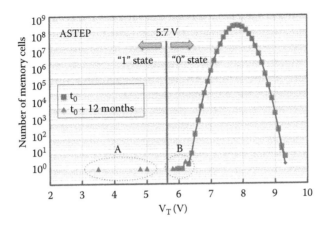

FIGURE 11.18
Comparison between the two distributions of V_T values measured at t_0 and at $t_0 + 12$ months for population of programmed memory cells exposed to natural radiation on the ASTEP platform. (Reprinted from Just, G. et al., Soft errors induced by natural radiation at ground level in floating gate flash memories, in *Proceedings of the IEEE International Reliability Physics Symposium*, 3D.4.1–3D.4.8, 14–18 April, Monterey, CA. © (2013) IEEE. With permission.)

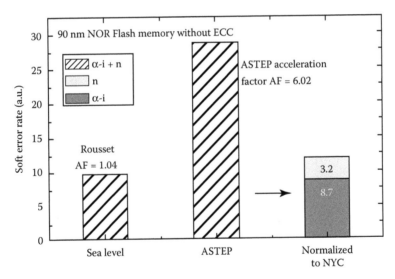

FIGURE 11.19
Summary of the SER deduced from data of Figure 11.17 for sea-level and ASTEP conditions. The two components of the SER are given for normalized New York City conditions. SER values are in a.u. for confidentiality reasons, but the order of magnitude of these values is a few hundred FIT per GBit. (Reprinted from Just, G. et al., Soft errors induced by natural radiation at ground level in floating gate flash memories, in *Proceedings of the IEEE International Reliability Physics Symposium*, 3D.4.1–3D.4.8, 14–18 April, Monterey, CA. © (2013) IEEE. With permission.)

Figure 11.19 shows the results of this SER extraction, considering the results in Figure 11.17 and the durations and memory capacities related to the different experiments. Global SER values of 9.7 and 28.8 a.u. are obtained for the Rousset (sea-level) and ASTEP (altitude) experiments, which leads to an estimation of α-i-SER = 3.2 and n-SER = 8.7 a.u.

These results demonstrate a very limited impact of atmospheric radiation on the total SER without ECC, typically in the range 10–100 FIT/GBit. With respect to all other internal failure mechanisms, the external natural radiation constraint is found to represent less than one third (27%) of the total SER. Note that all these SER values are found strictly equal to 0 if ECC is activated on the chips, due to the fact that only rare events, always corresponding to single-cell upsets, have been detected.

11.6 Experimental versus Simulation Results: Discussion

In this last section, we conclude by comparing these experimental results with predictive values obtained using the TIARA-G4 NVM simulation platform. Figure 11.20 summarizes this comparison for the different defined SER components.

A good agreement is found for the neutron-SER, taking into account all experimental and simulation uncertainties, in the first instance the relatively weak statistics of the experiment in terms of number of events detected. Indeed, despite the duration of the experiment (18 months) and the huge quantity of data to manipulate (the individual V_T evolution of more than 50 Gbits of memory cells has been stored and processed), the statistics of this first experiment remain relatively weak because of the extremely low rate of cell flips in this kind of memory.

For alpha SER, the discrepancy between the two values is striking. This confirms our initial precaution in naming the second extracted component of the SER (Figure 11.19) α-i-SER instead of classically α-SER, because, in the present case of FG devices, this component may be the result of other intrinsic failure mechanisms occurring in parallel inside the chips. From literature (Cappelletti and Golla 1999; Pavan et al. 2004; Van Houdt et al. 2007), we can invoke different intrinsic or extrinsic leakage current mechanisms though the dielectric layers present in the FG stack (TO, ONO, spacers). Intrinsic mechanisms that

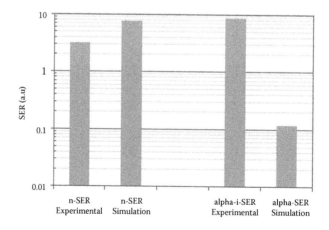

FIGURE 11.20

Comparison of the SER component values obtained by TIARA-G4 NVM simulation and from exposure to natural radiation in Rousset and ASTEP. (Reprinted from Just, G. et al., Soft errors induced by natural radiation at ground level in floating gate flash memories, in *Proceedings of the IEEE International Reliability Physics Symposium*, 3D.4.1–3D.4.8, 14–18 April, Monterey, CA. © (2013) IEEE. With permission.)

contribute to charge loss are field-assisted electron emission, thermionic emission, and electron detrapping. Extrinsic mechanisms are essentially oxide defects that can form conductive paths through a given dielectric. Whatever the mechanism or, eventually, the activation of several leakage paths, our results suggest that these electrical processes appear to be dominant in the observed failure rate with respect to the contribution of alpha-particle internal emission. This point will have to be carefully reevaluated in future work.

Another interesting point of comparison comes from the ratio of the numbers of upset cells to the numbers of cells for which V_T has shifted, but not enough to cross the limit of 5.7 V delimiting the two logical states. Although statistics are low for the data in Figure 11.18, the ratio (number of cells B/number of cells A) can be roughly evaluated as 50%. From simulation results with much larger statistics, this ratio is 40.7%, which is clearly in the same order of magnitude. Beyond the fact that this point consolidates the comparison between experiment and simulation, this result shows that the number of impacted cells with a final V_T ranging between the sense voltage value and the edge of the initial Gaussian distribution is approximately twice as large as the number of cells verifying the upset criterion.

References

Autran, J.L., Munteanu, D., Gasiot, G., and Roche, P. 2014. Computational modeling and Monte Carlo simulation of soft errors in flash memories. In J. Awrejcewicz (ed.), *Computational and Numerical Simulations*. Vienna: InTech.

Autran, J.L., Semikh, S., Munteanu, D., Serre, S., Gasiot, G., and Roche, P. 2012a. Soft-error rate of advanced SRAM memories: Modeling and Monte Carlo simulation. In M. Andriychuk (ed.), *Numerical Simulation—from Theory to Industry*. Vienna: InTech.

Autran, J.L., Serre, S., Munteanu, D., et al. 2012b. Real-time soft-error testing of 40 nm SRAMs. In *Proceedings of the IEEE International Reliability Physics Symposium*, pp. 3C.5.1–3C.5.9. 15–19 April, Anaheim, CA, IEEE.

Baumann, R. and Smith, E. 2001. Neutron-induced ^{10}B fission as a major source of soft errors in high density SRAMs. *Microelectronics Reliability* 41:211–218.

Butt, N.Z. and Alam, M. 2008. Modeling single event upsets in floating gate memory cells. In *Proceedings of the IEEE International Reliability Physics Symposium*, pp. 547–555. 27 April–1 May, Phoenix, AZ, IEEE.

Cappelletti, P. and Golla, C. 1999. *Flash Memories*. Norwell, MA: Kluwer Academic.

Cellere, G., Gerardin, S., Bagatin, M., et al. 2008. Neutron-induced soft errors in advanced flash memories. In *Proceedings of the IEEE International Electron Devices Meeting Technical Digest*, pp. 1–4. 15–17 December, San Francisco, CA, IEEE.

Cellere, G., Gerardin, S., Bagatin, M., et al. 2009. Can atmospheric neutrons induce soft errors in NAND floating gate memories? *IEEE Electron Device Letters* 30:178–180.

Cellere, G., Paccagnella, A., Visconti, A., and Bonanomi, M. 2006. Subpicosecond conduction through thin SiO_2 layers triggered by heavy ions. *Journal of Applied Physics* 99:074101.

Cellere, G., Paccagnella, A., Visconti, A., Bonanomi, M., and Candelori, A. 2004. Transient conductive path induced by a single ion in 10 nm SiO_2 layers. *IEEE Transactions on Nuclear Science* 51:3304–3311.

Gedion, M., Wrobel, F., Saigné, F., and Schrimpf, R.D. 2011. Uranium and thorium contribution to soft error rate in advanced technologies. *IEEE Transactions on Nuclear Science* 58:1098–1103.

Gerardin, S., Bagatin, M., Ferrario, A., et al. 2012. Neutron-induced upsets in NAND floating gate memories. *IEEE Transactions on Device and Materials Reliability* 12:437–444.

Gerardin, S., Bagatin, M., Paccagnella, A., et al. 2010. Scaling trends of neutron effects in MLC NAND flash memories. In *Proceedings of the IEEE International Reliability Physics Symposium*, pp. 400–406. 2–6 May, Anaheim, CA, IEEE.

Gerardin, S., Bagatin, M., Paccagnella, A., et al. 2013. Radiation effects in flash memories. *IEEE Transactions on Nuclear Science* 60:1953–1969.

Gerardin, S. and Paccagnella, A. 2010. Present and future non-volatile memories for space. *IEEE Transactions on Nuclear Science* 57:3016–3039.

JEDS89A. 2006. Measurement and reporting of alpha particle and terrestrial cosmic ray-induced soft errors in semiconductor devices, JEDEC Standard JESD89A, October 2006. Available at: http://www.jedec.org/download/search/jesd89a.pdf.

Just, G., Autran, J.L., Serre, S., et al. 2013. Soft errors induced by natural radiation at ground level in floating gate flash memories. In *Proceedings of the IEEE International Reliability Physics Symposium*, pp. 3D.4.1–3D.4.8. 14–18 April, Monterey, CA, IEEE.

Kim, J.H. and Choi, J.B. 2004. Long-term electron leakage mechanisms through ONO interpoly dielectric in stacked-gate EEPROM cells. *IEEE Transactions on Electron Devices* 51:2048–2053.

Lei, F. 2012. Quotid Atmospheric Radiation Model—QARM (previously Qinetiq Atmospheric Radiation Model). Available at: http://86.4.255.185:8080/qarm/.

Ma, T.M. and Dressendorfer, P.V. 1989. *Ionization Radiation Effects in MOS Device and Circuit*. New York: Wiley.

Martinie, S., Autran, J.L., Munteanu, D., Wrobel, F., Gédion, M., and Saigné, F. 2011. Analytical modeling of alpha-particle emission rate at wafer-level. *IEEE Transactions on Nuclear Science* 58:2798–2803.

Paccagnella, A., Bagatin, M., Cellere, G., and Gerardin, S. 2009. Flash memories and soft errors at ground level. In RADSOL Workshop, June 2009.

Pavan, P., Larcher, L., and Marmiroli, A. 2004. *Floating Gate Devices: Operation and Compact Modeling*. New York: Springer.

Puchner, H. 2011. Correlation of life testing to accelerated soft error testing. In *Third Annual IEEE-SCV Soft Error Rate (SER) 2011*, Workshop, San Jose, CA, IEEE.

Snyder, E.S., McWhorter, P.J., Dellin, T.A., and Sweetman, J.D. 1989. Radiation response of floating gate EEPROM memory cells. *IEEE Transactions on Nuclear Science* 36:2131–2139.

Van Houdt, J., Degraeve, R., Groeseneken, G., and Maes, H.E. 2007. Physics of flash memories. In J.E. Brewer and M. Gill (eds), *Nonvolatile Memory Technologies with Emphasis on Flash: A Comprehensive Guide to Understanding and Using NVM Devices*. Hoboken, NJ: Wiley.

12

SOI, FinFET, and Emerging Devices

12.1 Introduction

The microelectronics industry has experienced tremendous progress in the last 40 years, especially with regard to the evolution of the performance of the products (i.e., integrated circuits) and, at the same time, the drastic reduction of manufacturing costs by elementary integrated function. So far, this considerable growth of the semiconductor industry has been due to its technological capability to constantly miniaturize the elementary components of circuits, namely the metal-oxide-semiconductor field effect transistor (MOSFET), the basic building block of very-large-scale-integration (VLSI) integrated circuits. The continuous decrease of the silicon surface occupied by these elementary components has kept the speed of integration at the pace dictated by the famous *Moore's law*, which states that the number of transistors per integrated circuit doubles every 18–24 months (Moore 1965). However, in recent decades conventional bulk MOSFET scaling has encountered serious physical and technological limitations, mainly related to gate-oxide (SiO_2) leakage currents (Gusev et al. 2006; Taur et al. 1997), a large increase in parasitic short-channel effects, and dramatic mobility reduction (Fischetti and Laux 2001) due to highly doped silicon substrates necessarily used to reduce these short-channel effects. Technological solutions have been proposed in order to continue to use the *bulk solution* until the 32–38 nm nodes (ITRS 2013). Most of these solutions have introduced high-permittivity gate dielectric stacks (to reduce the gate leakage) (Houssa 2004), a midgap metal gate (to suppress the silicon-gate polydepletion-induced parasitic capacitances), and strained silicon channels (to increase carrier mobility) (Rim et al. 1998). However, in parallel to these efforts, alternative solutions to replace the conventional bulk MOSFET architecture have been proposed and studied in the recent literature. These options are numerous, and can be classified in general according to three main directions: (i) use of new materials in the continuity of the bulk solution, allowing MOSFET performance to be increased due to their dielectric properties (permittivity), electrostatic immunity (silicon-on-insulator materials), mechanical (strain), or transport (mobility) properties; (ii) a complete change of device architecture (e.g., multiple-gate devices, silicon-nanowire MOSFETs) allowing better electrostatic control, and, as a result, intrinsic channels with higher mobilities and currents; and (iii) exploitation of certain new physical phenomena that appear at the nanometer scale, such as quantum ballistic transport, substrate orientation, or modifications of the material band structure in devices/wires with nanometer dimensions (Haensch et al. 2006; Hiramoto et al. 2006).

The objective of this chapter is to provide a survey of the most recent research work in the domain of single-event effects (SEE) applied to advanced CMOS technologies, such as fully depleted silicon-on-insulator (FDSOI), FinFET, or triple-gate or multigate configurations based on ultrathin films or nanowire channels.

12.2 Silicon-on-Insulator (SOI) Technologies

12.2.1 SEE Mechanisms in SOI Technologies

The most widely used material for the fabrication of integrated circuits in microelectronics is bulk silicon. In MOS transistors made on bulk silicon (Figure 12.1a), also called bulk MOSFETs, only a very thin region (inversion channel), near the interface, is used for electronic conduction; the rest of the wafer, more than 99.9% of the substrate volume, is not used. Unlike the bulk technology, the silicon-on-insulator (SOI) technology uses only a thin silicon layer (called *silicon film*) carried by a more or less thick insulating layer, beneath which there is usually a silicon substrate (Figure 12.1b). Therefore, the SOI material is composed of a stack of two or three layers:

1. A monocrystalline silicon film, on which integrated circuits are fabricated and which is the unique *active* layer (which effectively participates in device operation)
2. An insulating layer, which may be either a thick completely insulating substrate (e.g., sapphire) or, more often, a thin insulating layer, made of silicon oxide and called *buried oxide*, that separates the silicon film from the underlying thick silicon substrate
3. A thick layer of silicon, called *substrate*, which contributes slightly to device operation and essentially acts as a mechanical support

The principle of operation of the SOI MOSFET (Figure 12.1b) is similar to that of the bulk MOSFET, with some differences resulting from its structure. The conduction can be controlled by two gates (front and back gates; Figure 12.1b), each creating a depletion region. According to the ratio between the silicon film thickness and the depth of the two existing

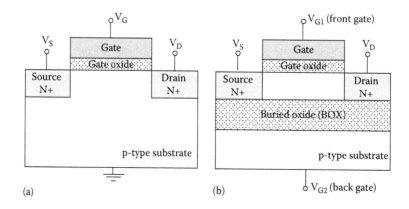

FIGURE 12.1
Schematic structure of the NMOS transistor fabricated on (a) bulk silicon and (b) silicon-on-insulator (SOI) wafers.

depletion regions, SOI MOSFETs can be classified into two categories: partially depleted and fully depleted. In partially depleted SOI transistors (PDSOI), the film thickness is large enough for the two depletion regions to be independent and there is still a neutral region in the silicon film, regardless of the two gates' biases. For relatively thin silicon films, the depletion region under the channel extends to the buried oxide and covers the entire silicon film when the gate is biased in strong inversion. The device is then called FDSOI. Therefore, the depletion charge is geometrically limited and no longer varies with the gate voltage, constituting a major difference from the conventional bulk MOSFET. A strong coupling between the gate and the inversion charge in the channel is then obtained, providing efficient control of the potential in the channel (especially when the ratio L/t_{Si} is higher than 4) and low power consumption.

SOI technology, and more precisely its older version, SOS (silicon-on-sapphire) technology, was originally proposed as a solution to the problem of bulk-device sensitivity to ionizing radiation (Cristoloveanu and Li 1995) and, in particular, to transient effects. As explained in the previous chapters, a transient effect arises, for example, when the semiconductor is crossed by high-energy particles. Electron–hole pairs are generated in the crossing region, inducing a very important transient current in the device (Colinge 1997). In a bulk MOSFET, this parasitic current can flow through the active area of the device, which may seriously affect its operation. On the contrary, in SOI devices the active region of the silicon film is completely isolated from the substrate, so the current induced by the ionizing particle has a much smaller influence on device performance. The first applications of the SOI technologies were therefore designed for high densities of ionizing radiation, such as encountered in military or civil nuclear applications. But, apart from their resistance to irradiation, it was observed that the source/substrate and drain/substrate capacities are four to seven times lower in SOI devices than in bulk MOSFETs (Colinge 1997). While in bulk transistors the source/substrate and drain/substrate capacities are those of reverse-biased junctions, in SOI devices these capacities are dominated by the capacity of the buried oxide under the source or drain regions, which is much lower (Kado 1997). Reduction of parasitic capacitances is reflected by an increase in circuit speed. Therefore, SOI MOSFETs exhibit significantly improved high-frequency performances compared with bulk transistors, which has attracted the interest of the scientific and industrial community.

It was shown later that the power dissipated in the SOI technology is greatly reduced compared with the bulk technology, and the SOI technology provides a higher density of integration. Indeed, the existence of the buried oxide in the SOI material enables the total separation of adjacent devices. This leads to the elimination of certain parasitic phenomena existing in bulk devices, such as the latchup effect, which consists in triggering the parasitic thyristor (illustrated in Figure 12.2a by two bipolar transistors) (Young and Burns 1988). As a result, a strong current occurs between the source and drain that may damage the device. The latchup triggering is related to the distance between the N+ and P+ regions of the two neighboring N-channel MOSFET (NMOS) and P-channel MOSFET (PMOS) transistors; to avoid this, it is necessary to maintain a certain limiting distance between these two regions. This leads to a lower integration density in bulk technologies. In the SOI technology, the N+ and P+ regions are completely isolated (Figure 12.2b), and breakdown between these two regions is impossible regardless of the distance that separates them; then the latchup effect is completely eliminated. In addition, the absence of wells allows the P+ and N+ regions of PMOS and NMOS transistors to be brought closer together. In consequence, the total separation of devices in the SOI technology allows a much higher integration density to be obtained than in the bulk technology (Morishita et al. 1999). The absence of wells also leads to the simplification of the interconnection levels. Due to all

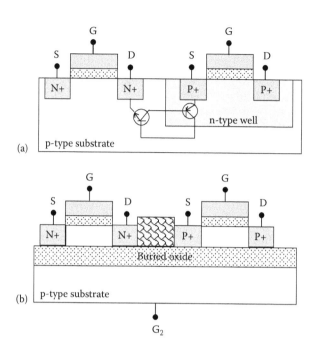

FIGURE 12.2
Schematic representation of CMOS structures fabricated on bulk silicon (a) and SOI (b). Illustration of the parasitic thyristor inducing the latchup phenomenon in bulk CMOS. Due to the total isolation between NMOS and PMOS transistors assured by the buried oxide, the latchup is completely eliminated in the SOI technology.

these advantages, the SOI technology has progressively became a very attractive candidate for integrated circuit fabrication.

Recently, due to the extremely rapid development of VLSI circuits, the size of CMOS devices has been greatly reduced. At the same time, because of the emergence and rapid expansion of portable applications, it has become imperative to design low-voltage/low-power VLSI circuits. SOI has emerged as the most appropriate technology for the deeply-submicron range, allowing the integration of VLSI circuits for low-voltage, low-power, and high-frequency applications. After more than three decades of research studies on materials and devices, SOI wafers have definitely entered into the mainstream of microelectronics. MOS transistors on SOI show reinforced immunity to short-channel effects and offer new possibilities for reducing device dimensions to the nanometer scale (sub-20 nm channel length).

Regarding sensitivity to ionizing radiation, charge collection in SOI devices is limited to the silicon film, which is very thin compared with bulk devices. This is why SOI devices are naturally resistant to single events. However, the unique configuration of SOI technology is responsible for a certain number of drawbacks, such as floating-body phenomena, which are not present in the bulk technology. Floating-body phenomena are parasitic effects specific to SOI devices, including:

1. Current transient effects, such as drain-current *overshoots* and *undershoots* (Munteanu et al. 1998; Munteanu and Cristoloveanu 1999)

2. Kink effect (Cristoloveanu and Li 1995)

3. Parasitic bipolar amplification (Ferlet-Cavrois et al. 2002b; Massengill et al. 1990; Musseau et al. 1994; Schwank et al. 2003)

These parasitic phenomena are due to the presence of the buried oxide, which prevents the evacuation of excess charges created by impact ionization (for the kink effect), by the sudden extension of the depletion region due to a voltage step (for the drain-current overshoot), or by an ionizing particle crossing the device (for the parasitic bipolar effect). In all these cases, excess charges are accumulated in the silicon film and considerably modify the electrostatic potential, inducing the parasitic phenomena described above. These phenomena do not occur in the bulk technology, since the charges are quickly evacuated via substrate contact.

The parasitic bipolar effect is an essential phenomenon for the sensitivity of SOI devices and circuits to single events (Massengill et al. 1990; Schwank et al. 2003). As will be explained later, SOI devices are not intrinsically immune to radiation environment due to the bipolar amplification phenomenon, even though they have sensitive volumes significantly smaller than those of bulk transistors.

12.2.1.1 Bipolar Amplification

The basic mechanism of bipolar amplification is as follows. The heavy-ion strike on the device creates electron–hole pairs in the silicon film. While minority carriers recombine quickly, the lifetime of majority carriers in the body region can be very long. Majority carriers that do not recombine can drift toward the source region and raise the body potential. Then, the source-to-body potential barrier is lowered, which triggers the lateral parasitic bipolar transistor inherent in the SOI transistor. The potential rise is maintained until the majority carriers are recombined. The bipolar current amplifies the collected charge and decreases the SEU/single-event transient (SET) immunity, especially at low LET (Ferlet-Cavrois et al. 2004). This effect is further enhanced by the impact ionization mechanism induced by the high electric field at the body–drain junction. The consequence is that SOI immunity to radiation is degraded; although SOI devices have a smaller sensitive volume than bulk silicon devices, this is counterbalanced by the enhanced bipolar amplification (Ferlet-Cavrois et al. 2002b; Massengill et al. 1990; Musseau et al. 1994; Schwank et al. 2003, 2004).

To reduce these bipolar effects, the most common technique involves the use of body ties (which connect the floating-body region to a fixed potential). The excess holes created by the ion strike no longer accumulate in the floating-body region because they are evacuated through the body contact. This considerably reduces the parasitic bipolar transistor effects. However, body ties do not completely eliminate the bipolar effect; a voltage drop exists along the body tie due to its finite resistance, and the reduction of bipolar effect is less effective. The ability of body ties to suppress the bipolar effect strongly depends on the location of the body tie in relation to the ion strike (Musseau et al. 2000). The farther the ion strike is from the body tie, the larger the effect of the parasitic bipolar transistor (Hite et al. 1992; Kerns et al. 1989; Massengill et al. 1990; Schwank et al. 2003).

Bipolar amplification can also occur in fully depleted transistor circuits. Previous experimental and theoretical studies have shown that, generally, FDSOI devices exhibit reduced floating-body effects and lower bipolar amplification of the collected charge than PDSOI devices (Ferlet-Cavrois et al. 2002a, 2005; Massengill et al. 1990). The bipolar transistor mechanism in fully depleted devices has been explained in Brisset et al. (1994) using MC simulations of 0.25 μm FDSOI circuits: after irradiation of an n-channel MOSFET biased in its off state, excess holes are accumulated in the channel (mainly near the gate oxide) and lower the potential barrier; then electrons diffuse from source to drain to maintain electrical neutrality. This mechanism is comparable to the bipolar transistor effect in PDSOI devices (Massengill et al. 1990; Musseau et al. 1994). Because bipolar amplification is less important for fully depleted than for partially depleted transistors, circuits based on fully depleted transistors are less sensitive to single-event upset than partially depleted circuits (Ferlet-Cavrois et al. 2002a).

The effect of the parasitic bipolar transistor in SOI devices is quantified using a metric called *bipolar gain*, β. The bipolar gain corresponds to the amplification of the deposited charge and is given by the ratio between the total collected charge, Q_{coll}, at the drain electrode and the deposited charge, Q_{dep}:

$$\beta = \frac{Q_{coll}}{Q_{dep}} \tag{12.1}$$

Bipolar amplification is a relevant factor for estimating device sensitivity to ionizing particles. Indeed, the charge collected at the drain contact, following the ion strike, results in a drain current transient, $I_D(t)$, which is then used to accurately calculate Q_{coll} as follows:

$$Q_{coll} = \int_0^t I_D(t)\,dt \tag{12.2}$$

The deposited charge in a SOI device is calculated as a function of the particle LET using Equation 12.3 (Ferlet-Cavrois 2004):

$$Q_{dep}\left(fC\right) = 10.3 \times LET\left(MeV\ cm^2\ mg^{-1}\right) \times t_{Si}\left(\mu m\right) \tag{12.3}$$

where t_{Si} is the silicon film thickness and 10.3 is a multiplication factor for silicon (calculated using the silicon density and the energy needed for creating an electron–hole pair in silicon [3.6 eV]) (Ferlet-Cavrois 2004). In this equation, a normal incident ion strike is considered and the LET is assumed to be constant along the ion path in the active silicon film.

12.2.1.2 Analysis of Radiation-Induced Drain-Current Transient in FDSOI MOSFETs

When a heavy ion strikes an SOI MOSFET operating in the off state, a drain-current transient is observed at the drain terminal. This type of transient has been thoroughly investigated in literature in both PDSOI and FDSOI transistors. A very detailed study of the transient current components was conducted by Kobayashi et al. (2006). To illustrate its conclusions, we simulated the behavior under irradiation of FDSOI devices with 50 nm gate lengths and thin films (11 nm-thick silicon films).

The schematic description of the SOI architectures considered in the three-dimensional (3D) simulation study is shown in Figure 12.3. Simulated devices are NMOS transistors (without body contact) fabricated in a FDSOI technology (Paillet et al. 2005). The ion strike was simulated using the Dessis HeavyIon module (Synopsys 2014). The simulated irradiation track has a Gaussian shape with narrow radius of 14 nm and a Gaussian time dependence, centered on 50 ps and with a characteristic width of 2 ps. More details about the simulated devices and simulation models will be given in Section 12.2.2. Figure 12.4 shows the time evolution of the drain current induced by a heavy ion crossing the device with two LET values. Three components of the total current in the transistor are highlighted in Figure 12.4: the electron current measured at the drain electrode (JeD), the electron current

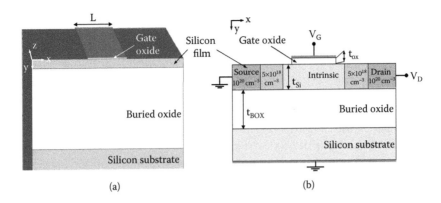

FIGURE 12.3
(a) Simulated 3D structure of Single-Gate FDSOI devices; (b) schematic description of a 2D cross-section (plane y–z) showing the main geometric parameters and doping levels in the silicon film.

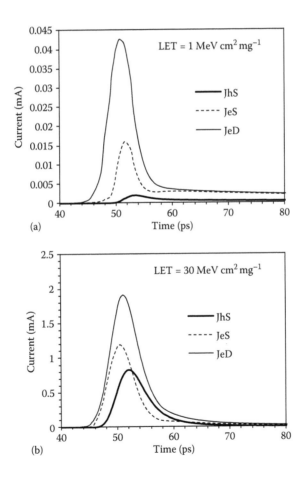

FIGURE 12.4
(a,b) Time variations of JeD (electron current at the drain electrode), JeS (electron current at source), and JhS (hole current at source) for two LET values. The FDSOI transistor with 50 nm gate length is biased in the off state at $V_D = 0.7$ V.

at the source (JeS), and the hole current at the source (JhS). The other components of the current are too small and can be neglected. The electron current measured at the drain is identical to the drain current in the transistor, that is, $I_D(t) = JeD(t)$. Figure 12.4 shows that the electron current increases immediately after the ion strike and slowly decreases thereafter. For a low LET value (e.g., LET = 1 MeV cm² mg⁻¹ in Figure 12.4a), the drain-current tail is substantially equal to the electron current to the source. This indicates that the current is controlled by the electron flow injected by the source due to the parasitic bipolar current (activated by the floating-body phenomenon). This result is in perfect agreement with the conventional description of drain-current transient induced by an ion strike in SOI transistors. On the other hand, at high LET values, this description is no longer valid. In the case of a high LET (e.g., LET = 30 MeV cm² mg⁻¹ in Figure 12.4b), the current tail is much larger than the electron current at the source. This amplification is not due to impact ionization: we have verified by simulation that, when the impact ionization model is disabled, the transient current is identical to that simulated when the impact ionization is active. This shows that there is an additional component at high LET that explains the width of the current transient. In order to reveal this component, the difference between JeD and JeS is plotted for LET = 30 MeV cm² mg⁻¹ in Figure 12.5a, as shown by Kobayashi

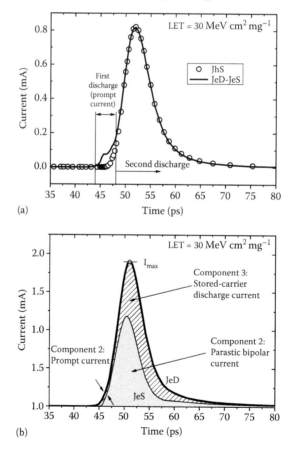

FIGURE 12.5

(a,b) Time variations of JeD (electron current at the drain) and JeS (electron current at the source) for LET = 30 MeV cm² mg⁻¹. Illustration of the different components of the drain-current transient induced by a heavy ion striking the device. The current JeD is identical to the drain current (I_D).

et al. (2006). In this figure, the hole current at the source, JhS, is also shown. The time variation of the difference JeD–JeS can be divided into two parts: a first electron discharge (due to the *prompt* component of the transient) and a second electron discharge, which is identical to the hole current at source. From these observations and detailed inventory of deposited charges in the transistor, it has been shown in Kobayashi et al. (2006) that this additional component is due to the discharge of a portion of the deposited electrons that are stored in the high-injection condition body to maintain quasi-neutrality. Therefore, the transient drain current has three components (Figure 12.5b): (1) a prompt component due to the discharge of electrons immediately after the ion strike (*first discharge component* in Figure 12.5a), (2) a parasitic bipolar current (*second discharge* in Figure 12.5a) with a slow decay, due to floating-body effects, and (3) a tail component due to the electrons stored in the high-injection regime body (*stored-carrier discharge current* in Figure 12.5b). This additional third component increases considerably both the amplitude and the width of the current transient (Kobayashi et al. 2006). Figure 12.6a and b show the impact of the third component due to the electrons stored in the substrate on the transient peak and width at 50%. When LET increases, the influence of this component increases significantly. For example, the peak current increases by 0.72 mA and the pulse is broadened by 2.93 ps for

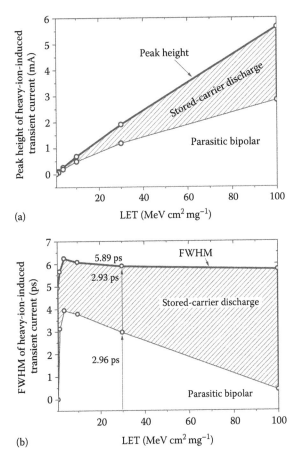

FIGURE 12.6
(a) Drain-current transient peak as a function of LET. (b) Pulse widths (FWHM) plotted as a function of LET. The contribution of each component of transient current is indicated.

an LET value of 30 MeV cm² mg⁻¹. The pulse width (which is a key feature of the propagation of current transient induced by the ion strike in logic circuits) is strongly enhanced by this new component, especially at high LET. In Kobayashi et al. (2006), the authors have also analyzed the effects of device downscaling on this new transient component. It was shown that the third component of the current transient must be analyzed carefully, especially for future ultrashort SOI devices, for which this component is likely to become very important.

12.2.2 3D Simulation Study of Radiation Response of 50 nm FDSOI Devices

The transient response to radiation of FDSOI devices has been addressed in literature in numerous studies. For example, in Munteanu et al. (2006), the drain-current transient and the bipolar amplification of FDSOI MOSFETs with 50 and 80 nm gate length submitted to heavy-ion irradiation has been studied using both measurements and 3D numerical simulations. The tested devices are floating-body (without body contacts) NMOS transistors fabricated with a single-gate FDSOI technology (Paillet et al. 2005). The transistors have been processed with a midgap TiN gate, and the silicon film is nearly intrinsic (p-type, 10^{15} cm⁻³). The thickness of the silicon film and of the buried oxide is 11 and 100 nm, respectively, and the equivalent gate-oxide thickness is close to 1.8 nm. Three-dimensional numerical simulation results have been compared with experimental data measured by heavy ion experiments performed at GANIL (Grand Accélérateur National d'Ions Lourds, Caen, France). A detailed description of the experimental setup and measured data can be found in Munteanu et al. (2006). Numerical simulations were conducted in order to obtain more detail of the physical quantities (such as the concentration of carriers in the channel, the electrostatic potential, etc.) that are involved in device operation and to extrapolate experimental results when the device dimensions are reduced. The schematic description of the 3D SOI architecture considered in the simulation is represented in Figure 12.3. For these ultrathin devices with silicon film thicknesses around and below 10 nm, quantum confinement phenomena must be considered in the numerical simulation of the devices, as shown in Munteanu et al. (2006). Quantum simulation results are thoroughly analyzed in that paper and compared each time with the *classical* case (i.e., without quantum confinement effects), in order to evidence the quantum confinement effects on the device transient response when a heavy ion is crossing the device. Several important results are described below, in particular the impact of quantum effects on the 3D carrier concentration distribution, drain current transient, and bipolar amplification.

12.2.2.1 Calibration of the Quantum Model

Three-dimensional numerical simulations have been performed with the 3D Synopsys tools (Dessis module) (Synopsys 2014), in which quantum confinement of carriers is taken into account using the Density Gradient model, described previously in Section 8.2.2.1. As stated in Chapter 8, it has been shown in Wettstein et al. (2002) that the Density Gradient model can accurately account for quantum carrier confinement in single-gate SOI and double-gate devices with an appropriate calibration step of the fit factor γ (see Equations 8.11 and 8.12). In Munteanu et al. (2006), the calibration of the Density Gradient model was performed using the exact solution of the Schrödinger–Poisson system of equations given by a homemade code called Balmos3D (Munteanu and Autran 2003). This code is a homemade quantum simulator for ultrathin devices, written in Fortran, which self-consistently solves the 1D Schrödinger equation and the 3D Poisson equation on a 3D mesh. The solution of

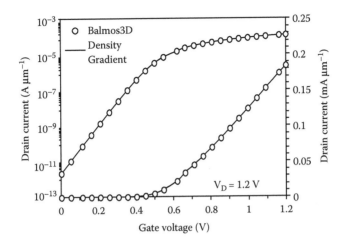

FIGURE 12.7
Calibration of the Density Gradient model on the exact solution of the Schrödinger–Poisson system of equations (Balmos3D code) for 50 nm single-gate FDSOI MOSFET.

this system of equations is coupled with the diffusion-drift transport equation (or with a ballistic transport equation) in the channel. In Munteanu et al. (2006), the Density Gradient model calibration was carried out on devices with 50 and 80 nm gate lengths and with 11 nm-thick silicon films. For each device, the characteristics $I_D(V_G)$ were calculated with Balmos3D. The same device was then simulated with the Synopsys 3D simulator using the Density Gradient model. The factor γ was finely tuned in order to obtain a perfect match between the characteristics simulated with Synopsys and the data calculated by Balmos3D, as shown in Figure 12.7.

12.2.2.2 Calibration of Simulation Models on Experimental Data

In the simulations presented in Munteanu et al. (2006), a certain number of physical models have been considered (in addition to the Density Gradient model). These models include the Shockley–Read–Hall (SRH) and Auger recombination models, as well as the Fermi–Dirac carrier statistics. In the SRH recombination model, carrier lifetimes depend on the doping level (Munteanu et al. 1998; Synopsys 2014). The model of the effective intrinsic density includes a doping-dependent band-gap narrowing (Slotboom's model) (Synopsys 2014) and a lattice temperature-dependent band gap. The carrier transport model used in the simulation is the hydrodynamic model, which includes the energy-balance equations for electrons, holes, and lattice. We also considered models for impact ionization and carrier mobility depending on the carrier energy calculated by the hydrodynamic model. The mobility model also takes into account the dependence of mobility on normal electric field (through Lombardi's model) (Synopsys 2014), lattice temperature, and channel doping. As explained in Chapter 8 (Section 8.2.3.1), the models used in the numerical simulator contain a number of empirical parameters that must be absolutely calibrated on experimental data in order to simulate device operation with maximum accuracy. Model parameters used in Munteanu et al. (2006) (particularly carrier mobility) were calibrated to obtain the best match between the simulated quantum drain current and experimental data. Figure 12.8 shows the result of this calibration step performed on FDSOI devices with 50 nm gate length.

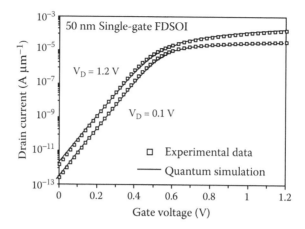

FIGURE 12.8
Calibration of the simulated quantum-mechanical drain current on experimental data in 50 nm single-gate FDSOI devices. (Reprinted from Munteanu, D. et al., Investigation of quantum effects in ultra-thin body single- and double-gate devices submitted to heavy ion irradiation, *IEEE Trans. Nucl. Sci.* 53, 3363–3371. © (2006) IEEE. With permission.)

12.2.2.3 Transient Simulation Details

The calibrated physical parameters have been further used in Munteanu et al. (2006) for the simulation of drain-current transients produced by an ion strike on the sensitive regions of the device. The drain-current transients have been simulated in two cases: the classical case (i.e., without quantum effects) and the quantum case (using the Density Gradient model with the fit factor γ as calibrated on Balmos3D). The simulated transistors were biased in the off state, which is the most sensitive case.

The ion strike was simulated using the Dessis HeavyIon module (Synopsys 2014). The simulated irradiation track has a Gaussian shape with a narrow radius of 14 nm and a Gaussian time dependence, centered on 50 ps and with a characteristic width of 2 ps. This small radius value was chosen in order to be comparable with the actual radii used in the experiment. In a first step, an angle of incidence of the ion strike of 60° was also considered in Munteanu et al. (2006) in order to mimic the heavy-ion experiment. The deposited charge was calculated considering the Gaussian distribution and the 3D geometry of the Si film. The deposited carriers are rapidly transported (mainly by drift and diffusion mechanisms) (Munteanu and Autran 2008) and collected by the drain contact. Part of these carriers can be recombined by carrier recombination mechanisms; the deposited charge can also be amplified by the parasitic bipolar amplification mechanism, as explained in Section 12.2.1.1. The collected charge at the drain terminal was obtained by the integration of the drain current over the transient duration (Equation 12.2), and the bipolar gain was finally calculated as the ratio between collected and deposited charges (Equation 12.1).

12.2.2.4 Transient Simulation Results: Impact of Quantum Effects on Radiation Response

Simulations performed in Munteanu et al. (2006) have shown that, as expected, the maximum of the inversion charge is located at the silicon/gate-oxide interface. This is not the case for quantum simulations, where the maximum charge is no longer at the interface, but is moved into the silicon film at a depth of several nanometers relative to the interface. The total inversion charge is lower in the quantum case than in the classical case; thus, the threshold voltage increases and the drain-current value in the off state is lower in the quantum case than in the classical case. Figure 12.9 shows 3D distributions of the carrier density in the classical and quantum cases at different times before and after the ion strike. The distributions in the quantum case are different from the classical case, due to the quantum carrier confinement.

However, this effect is only slightly reflected in the drain-current transients induced by the ion strike, shown in Figure 12.10 for different LET values. Quantum transients have current prompt components and transient tails similar to the classical case. Simulations presented in Munteanu et al. (2006) show that the transients are almost identical at very high LET values, where the charge generation is significant and masks the impact of the different carrier distributions in the channel between the quantum and classical cases.

Quantum drain-current transients were normalized in Figure 12.11 with respect to their maximum value and compared with the initial Gaussian time-dependent charge generation. Figure 12.11 shows that the transient current quantum drain for the low-injection regime (LET = 0.1 MeV cm² mg⁻¹) is almost synchronous with the initial charge generation, which indicates a weak bipolar amplification. At high LET, the prompt component of the drain-current transient is shifted to the right and the current pulse is visibly wider than the initial charge generation, which reflects an important bipolar gain.

The simulations carried out in Munteanu et al. (2006) also showed that the transient quantum drain currents have very small durations, shorter than those obtained in experimental measurements. For FDSOI devices with 50 nm gate length, the total duration of the transient is 15 ps at 10% of the peak current for an LET of 30 MeV cm² mg⁻¹. However, these values are consistent with transient durations obtained by simulation in Ferlet-Cavrois et al. (2005). Finally, the bipolar amplification values in the classical and quantum cases are compared in Figure 12.12. The bipolar gain is lower in the quantum case, mainly due to the lower off-state current (Munteanu et al. 2006). For an 80 nm gate-length device and for

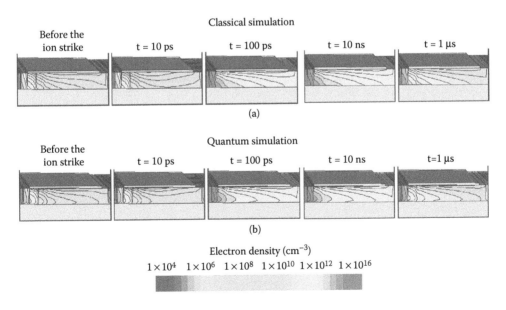

FIGURE 12.9
3D electron density distribution in the silicon film in 50 nm FDSOI MOSFETs (V_G = 0 V, V_D = 1.2 V) at different times before and after the ion strike: (a) classical simulation (without quantum effects) and (b) quantum simulation using the Density Gradient model calibrated on the exact solution of the Schrödinger–Poisson system of equations.

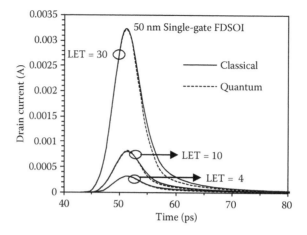

FIGURE 12.10

Drain-current transient at different LET values in 50 nm single-gate FDSOI. LET values are expressed in MeV cm² mg⁻¹. (Reprinted from Munteanu, D. et al., Investigation of quantum effects in ultra-thin body single- and double-gate devices submitted to heavy ion irradiation, *IEEE Trans. Nucl. Sci.* 53, 3363–3371. © (2006) IEEE. With permission.)

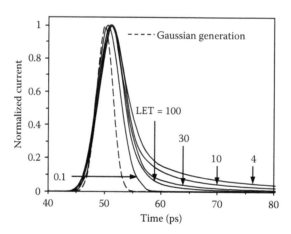

FIGURE 12.11

Normalized drain current versus time in the quantum case at different LET values expressed in MeV cm² mg⁻¹. (Reprinted from Munteanu, D. et al., Investigation of quantum effects in ultra-thin body single- and double-gate devices submitted to heavy ion irradiation, *IEEE Trans. Nucl. Sci.* 53, 3363–371. © (2006) IEEE. With permission.)

an ion strike near the drain with an LET = 30 MeV cm² mg⁻¹, the bipolar gain is 2.8 in the classical case and about 2.62 in the quantum case. This last result is in excellent agreement with the experimental results (bipolar gain of 2.6 in the worst case).

12.2.3 SEU Sensitivity of FDSOI SRAM Cells

Scaling trends concerning the sensitivity to SEU in SOI static memory cells and the propagation of digital single-event transients (DSETs) in SOI circuits have been addressed (Gaillardin et al. 2007) using mixed-mode simulations. In this approach, all devices (FDSOI NMOS and PMOS transistors with 50 nm gate length and 3 μm gate width) have been modeled

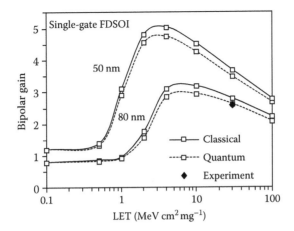

FIGURE 12.12
Bipolar gain in FDSOI devices for two gate lengths: 50 and 80 nm. The experimental bipolar gain obtained at $LET = 30$ MeV cm^2 mg^{-1} on the 80 nm FDSOI device is also reported. The angle of incidence in both experiment and simulation is 60°. (Reprinted from Munteanu, D. et al., Investigation of quantum effects in ultra-thin body single- and double-gate devices submitted to heavy ion irradiation, *IEEE Trans. Nucl. Sci.* 53, 3363–3371. © (2006) IEEE. With permission.)

in full 3D, then connected, and only the OFF NMOS is struck by an ionizing particle. The SEU-threshold LET of memory cells (LET$_{TH\text{-}SEU}$) and the DSET-critical LET for unattenuated propagation in a chain of inverters (LET$_{TH\text{-}SET}$) were analyzed. The SEU-threshold LET was determined by varying the ion-strike LET until the static memory cell was observed to upset. The authors added the LET$_{TH\text{-}SEU}$ obtained for 50 nm FDSOI technology to data presented in Dodd et al. (2004) on former SOI technologies. The results obtained in Gaillardin et al. (2007) showed that the threshold LET of 50 nm devices stands at a similar value to that of the previous technology node (LET$_{TH\text{-}SEU}$ = 2 MeV cm^2 mg^{-1}). As explained in Gaillardin et al. (2007), the reduced sensitive volume associated with the efficient electrostatic potential control provided by the gate in the body mitigates the effects induced by an ionizing-particle hit in this advanced 50 nm FDSOI technology. The LET$_{TH\text{-}SET}$ for unattenuated transient propagation in a chain of inverters was also studied by Gaillardin et al. (2007) as a function of device scaling. The results obtained for the FDSOI 50 nm chain of inverters were in good agreement with data from the literature (Dodd et al. 2004). It was shown that, contrary to the SEU-threshold LET, LET$_{TH\text{-}SET}$ decreases with technology scaling (Gaillardin et al. 2007). The speed of these circuits increases with size reduction and is sufficient to propagate shorter transients (Benedetto et al. 2004; Mavis and Eaton 2002) (<100 ps wide in sub-0.1 μm technologies) that become indistinguishable from normal circuit signals (Gaillardin et al. 2007). It is important to note that results concerning the 50 nm FDSOI devices in Gaillardin et al. (2007) were obtained by simulation and need to be confirmed experimentally. More recent results obtained for FDSOI 28 nm SRAMs have been reported in Roche et al. (2013).

12.3 Multiple-Gate Devices

As explained in the introduction to this chapter, one of the solutions envisaged to replace the bulk technology is to radically change the transistor architecture by introducing additional gates to improve the electrostatic control of the channel, which may allow MOSFET scaling to continue. Multiple-gate transistors (Figure 12.13) are original and very promising devices to fulfill the conditions imposed by the ITRS roadmap for MOSFET miniaturization at the decananometer scale (Park and Colinge 2002). Multiple-gate devices have received particular interest from the scientific community in the last 15 years, and an impressive number of theoretical and experimental studies have been published in the literature. Numerous multiple-grid architectures have been proposed and fabricated, the best known being listed below:

1. Planar double-gate (DG) (Frank et al. 1992; Harrison et al. 2003), vertical double-gate
2. Triple-gate (or tri-gate) (Guarini et al. 2001; Park and Colinge 2002), FinFET (Choi et al. 2001; Kedzierski et al. 2002)
3. Omega-gate (Ω-gate) (Park et al. 2001), Pi-gate (π-gate) (Yang et al. 2002), Δ-channel SOI MOSFET (Jiao and Salama 2001), DELTA transistor (Hisamoto et al. 1989)

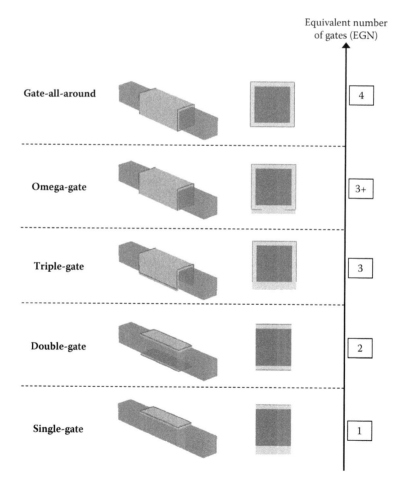

FIGURE 12.13
Schematic description of the single-gate, double-gate, triple-gate, omega-gate, and gate-all-around structures and their main geometric parameters. The devices are classified as a function of the "Equivalent Gate Number" (EGN). The schematic cross sections in the (y–z) plane are also shown.

4. Gate-all-around (GAA) (Colinge et al. 1990; Park and Colinge 2002), rectangular or cylindrical nanowires (Jiménez et al. 2004)

It has been shown that, when the *number of equivalent gates* (EGN) increases, the control by the gate of the electrostatic potential in the silicon film is considerably improved. This implies that short-channel effects are gradually reduced when increasing the number of gates, from single-gate to *four-gate* devices (i.e., GAA architecture, where the gate electrode is wrapped around the entire channel; Figure 12.13). Therefore, multiple-gate architectures show superior performance compared with single-gate devices in terms of short-channel effects and power dissipation. The good control of short-channel effects due to their geometric construction allows constraints on film doping level to be relaxed. Thus, in multi gate devices, there is no need to use high doping levels in the channel to reduce short-channel effects, which are already limited by the existence of multiple gates. These devices may be designed with intrinsic channels, which is very beneficial for increasing the channel mobility. The absence of doping in the channel also eliminates the problem of doping fluctuations, which is a major issue in bulk transistors. Finally, the condition of *volume inversion* (Yang and Fossum 2005), which occurs in multiple-gate devices in the subthreshold regime, can be beneficial with regard to carrier mobility and source–drain transport.

12.3.1 Impact of Quantum Effects

When the dimensions of multiple-gate MOSFET are reduced, it becomes necessary to use increasingly thinner silicon films in order to maintain the benefits of these structures in terms of short-channel effects. For these very thin films, the quantum confinement of carriers must be taken into account in the device simulation. Such quantum confinement effects induce strong subband splitting and carrier confinement in the narrow potential well formed by the silicon film (Taur and Ning 1998).

Carrier confinement is not the same in all multiple-gate structures, since it strongly depends on gate geometry. This is explained in Figure 12.14, which shows the quantum confinement directions in three different generic configurations: single-gate FDSOI, double-gate, and GAA. In single-gate or double-gate structures, carriers are confined in a single direction (y), vertically, perpendicular to both the gate electrode and the source–drain axis. In multiple-gate structures having at least three gates, the quantum confinement occurs in two directions (y and z) perpendicular to both the gate electrodes and the source–drain axis. Therefore, the carrier confinement and its effects are stronger in multi-gate devices with EGN ≥ 3 than in single- or double-gate configurations.

Figure 12.14 also shows the impact of quantum effects on the electron density extracted along a cutline parallel to the confinement direction y. In this figure, all devices are biased in strong inversion regime. Similar to the bulk device (see Section 8.2.2.1), in single-gate FDSOI devices carriers are confined in a very narrow triangular well potential formed at the Si/SiO₂ interface.

In the case of a double-gate structure, carriers are confined in a rectangular potential well formed between the two Si/SiO₂ interfaces. The size of this potential well is controlled by the film thickness, which then becomes a key parameter in the analysis of quantum effects. Figure 12.14 shows that in double-gate devices the classical electron density is at a maximum at the two interfaces. In the quantum case, the density profile shows two maxima situated within the silicon film at a depth of several nanometers from the interface. Figure 12.14 is typical for double-gate devices with thin silicon film (Majkusiak et al. 2002). Due to this distribution of the inversion charge, the quantum drain current is always

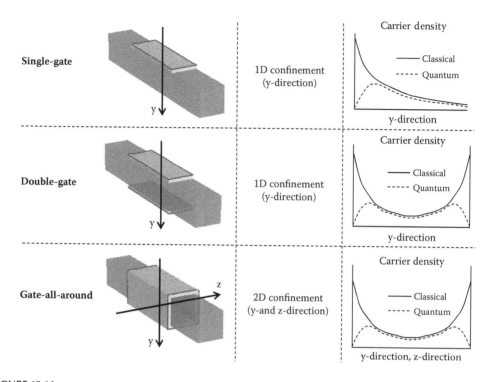

FIGURE 12.14

Schematic representation of the quantum-mechanical carrier confinement in single-gate, double-gate, and gate-all-around structures. The quantum confinement directions and the profiles of the carrier density in a cutline along the film thickness are also reported for both classical (i.e., without quantum confinement) and quantum cases.

divided into two separate channels, but they are no longer situated at the interface as in the classical case. As in the case of single-gate FDSOI devices (see Section 12.2.2.4), the total quantum inversion charge is lower than the classical inversion charge for a given polarization. As a result, the drain-current characteristics as a function of the gate voltage are shifted to the higher gate voltages in the quantum case compared with the classical case. Therefore, the quantum off-state current of double-gate devices is lower than the classical one (Munteanu et al. 2006).

Carrier-confinement effects are stronger as the silicon film thickness decreases, because band splitting is inversely proportional to the square of the film thickness (Munteanu et al. 2006). When the film thickness of double-gate devices is greatly reduced, the two electron-density maxima become superposed in the middle of the silicon film. As a result, the electrons form a single conduction channel, located in the middle of the film. The ratio between the quantum and classical off-state currents calculated in Munteanu et al. (2006) confirms that the quantum confinement is strengthened when the film thickness is reduced. This ratio is about 1.3 for an 11 nm-thick silicon film and increases to 3.4 when the film thickness is reduced to 5 nm. The quantum off-state current is more than three times lower than the classical current in a double-gate device with very thin film.

In GAA structures, carriers are confined in a double rectangular potential well (along the y and z axes); the movement of carriers is not free in the z direction (as in the case of single-gate and double-gate devices), but their energy is quantized as in the y direction. This considerably increases quantum effects in the GAA structure compared with the double-gate configuration. The quantum electron density in the z direction is no longer maximal at the interface but has two peaks located inside the silicon film, similarly to the carrier density in the y direction. Therefore, the total quantum inversion charge in the GAA structure is smaller than that of a double-gate device. In the GAA structure, not only the silicon film thickness but also the gate width, W, are essential parameters that control the intensity of quantum effects. When one or both dimensions decrease, quantum effects are considerably enhanced.

Figure 12.15 shows 3D electron-density distributions calculated in the classical and quantum cases for three structures: double-gate, triple-gate, and GAA. The electron densities in a 2D cross section in the plane (y–z) are also shown for each structure. In this figure, the devices are biased in the off state. Quantum effects obviously modify the 3D carrier distribution in the channel, since more carriers are concentrated in the middle of the silicon film in the quantum case than in the classical case. The quantum inversion charge is lower than the classical inversion charge for the same bias, which leads to a reduction of the drain current in the quantum case with respect to the classical case (Munteanu et al. 2007).

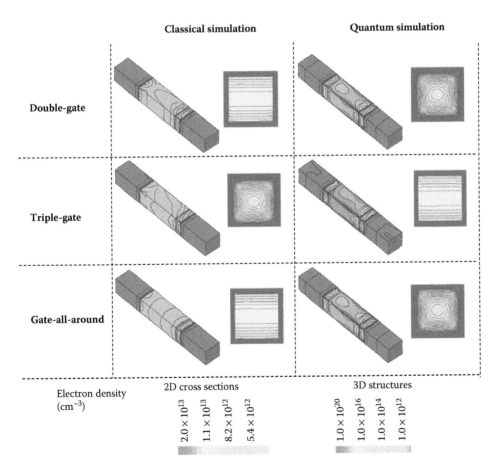

FIGURE 12.15
Electron density in classical and quantum cases in 3D structures and 2D cross sections (plane y–z) for double-gate, triple-gate, and gate-all-around devices.

12.3.2 Transient Response of Multiple-Gate Devices

The transient response of multiple-gate devices submitted to heavy ion irradiation has been investigated in literature by both experimental (Gaillardin et al. 2007) and 3D numerical simulation (Munteanu et al. 2006, 2007). These previous studies demonstrate that multiple-gate devices show better radiation hardness than single-gate FDSOI transistors, particularly due to enhanced control of body potential and reduction of floating-body effects. Simulation studies also show that the transient and charge collection are very fast in multiple-gate devices due to the small active volumes, which allow all the excess charge to be quickly evacuated (Munteanu et al. 2007). However, these very fast transients are specific to intrinsic devices and can be degraded by extrinsic elements related to the fabrication process, which may not be mature in multiple-gate technologies. An example is shown in Gaillardin et al. (2007), where the radiation-induced current transient of Omega-gate devices has been found to be noticeably longer than that of single-gate FDSOI devices due to the technological process, which includes long resistive access to source and drain electrodes. If the length of this region increases, the carriers produced in the body need more time to reach the drain electrode (Gaillardin et al. 2007). Consequently, the current transient is broader, with a large width at half maximum (FWHM) value of ~64 ps, and the transient tail is longer. Consequently, these devices collect a higher total charge than planar FDSOI devices with similar gate lengths (Gaillardin et al. 2007). This long transient is not an intrinsic characteristic of omega-gate devices. Thus, the length of the highly doped access area that connects the channel to the drain (source) electrode is a key issue for the radiation-induced current transient, and so access regions have to be carefully optimized in order to conserve the intrinsic radiation hardness of omega-gate devices (Gaillardin et al. 2007).

In order to illustrate the transient response of multiple-gate devices, we outline in the following the most important simulation results obtained by Munteanu et al. (2006, 2007). The transient responses of double-gate and FDSOI MOSFET are compared in Section 12.3.2.1, and the impact of quantum confinement effects is widely discussed and illustrated. In Section 12.3.2.2, different multiple-gate structures are compared in terms of sensitivity to radiation, taking into account the impact of quantum effects.

12.3.2.1 Double-Gate versus FDSOI MOSFET

The impact of quantum confinement on the radiation response of planar double-gate devices has been studied by 3D quantum numerical simulation in Munteanu et al. (2006) and compared with that of FDSOI devices. The double-gate structure simulated in Munteanu et al. (2006) is schematically described in Figure 12.16. The FDSOI devices are those simulated in Section 12.2.2 (see Figure 12.3). The simulation models are the same as in Sections 12.2.2.2 and 12.2.2.3.

Figure 12.17 shows the time variation of the electron density before and after the ion strike in the classical and quantum case for a double-gate transistor with $t_{Si} = 11$ nm. In the quantum case, the carrier density is more centered in the middle of the film, and, at the same time, the total quantum inversion charge is lower than the classical charge. Figure 12.17 also compares the quantum electron densities for $t_{Si} = 11$ and $t_{Si} = 5$ nm. This figure confirms that, when the film thickness is reduced, quantum effects are stronger and carriers are more concentrated in the middle of the silicon film.

Bipolar gains for double-gate structures with two film thicknesses (5 and 11 nm) calculated in the classical and quantum cases are shown in Figure 12.18. As explained

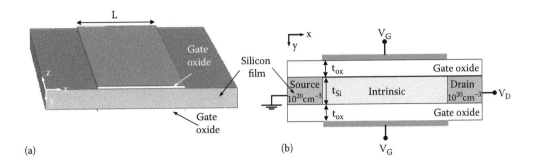

FIGURE 12.16
(a) Simulated 3D structure of double-gate MOSFETs and (b) schematic representation of a 2D cross section (plane x–y) showing the main geometric parameters and doping levels in the silicon film.

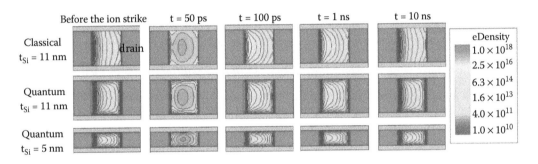

FIGURE 12.17
Classical and quantum electron density (expressed in cm^{-3}) in a vertical cross section along the source–drain axis (x–y plane) in the silicon film of 50 nm double-gate devices with 11 and 5 nm-thick channels. LET = 10 MeV cm^2 mg^{-1}. (Reprinted with permission from Munteanu, D. et al., Investigation of quantum effects in ultra-thin body single- and double-gate devices submitted to heavy ion irradiation, *IEEE Trans. Nucl. Sci.*, 53, 3363–3371. © (2006) IEEE.)

before, bipolar amplification is specific to PDSOI devices, but also exists in FDSOI (Ferlet-Cavrois et al. 2002b, 2004) and double-gate transistors (Munteanu et al. 2006). The results of Figure 12.18 show that, in the classical case, the bipolar gain is lower in the double-gate configuration than in the single-gate structure (see results reported in Section 12.2.2.4). This is mainly due to better electrostatic control of the channel in the double-gate configuration, which reduces floating-body effects (Munteanu et al. 2006). The bipolar gain is lower for a 5 nm-thick film than for a transistor with a film thickness of 11 nm, due to both a lower deposited charge and weaker floating-body effects in the thinner film. These results also show that the difference between classical and quantum bipolar gain is significantly higher in the double-gate than the single-gate structure. Unlike single-gate devices, in double-gate transistors the quantum bipolar gain exceeds the classical bipolar gain at medium and low LET. These differences between bipolar gains of double-gate and single-gate devices can be explained by: (i) a lower off-state current in the quantum case and (ii) a different distribution of carriers in the quantum case, which modifies the recombination rate at the source (Munteanu et al. 2006). Finally, it should be noted that the bipolar gain is quite low in double-gate devices, probably partly due to the Gaussian shape of the ion track considered in simulation.

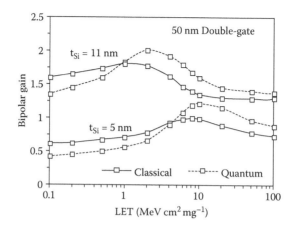

FIGURE 12.18
Bipolar gain as a function of LET in 50 nm double-gate devices for different silicon film thicknesses. (Reprinted from Munteanu, D. et al., Investigation of quantum effects in ultra-thin body single- and double-gate devices submitted to heavy ion irradiation, *IEEE Trans. Nucl. Sci.* 53, 3363–3371. © (2006) IEEE. With permission.)

12.3.2.2 Comparison of Transient Response of Different Multiple-Gate Devices

Three-dimensional quantum simulation has been used in Munteanu et al. (2007) for investigating the drain-current transient produced by the ion strike in multiple-gate nanowire MOSFETs with ultrathin channels (≤10 nm). In Munteanu et al. (2007), the gate length has been reduced from 32 to 20 nm and the corresponding film thickness has been reduced from 10 to 5 nm (Table 12.1). Four different multiple-gate configurations have been considered: double-gate, triple-gate, omega-gate, and GAA. The description of the 3D architectures considered in the simulation and the definition of their geometric parameters are presented in Figure 12.13 and Table 12.1. All simulated devices have an intrinsic channel and midgap gate, and the thickness of the buried oxide is 100 nm. The supply voltage is 0.8 V for structures with 32 and 25 nm channel lengths and 0.7 V for the device with 20 nm channel length (Munteanu et al. 2007).

Similarly to the study of FDSOI devices presented in Section 12.2.2.1, a calibration step of the Density Gradient model on Balmos3D has been performed in Munteanu et al. (2007) for obtaining the fit factor γ. The other models used in simulation are the same as in Section 12.2.2. $I_D(V_G)$ characteristics for the different multi gate structures with 32 nm gate length have been simulated in Munteanu et al. (2007) in the classical and quantum cases. The results show that increasing the EGN reduces the off-state current and improves the subthreshold slope. Indeed, the following values for the subthreshold slope S were obtained in Munteanu et al. (2007) for the simulated structures: S=70 mV dec^{-1} for double-gate, S=68.5 mV dec^{-1} for triple-gate, S=65 mV dec^{-1} for omega-FET, and S=61.5 mV dec^{-1} for GAA. This is due to better electrostatic control of the channel by the gate when the equivalent number of gates is increased, reducing short-channel effects and therefore improving the subthreshold slope. At the same time, the on-state drain current increases with the equivalent number of gates, due to the existence of multiple-channel conduction.

As in the previous simulations, the quantum drain current is lower than the classical drain current because the total inversion charge is reduced in the quantum case. Another result shown in Munteanu et al. (2007) is that the difference between classical and quantum off-state currents increases if one moves from a double-gate configuration to multiple-gate structures with EGN ≥ 3. To illustrate this effect, the ratio between classical and quantum

TABLE 12.1

Bipolar Gain in Multiple-Gate Nanowire MOSFETs as Obtained by
Quantum 3D Numerical Simulation

Structure	Gate Length (nm)	$W = t_{Si}$ (nm)	V_D (V)	EGN	I_{offcl}/I_{offq}	Bipolar Gain
Double-Gate	32	10	0.8	2	1.91	1.79
Triple-Gate				3	2.24	1.49
Omega-Gate				3+	2.32	1.3
GAA				4	2.36	1.18
Double-Gate	25	8	0.8	2	2.02	1.53
Triple-Gate				3	2.30	1.21
Omega-Gate				3+	2.48	0.94
GAA				4	2.67	0.82
Double-Gate	20	5	0.7	2	3.01	0.82
Triple-Gate				3	3.66	0.78
Omega-Gate				3+	3.86	0.73
GAA				4	4.11	0.72

Source: Data from Munteanu, D. et al., *IEEE Trans. Nucl. Sci.*, 54, 994–1001,
2007.

off-state currents was calculated and reported in Table 12.1 for the four simulated structures. For a given gate length, this ratio increases with the equivalent number of gates.
This effect can be explained by the dimensionality of the quantum confinement. In the
double-gate configuration, carriers are confined in one direction (y), while in triple-gate,
omega-FET, and GAA configurations carriers are confined in two directions (y and z),
which significantly enhances quantum effects compared with double-gate devices.

The number of equivalent gates is also a key parameter for radiation-induced response
of multiple-gate devices. Figure 12.19 shows a collection of drain-current transients in different multiple-gate devices obtained by 3D numerical simulation. These results indicate

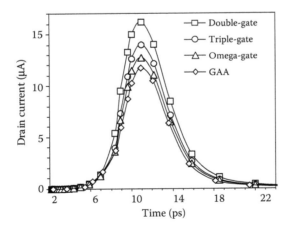

FIGURE 12.19
Classical simulation of drain-current transients induced by an ion strike vertically (y direction) in the middle
of the silicon film in 32 nm gate-length double-gate, triple-gate, omega-gate, and GAA devices. LET = 2 MeV
cm² mg⁻¹.

that the peak value of the drain-current transient is reduced when EGN increases; in this case, the channel is better controlled by the gate and the floating-body effects are strongly reduced. The drain-current transient tail is logically shorter when going from double-gate to GAA devices.

The time evolution of the 3D electron density in the silicon film when an ion vertically strikes the middle of the channel of 32 nm double-gate, triple-gate, and GAA devices has also been investigated in Munteanu et al. (2007). The irradiation track has been simulated in a vertical incidence with a Gaussian shape (radius = 14 nm) and with a Gaussian time dependence, centered on 10 ps (characteristic width = 2 ps). Figure 12.20 shows the electron-density distribution in a vertical cross section (y–z plane) in the middle of the channel. This figure confirms that in the quantum case the maximal value of the electron density is no longer located at the interface (as is the case in the classical approach) but into the silicon film. For the three devices, the quantum electron charge is centered in the middle of the film and the electron density has lower values than in the classical case. In off-state bias condition, the carrier conduction in all devices is mainly dominated by the volume-inversion phenomenon: carriers flow from source to drain over the entire thickness of the silicon film. In the quantum case, the volume-inversion phenomenon is reinforced because the quantum carrier density becomes more centered in the middle of the

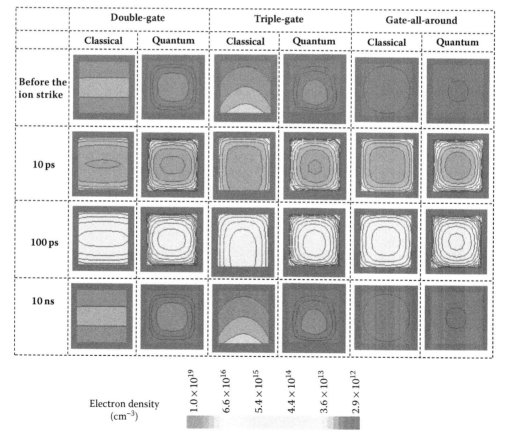

FIGURE 12.20
Classical and quantum electron density in a vertical cross section (y–z plane) in the middle of the channel of 32 nm double-gate, triple-gate, and gate-all-around at different times before and after the ion strike.

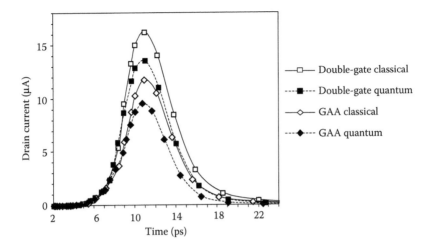

FIGURE 12.21
Comparison between classical and quantum simulation of drain-current transients induced by an ion strike vertically (y direction) in the middle of the silicon film in double-gate and GAA MOSFETs. LET = 2 MeV cm² mg⁻¹.

film (Figure 12.20). This latter effect is enhanced when EGN increases from 2 (for double-gate) to 4 (for GAA). The drain current in the classical and quantum cases is shown in Figure 12.21 for double-gate and GAA devices. The peak current is higher in the classical case than in the quantum case for both devices.

Simulation results in Munteanu et al. (2007) showed that bipolar amplification decreases increasing EGN increases due to reduced floating-body effects. However, the difference between the different architectures of multiple-gate devices vanishes at high LET (Table 12.2), as the charge generated by the ion strike is high enough to collapse the electric field at the body-to-drain junction and to reduce the bipolar amplification. The classical bipolar amplification of these multiple-gate devices, reported in Table 12.2, is very low compared with FDSOI devices, because in multiple-gate devices the control of the channel by the gates is naturally reinforced and strongly reduces the floating-body effects.

The quantum bipolar gain, shown in Figure 12.22, is lower than the classical one, except at very high LET. As explained in Munteanu et al. (2007), two phenomena, with opposite effects on the bipolar gain, are to be considered to explain these results: (i) the lower quantum off-state current leads to a lower quantum bipolar amplification; (ii) in the quantum case, the electron density is lower than the classical one, leading to a slower recombination process (reflected in a longer transient tail) and then to a higher collected charge. Depending on the injection regime, one phenomenon or the other prevails.

Simulation results reported in Munteanu et al. (2006) showed that in double-gate devices the effects of the carrier quantum confinement become more significant when the silicon film is thinner. The same behavior is found in all multiple-gate devices with EGN ≥ 3. As explained previously, this is due to the energy subband splitting, which is directly proportional to the reverse of the square of the potential well dimension (equal to the film thickness). The ratio between the classical and quantum off-state currents (Table 12.1) as a function of silicon film thickness confirms that the quantum confinement is strongly enhanced when the film thickness is reduced. The collected charge and the bipolar gain (Table 12.3) are lower for thinner channels because floating-body effects are reduced when the film thickness is reduced.

TABLE 12.2

Bipolar Gain Simulated in the Classical Case at Different LET Values for 32 nm Gate-Length Multiple-Gate Nanowire MOSFETs

Structure	LET (MeV cm² mg⁻¹)	Bipolar Gain
Double-Gate	0.1	3.12
	1	2.28
	10	1.94
	100	1.40
Triple-Gate	0.1	2.69
	1	1.96
	10	1.82
	100	1.40
Omega-Gate	0.1	1.80
	1	1.62
	10	1.72
	100	1.40
GAA	0.1	1.65
	1	1.59
	10	1.71
	100	1.40

Source: Data from Munteanu, D. et al., *IEEE Trans. Nucl. Sci.*, 54, 994–1001, 2007.

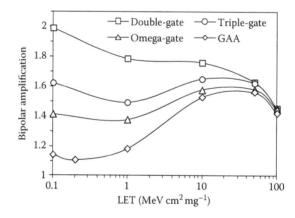

FIGURE 12.22
Bipolar gain simulated in the quantum case as a function of LET in 32 nm gate-length multiple-gate MOSFET.

Finally, the quantum bipolar gain for multiple-gate devices scaled down to 20 nm gate length and 5×5 nm silicon film cross section was also predicted in Munteanu et al. (2007). The difference between the four architectures is reduced for the 20 nm gate length devices compared with the 25 and 32 nm ones, due to the very thin square-wire cross section ($t_{Si} = W = 5$ nm). When the cross section is decreased, the influence of the gate configuration is attenuated and the values of the bipolar gain for the different structures are almost the same. This behavior can be explained by the fact that, around 5 nm and below, the combination of gate electrostatic control and quantum-mechanical confinement leads

TABLE 12.3

Bipolar Gain Simulated in the Quantum Case for Several LET Values in 32 nm GAA MOSFET with Different Silicon Film Thickness

t_{Si} (nm)	LET (MeV cm^2 mg^{-1})	Bipolar Gain
10	0.1	1.14
	1	1.18
	10	1.53
	100	1.42
8	0.1	0.93
	1	0.83
	10	1.37
	100	1.42
5	0.1	0.82
	1	0.78
	10	1.17
	100	1.50

Source: Data from Munteanu, D. et al., *IEEE Trans. Nucl. Sci.*, 54, 994–1001, 2007.
Note: Gate width W = 10 nm.

to similar carrier-density distributions in the film for all gate configurations (Munteanu et al. 2007). At this ultimate scale of integration, it should be expected that the sensitivity of all multiple-gate nanowire architectures (EGN ≥ 2) to heavy-ion irradiation becomes equivalent.

12.3.3 Radiation Hardness of Circuits Based on Multiple-Gate Devices

Concerning the radiation hardness of circuits based on multiple-gate devices, only a few studies are available in literature. Recently, Seifert et al. (2012) reported on radiation-induced SER measured on SRAM logic circuits based on 22 nm triple-gate device technology. This work shows that the SER of SRAM and sequential elements is reduced in the order of 1.5 to 4 times in 22 nm triple-gate compared with 32 nm planar devices for high-energy neutrons and protons (Seifert et al. 2012). The SER reduction is even larger for alpha particles (10× lower), which shows that alpha-particle SER has become negligible in the investigated technology (Seifert et al. 2012). This study confirms the benefits of multiple-gate devices in terms of radiation hardness compared with planar bulk technology (Seifert et al. 2012).

12.4 Bulk and SOI FinFET

FinFET devices can be manufactured on both SOI and bulk substrates (Huang et al. 2001; Nowak et al. 2004). However, many disadvantages characterize SOI wafers compared with bulk wafers, such as the problems of self-heating, cost, density of defect, and so on. Bulk substrates, besides eliminating most of the disadvantages of SOI, are compatible with existing planar CMOS technology processes and reduce manufacturing costs considerably. This explains the great interest in fabricating FinFETs on bulk wafers (Manoj et al. 2008). Figure 12.23a shows the 3D structure of a typical bulk FinFET; the comparison between bulk and SOI FinFET architectures is illustrated in Figure 12.23b.

Concerning radiation effects, the performance of bulk FinFET compared with SOI FinFET is significantly mitigated, as illustrated in the recent publications discussed below. The transient responses of bulk and SOI FinFETs have been studied by El-Mamouni et al. (2011b) using top-side single-photon absorption (SPA). In this work, it was reported that the charge collection of PMOS SOI FinFETs is lower than that of similar bulk devices. Transients measured on bulk FinFET devices were found to be larger and longer than those of SOI FinFETs. SOI FinFETs also have a smaller volume of collection on bulk than their counterparts due to the buried oxide layer, which isolates the active layer from the substrate. Consequently, the collection volume of a SOI FinFET transistor is limited to the fin, while the collection volume of a bulk FinFET also extends into the well region.

Additional measurements using laser testing have been reported in El-Mamouni et al. (2011a). The authors showed that charge collection in 130 nm bulk p-channel FinFETs strongly depends on the structure of the drain region. In this paper, it is also reported that, in the devices tested, charge collection in the drain region masks the contributions of the fins to the charge-collection process (El-Mamouni et al. 2011a). The drain/substrate p-n junction efficiently collects the charge generated in the substrate (El-Mamouni et al. 2011a).

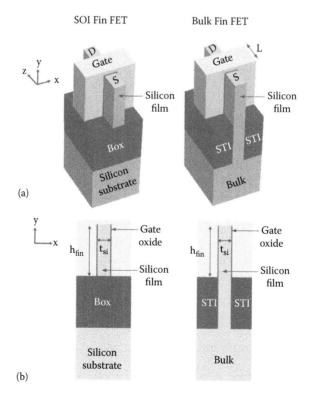

FIGURE 12.23
(a,b) Schematic description of 3D structures of SOI FinFET and bulk FinFET and 2D cross sections (plane x–y) showing the main geometric parameters.

Carriers generated in the substrate below the drain junction can also diffuse to the drain junction, where they are collected. Heavy-ion-induced charge-collection measurements in n-channel bulk FinFETs showed that the greatest amount of collected charge occurs for strikes in the drain region (El-Mamouni et al. 2011a). Device scaling affects the geometry of the fin, but the drain (and source) region may dominate charge collection (El-Mamouni et al. 2011a).

Finally, heavy-ion-induced charge collections obtained from sub-70 nm bulk and SOI FinFETs with both conventional and reduced-area drain regions (saddle layout) have been recently reported by El-Mamouni et al. (2012), as well as drain-current transient measured on these devices. The amount of charge collected in bulk FinFETs with saddle contacts is at least 17% less than that collected in bulk FinFETs with dumbbell contacts (El-Mamouni et al. 2012). This result is extremely important when values of critical charge are in the order of 1–10 fC, as shown in Ball et al. (2010). The shunt effect plays a key role in the charge-collection process in the investigated bulk and SOI FinFETs (El-Mamouni et al. 2012). The small feature size allows the ion track to affect the whole channel region. SOI FinFETs exhibit higher tolerance to SEEs in comparison to their bulk counterparts. The results presented here suggest that improved designs of the drain region will significantly increase bulk FinFET SEE tolerance (El-Mamouni et al. 2012). Charge collection in SOI FinFETs with dumbbell and saddle contacts shows a strong dependence on the substrate bias, with the highest amount of charge collected when the substrate is negatively biased (El-Mamouni et al. 2012). This effect can decrease the SEE tolerance of SOI devices (El-Mamouni et al. 2012).

12.5 Multichannel Architectures with Multiple-Gate Devices

In order to further enhance the ratio of on-state current to off-state current ratio (I_{ON}/I_{OFF}) and to achieve a higher current drivability compared with conventional single-channel devices, 3D vertically stacked nanowire MOSFETs with multiple-gate operation, also called multichannel nanowire MOSFETs (MC-NWFETs), have been proposed (Ernst et al. 2006; Bernard et al. 2007). MC-NWFETs (Figure 12.24) combine the advantages of excellent control of short-channel effects with a high on-state current due to a multiple-gate architecture and the 3D integration of vertically stacked channels. GAA devices with ultrathin and narrow channels (about 10 nm) are seen as the ideal architecture for off-state current control of sub-10 nm gate lengths (Ernst et al. 2006). Meanwhile, the current density per surface of such a device is limited by the lithography pitch, which dictates the distance between nanowires. The current density can be improved by the vertical integration of GAA devices. Thanks to the integration of vertically stacked channels, a fivefold increase in current density per layout surface can be achieved compared with planar transistors with the same gate stack (Ernst et al. 2006).

The transient response of MCFETs submitted to heavy-ion irradiation has been investigated in Munteanu and Autran (2009b) using 3D numerical simulation. The simulated MC-NWFETs have a 3×3 nanowire matrix containing square-cross-section nanowires. The description of the 3D architecture considered here and the definition of the geometric parameters are shown in Figure 12.24. The MC-NWFET matrix is composed of three parallel transistor stacks, each stack containing three vertically stacked nanowire devices (two GAA and one FinFET). The individual nanowire MOSFETs are designed with a 32 nm gate

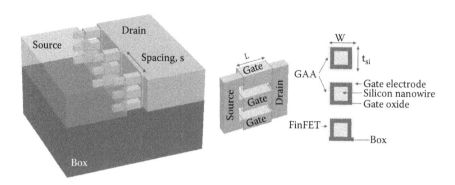

FIGURE 12.24
Schematic description of the 3D simulated MCFET structures and their main geometric parameters. The MCFET matrix is composed of three parallel transistor stacks, each stack containing three vertically stacked nanowire devices (two GAA and one FinFET). All nanowires have silicon film with square section ($t_{si} = W = 10$ nm). For a better view of the nanowires, the gate material, spacers, isolation oxide, and a part of the source and drain regions are not shown. (Reprinted from Munteanu, D., Autran, J.L., Transient response of 3D multi-channel nanowire MOSFETs submitted to heavy ion irradiation: A 3D simulation study. *IEEE Trans. Nucl. Sci.* 56, 2042–2049. © (2009) IEEE. With permission.)

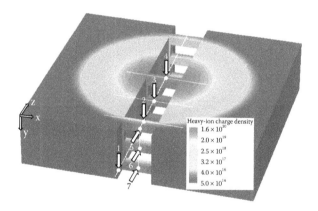

FIGURE 12.25
3D profile of the heavy-ion charge density for ion-strike location "4." The positions (*arrows*) of the ion strike considered in this work are also shown; the ion strikes in the middle of the channel (between the source and drain regions). For a better view of the nanowires, the gate material, spacers, and isolation oxide are not shown. (Reprinted from Munteanu, D., Autran, J.L., Transient response of 3D multi-channel nanowire MOSFETs submitted to heavy ion irradiation: A 3D simulation study. *IEEE Trans. Nucl. Sci.* 56, 2042–2049. © (2009) IEEE. With permission.)

length and a square cross section, with $t_{si} = W = 10$ nm, and a 3 nm-thick gate oxide. An intrinsic silicon film and a midgap gate are included. Three lateral spacings between the nanowire stacks are considered: 100, 75, and 50 nm.

Three-dimensional numerical simulations have been performed following the same procedure as previously described in Sections 12.2.2.2 and 12.2.2.3. Two characteristic radii have been considered for the spatial Gaussian dependence of the ion track: 50 and 20 nm. The time distribution of the ion track had a Gaussian shape, centered on 10 ps, and a characteristic width of 2 ps. The ion struck the middle of the channel between the source and the drain and perpendicular to the gate electrode, as shown in Figure 12.25. The different ion-strike locations considered in Munteanu and Autran (2009b) are schematically presented in Figure 12.25. Ion-strike locations "1," "2," "3," and "4" are parallel to the y axis (perpendicular to the x–z plane). The lateral spacing between locations "1," "2," "3," and "4" was equal to s/2. The 3D profile of heavy-ion charge density generated in the structure is shown in Figure 12.25 for the ion-strike location "4." The MC-NWFET was biased in the off state ($V_G=0$ V), and the drain terminal was constantly biased at 0.8 V.

In Munteanu and Autran (2009b), the evolution in time after the ion strike of electrostatic potential and electron density in the nine individual devices of the MC-NWFET matrix was analyzed. In addition, the authors investigated in detail the influence of the ion-strike parameters (location, direction, and radius of the ion track) and the lateral spacing between the nanowire stacks on the current transient and charge collection. The results showed that the drain-current transients and collected charge strongly depend on the ion-strike location, direction, and track radius. An example of drain-current transients produced by the ion strike is plotted in Figure 12.26 for the ion strike locations "1" to "4" and

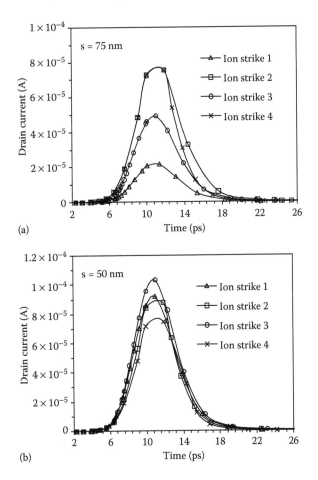

FIGURE 12.26
Drain-current transients induced by an ion striking vertically (parallel to y axis) in the middle of the structure. Two lateral spacings are simulated, (a) s=75 nm and (b) s=50 nm, and the four locations "1" to "4" are considered for the ion strike. The MCFET is off-state biased ($V_G=0$ V, $V_D=0.8$ V). The ion LET is 10 MeV cm² mg⁻¹ and the ion track radius is 50 nm. (Reprinted from Munteanu, D., Autran, J.L., Transient response of 3D multi-channel nanowire MOSFETs submitted to heavy ion irradiation: A 3D simulation study. *IEEE Trans. Nucl. Sci.* 56, 2042–2049. © (2009) IEEE. With permission.)

for s = 75 and s = 50 nm. The collected charge corresponding to these current transients is shown in Figure 12.27. The lateral spacing between adjacent nanowire stacks is found to be a key parameter in the analysis of the worst-case location of the ion strike. It was shown that, for a large lateral spacing between stacks compared with the ion-track radius, strikes centered on any nanowire produce almost the same current transients. In this case, the transient peak is higher than that obtained for a strike between nanowires on the isolation oxide. On the contrary, for a small lateral spacing, comparable to the ion-track radius, the highest current peak is obtained for a strike between the nanowires. However, the highest collected charge is obtained for the strike on the nanowire situated on the center of the MC-NWFET matrix.

The results presented in Munteanu and Autran (2009b) also showed that the charge collection is very fast in a MC-NWFET for all ion-strike parameters and configurations. This is due to the multiple-gate devices, which have small active volumes that allow all the excess charge to be quickly evacuated. The MC-NWFET simulated in Munteanu and Autran (2009b) had a total transient duration of 10 ps at 10% of the peak value for LET = 10 MeV cm^2 mg^{-1},

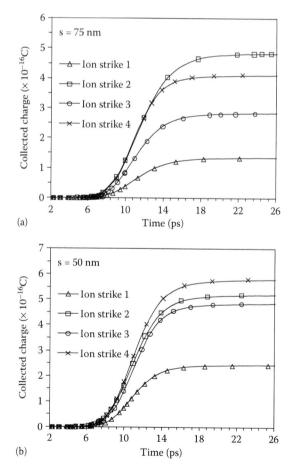

FIGURE 12.27
(a,b) Collected charge for the drain-current transients presented in Figure 12.26. (Reprinted from Munteanu, D., Autran, J.L., Transient response of 3D multi-channel nanowire MOSFETs submitted to heavy ion irradiation: A 3D simulation study. *IEEE Trans. Nucl. Sci.* 56, 2042–2049. © (2009) IEEE. With permission.)

which is almost identical to that of an individual GAA device, but lower than that obtained in simulation of single-gate FDSOI devices. From these results, one could expect better immunity to single-event phenomena of MC-NWFETs compared with other conventional structures, such as FDSOI devices. This will probably have a consequence for the behavior under irradiation of circuits based on these devices. However, single-device behavior is not enough to determine the circuit sensitivity to single events, because this also depends on the load capacitance. Since the single-event transients of MC-NWFETs are high bandwidth, they are very sensitive to inductive and reactive capacitance (i.e., node loading) in the circuit. More detailed study concerning this point is needed to exactly quantify the sensitivity of MC-NWFETs-based circuits to single events.

12.6 Multiple-Gate and Multichannel Devices with Independent Gates

In spite of the excellent electrical performance due to their multiple conduction surfaces, conventional multiple-gate MOSFETs or MC-NWFETs provide only three-terminal (3T) operation, because devices are designed with a single-gate electrode (surrounding the channel) or, in the case of double-gate devices, the two gates are tied together. Planar double-gate, FinFET, and MC-NWFET structures with independent gates have been recently proposed (Yang and Fossum 2005; Masahara et al. 2005; Pei and Kan 2004; Autran and Munteanu 2008; Mathew et al. 2004; Moreau et al. 2008; Zhang et al. 2005), which makes possible a four-terminal (4T) operation. These devices offer novel possibilities, such as dynamic threshold voltage control by one of the two gates, transconductance modulation, and signal mixer, in addition to the conventional switching operation. Thus, 4T-dual-gate field effect transistor (DGFET), 4T-FinFET, and 4T-MC-NWFET are promising for future high-performance and low-power-consumption integrated circuits.

The radiation response of independent-gate devices has been studied by 3D numerical simulation in Munteanu and Autran (2009a, 2012b) and compared with their 3T-device counterparts. In Munteanu and Autran (2009a), 3D numerical simulation was used to investigate the sensitivity to single events of 4T-DGFET devices compared with that of conventional 3T-DGFET. The impact of the second gate bias on the transient response and bipolar amplification of the device submitted to heavy-ion irradiation was particularly addressed. The results showed that the bipolar amplification is higher in independent-gate devices for both positive and negative back-gate biases. The response to heavy-ion irradiation of FinFET and MC-NWFET operated with independent gates is investigated by 3D numerical simulation in Munteanu and Autran (2012b); in the following we detail the main results of this study.

12.6.1 Simulation Details

Figure 12.28 schematically defines the 3D architectures analyzed in Munteanu and Autran (2012b) and the main geometric parameters of the simulated devices. FinFET devices had the following parameters: an intrinsic fin body with a 10 nm-thick film thickness (or fin width, t_{Si}), a fin height $h_{fin} = 62$ nm, a gate length of 32 nm, and 3 nm-thick front and back-gate oxides. MC-NWFET structures, in both 3T-MC-NWFET and 4T-MC-NWFET configurations, were based on the fabricated devices reported in Dupre et al. (2007). These devices had three vertically stacked nanowire devices, as shown in Figure 12.28. The 3T-MC-NWFET

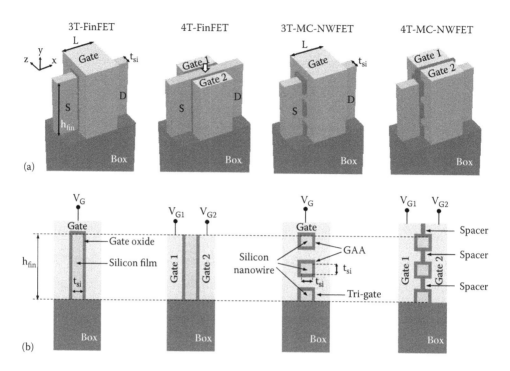

FIGURE 12.28
(a) Schematic description of the 3D simulated FinFET and MC-NWFET structures with three and four terminals considered in this work. The main geometric parameters used in simulation are also defined (S = source region and D = drain region). The position of the ion strike is also indicated by an *arrow*; the ion strikes vertically in the middle of the channel (between the source and drain regions). (b) Schematic cross-sections in the y–z plane, showing the bias conditions specific to each device. For MC-NWFET structures, this figure also indicates the particular devices which compose the nanowire stack. (Reprinted from Munteanu, D., Autran, J.L., Simulation analysis of bipolar amplification in independent-gate FinFET and multi-channel NWFET submitted to heavy-ion irradiation. *IEEE Trans. Nucl. Sci.* 59, 3249–3257. © (2012) IEEE. With permission.)

contains two 3T-GAA devices and one 3T-Trigate MOSFET. The 4T-MC-NWFET contains two 4T-ΦFETs (Dupre et al. 2007) and one 4T-FinFET. The individual nanowire MOSFETs have been designed with a gate length of 32 nm, square cross section with $t_{Si} = 10$ nm, and 3 nm-thick gate oxide. An intrinsic silicon film was also considered for each individual nanowire. For all devices (FinFET and MC-NWFET), the channel doping (intrinsic) as well as the source and drain doping (n type) were uniform. Abrupt doping profiles were used between the channel and the highly doped source and channel regions. Midgap gates were used for all devices. In 3T-devices, the gate was biased at V_G, and in 4T-devices the gates were biased independently at V_{G1} (front gate or gate 1) and V_{G2} (back gate or gate 2). The drain voltage was kept constant at $V_D = 0.8$ V, and the source voltage was always $V_S = 0$ V.

Three-dimensional numerical simulations were performed as previously explained in Sections 12.2.2.2 and 12.2.2.3. The ion strike was modeled using a carrier-generation function, which has a Gaussian radial distribution with a characteristic radius of 20 nm and a time distribution with a Gaussian shape, centered on 10 ps and with a characteristic width of 2 ps. Devices were biased in the off state (which means that $V_G = 0$ V in three terminal devices and $V_{G1} = 0$ V in four terminal devices). The ion was considered to strike in the middle of the channel (Figure 12.28), and the ion strike was simulated in vertical incidence (parallel to the y axis). The deposited charge was calculated considering the Gaussian distribution of the ion track and the 3D geometry of the silicon film: the fact that the ion track is not completely contained in the active device has been carefully taken into account in the calculation of the deposited charge.

12.6.2 FinFET Devices

12.6.2.1 Static Characteristics

The steady-state drain-current characteristics in 3T-FinFET and 4T-FinFET simulated in Munteanu and Autran (2012b) are plotted in Figure 12.29. This figure shows that the drain current is deeply modified by the bias of the second gate. The off-state current of the 4T-FinFET is lower than that of the 3T-FinFET for negative V_{G2} and higher for positive V_{G2}. The off-state current in 3T-FinFET and 4T-FinFET biased at $V_{G2}=0$ V is almost the same, but the subthreshold swing (SS) (inverse of the subthreshold slope) is lower (better) in 3T-FinFET. The SS is always lower in 3T-FinFET than in 4T-FinFET (for all V_{G2}) due to a better control of short-channel effects in 3T-FinFET. The curves presented in Figure 12.29 show that the bias applied to the back gate (at V_{G2}) modifies not only the drain current but also all the key electrical parameters in the regime below the threshold, such as threshold voltage, off-state current, and SS. This is due to the progressive increase of the electron density at the back interface when the back interface switches from an accumulation regime (for a negative V_{G2}) to a strong inversion regime (for a high positive V_{G2}). Thus, when V_{G2} increases, threshold voltage decreases, SS is degraded, and off-state current strongly increases. For a V_{G2} higher than the threshold voltage, the back-interface channel turns on and two conduction channels exist simultaneously in the device, which additionally enhances (degrades) the off-state current.

12.6.2.2 Transient Simulation Results

The time variations of the electron-density profile in 4T-FinFET after the ion strike were studied and illustrated in Munteanu and Autran (2012b). These results showed that the ion

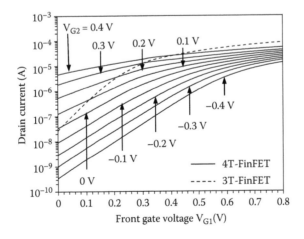

FIGURE 12.29
Drain-current characteristics as function of V_{G1} for 4T-FinFET with different back-gate biases V_{G2}. The drain current vs. V_G of 3T-FinFET is also reported for comparison. For all curves, $V_D=0.8$ V and $V_S=0$ V. (Reprinted from Munteanu, D., Autran, J.L., Simulation analysis of bipolar amplification in independent-gate FinFET and multi-channel NWFET submitted to heavy-ion irradiation. *IEEE Trans. Nucl. Sci.* 59, 3249–3257. © (2012) IEEE. With permission.)

strike strongly perturbs the electron-density profiles in all devices. It was demonstrated that the electron-density distribution at t = 10 ps (maximum of the deposited charge) is considerably modified as compared with the density profile in steady-state conditions; for all devices, the electron density at t = 10 ps is much higher than that at steady state, due to the high charge density deposited in the film by the ion. After t = 10 ps, the structure begins to return to the steady state, and then the electron density begins to progressively decrease. However, as shown in Munteanu and Autran (2012b), at t = 100 ps after the ion strike, no structure has yet reached the steady state. To better visualize the time variation of the main internal parameters and to estimate the time required for return to steady state, Figure 12.30a and b plot the 1D potential profile in a cutline along the x axis in the middle of the channel for 3T-FinFET and 4T-FinFET ($V_{G2} = -0.4$ V), respectively. The potential profile at different times has been considered, including the steady-state potential profile. These figures show that the potential reaches the initial (steady-state) profile much more quickly in 3T-FinFET than in 4T-FinFET operating under $V_{G2} = -0.4$ V. The

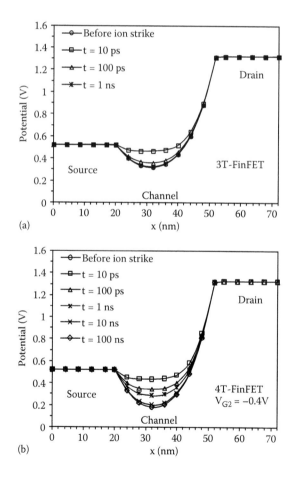

FIGURE 12.30
(a,b) 1D potential profiles in a cutline along the x axis in the middle of the channel at different times before and after the ion strike. The ion LET is 0.1 MeV cm² mg⁻¹. The symbols of the curve "Before ion strike" are superposed with those of the curve "100 ns" in Figure 12.30b. (Reprinted from Munteanu, D., Autran, J.L., Simulation analysis of bipolar amplification in independent-gate FinFET and multi-channel NWFET submitted to heavy-ion irradiation. *IEEE Trans. Nucl. Sci.* 59, 3249–3257. © (2012) IEEE. With permission.)

time needed to return to equilibrium is about (lower than) 1 ns for 3T-FinFET, whereas the 4T-FinFET has not yet reached the steady state at t = 10 ns (Munteanu and Autran 2012b).

In the same study, the collected charge in 3T-FinFET was compared with that of 4T-FinFET; this is an interesting analysis because the particular electrostatic conditions related to the 4T devices could influence the charge collection. The time variation of the collected charge induced by the ion strike is plotted in Figure 12.31 for LET = 0.1 MeV cm^2 mg^{-1}. To calculate the collected charge, the nominal steady-state leakage current (no ion track) is subtracted from the drain current (with ion track), so that the curves of the collected charge saturate at high time values. Figure 12.31 shows that the collected charge is higher in 4T-FinFET than in 3T-FinFET. A possible explanation, involving carrier recombination, is given in Munteanu and Autran (2012b) and will be detailed in Section 12.6.4. It is important to note here that the collected charge shown in Figure 12.31 was obtained from very small transient currents (comparable to the nominal leakage current) that persist for a relatively long time. This small, temporary leakage current increase is unlikely to be relevant to single-event effects. However, extending the time axis in Figure 12.31 far enough for the curves to saturate facilitates the understanding of the physical mechanisms of the charge-collection process.

The parasitic bipolar amplification of each structure calculated in Munteanu and Autran (2012b) is plotted in Figure 12.32a and b as a function of LET, for negative and positive V_{G2}, respectively. The bipolar gain for 3T-FinFET is also shown to facilitate the comparison. The bipolar gain curves exhibit a particular shape compared with that usually obtained for single-gate FDSOI MOSFETs (Ferlet-Cavrois et al. 2004, 2005; Munteanu et al. 2006). Indeed, even at very low LET, the low-injection regime of the parasitic bipolar transistor is not visible on the curves for either 3T-FinFET or 4T-FinFET. Only the second part of the characteristic curve, corresponding to the high-injection regime, is present for these

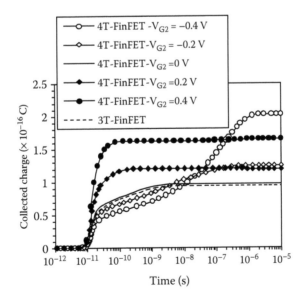

FIGURE 12.31

Collected charge transient in 3T-FinFET and 4T-FinFET. The transistors are biased in off state (V_G = 0 V for 3T-FinFET and V_{G1} = 0 V for 4T-FinFET). V_D = 0.8 V and V_S = 0 V. LET = 0.1 MeV cm^2 mg^{-1}. (Reprinted from Munteanu, D., Autran, J.L., Simulation analysis of bipolar amplification in independent-gate FinFET and multi-channel NWFET submitted to heavy-ion irradiation. *IEEE Trans. Nucl. Sci.* 59, 3249–3257. © (2012) IEEE. With permission.)

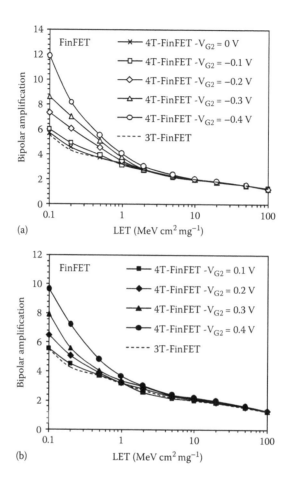

FIGURE 12.32
(a,b) Bipolar amplification as function of LET in 3T-FinFET and 4T-FinFET at different back-gate biases. (Reprinted from Munteanu, D., Autran, J.L., Simulation analysis of bipolar amplification in independent-gate FinFET and multi-channel NWFET submitted to heavy-ion irradiation. *IEEE Trans. Nucl. Sci.* 59, 3249–3257. © (2012) IEEE. With permission.)

devices. This is due to the very small dimensions of these transistors, which induce a particular sensitivity to the charge density deposited by the ion.

As expected, the bipolar gain has the same behavior as the collected charge: the gain is always higher in 4T-FinFET than in 3T-FinFET. This behavior is in perfect agreement with previous simulation results in Munteanu and Autran (2009a). Figure 12.32 also shows that, for LET > 5 MeV cm² mg⁻¹, the bipolar gain of all devices becomes the same. This is probably due to the high generated charge, which reduces the electric field at the body-to-drain junction, the impact ionization, and the bipolar amplification. The huge generated charge in this case masks the influence of the electrostatic conditions of the device.

12.6.3 MC-NWFET Devices

12.6.3.1 Static Characteristics

The 3D profiles of electrostatic potential simulated in Munteanu and Autran (2012b) for 3T-MC-NWFET and 4T-MC-NWFET are shown in Figure 12.33a. 3T-MC-NWFET is biased at $V_G = 0$ V and 4T-MC-NWFET is biased with $V_{G2} = -0.2$ and $V_{G2} = 0.2$ V ($V_{G1} = 0$ V). This fig-ure shows that, as expected, the potential is symmetrical with respect to the silicon/oxide interfaces in 3T-MC-NWFET. This behavior is common to all multiple-gate devices biased with 3T operation (Munteanu et al. 2005). In 4T-MC-NWFET the potential is modulated by the back-gate bias, which breaks the symmetry with respect to the silicon/oxide interfaces. This is confirmed in Figure 12.33b, where the electron density in a 2D cross section (plane y–z, Figure 12.28) is plotted for 4T-MC-NWFET biased in strong inversion at the front inter-face ($V_{G1} = 0.8$ V) and two different biases on the back gate. 3T-MC-NWFET biased in strong inversion ($V_G = 0.8$ V) is also shown for comparison. In 3T-MC-NWFET, the electron den-sity is symmetrical with respect to the silicon/oxide interfaces. The symmetry is broken in 4T-MC-NWFET when the second gate bias is different from the first gate bias. This figure indicates that the back-gate bias has a considerable influence on the internal parameters of 4T-MC-NWFETs. Figure 12.33 also shows that the electron density and the potential at the second gate interface are lower in 4T-MC-NWFET than in 3T-MC-NWFET. The electron

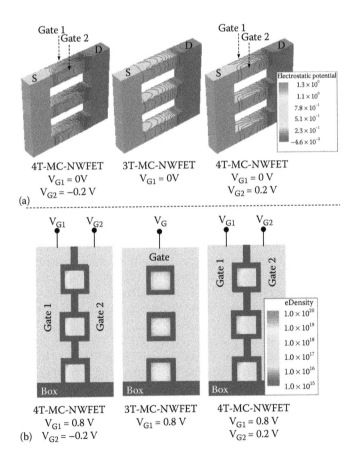

FIGURE 12.33
(a) 3D electrostatic potential profile in 3T-MC-NWFET and 4T-MC-NWFET at different V_{G2}. For a better view, the gate material, oxide gate, and spacers are not shown. $V_G = 0$ V for 3T-FinFET and $V_{G1} = 0$ V for 4T-FinFET (S = source region and D = drain region). (b) 2D electron-density profile in a vertical cross section (parallel to y–z plane, Figure 12.28) in the middle of the channel. $V_D = 0.8$ V. (Reprinted from Munteanu, D., Autran, J.L., Simulation analysis of bipolar amplification in independent-gate FinFET and multi-channel NWFET submitted to heavy-ion irradiation. *IEEE Trans. Nucl. Sci.* 59, 3249–3257. © (2012) IEEE. With permission.)

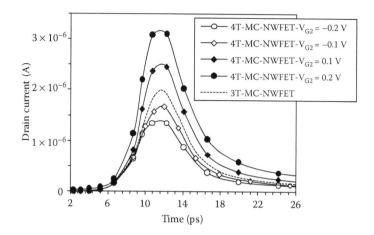

FIGURE 12.34
Drain-current transients in 3T-MC-NWFET and 4T-MC-NWFET. The transistors are biased in off state ($V_G = 0$ V for 3T-MC-NWFET and $V_{G1} = 0$ V for 4T-MC-NWFET). LET = 0.1 MeV cm² mg⁻¹. (Reprinted from Munteanu, D., Autran, J.L., Simulation analysis of bipolar amplification in independent-gate FinFET and multi-channel NWFET submitted to heavy-ion irradiation. *IEEE Trans. Nucl. Sci.* 59, 3249–3257. © (2012) IEEE. With permission.)

density (as well as the potential) at the back interface increases when V_{G2} increases from negative to positive values.

12.6.3.2 Transient Simulation Results

The time variations of the drain current induced by the ion strike, as simulated in Munteanu and Autran (2012b), are plotted in Figure 12.34 for LET = 0.1 MeV cm² mg⁻¹. The drain-current peak in 4T-MC-NWFET is higher than in 3T-MC-NWFET for positive V_{G2} and lower for negative V_{G2}. This behavior is explained as follows. Previous work (Munteanu and Autran 2008) has shown that the off-state current has a significant influence on the drain-current peak. When the off-state current increases, the drain-current peak increases. For the devices considered in the present analysis, the back-gate bias induces a strong variation in the threshold voltage and in the off-state current. Thus, when V_{G2} decreases, the threshold voltage increases, which leads to the decrease of the off-state current and of the peak of the drain-current transient.

As for the FinFET, the bipolar amplification (Figure 12.35) is higher in 4T-MC-NWFET (for both negative and positive second gate bias) than in 3T-MC-NWFET (Munteanu and Autran 2012b). This is due to more important floating-body effects in 4T operation than in 3T operation, induced by less effective control by the gate of the channel potential in 4T configuration. As explained in Munteanu and Autran (2012b), this behavior is consistent with values of the SS, which were higher in 4T-MC-NWFET (for all back-gate biases) than in 3T-MC-NWFET.

12.6.4 Comparison between FinFET and MC-NWFET Devices

In this last section, we consider the discussion detailed in Munteanu and Autran (2012b) concerning the comparison between FinFET and MC-NWFET devices in terms of bipolar amplification in 3T devices and bipolar gain variation as a function of V_{G2} in 4T devices. The bipolar gain of FinFET and MC-NWFET is also compared with results obtained in previous work for similar devices.

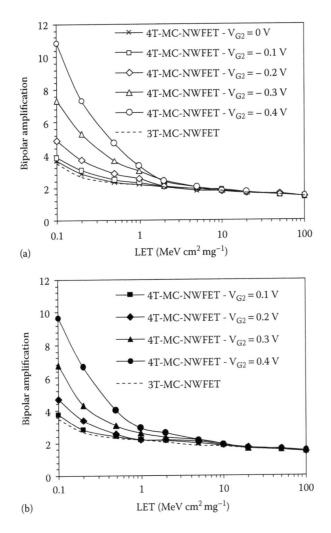

FIGURE 12.35
(a,b) Bipolar gain versus LET in 3T-MC-NWFET and 4T-MC-NWFET. (Reprinted from Munteanu, D., Autran, J.L., Simulation analysis of bipolar amplification in independent-gate FinFET and multi-channel NWFET submitted to heavy-ion irradiation. *IEEE Trans. Nucl. Sci.* 59, 3249–3257. © (2012) IEEE. With permission.)

The simulations in Munteanu and Autran (2012b) showed that, for a given LET, FinFET devices collect more charge than MC-NWFET, in both 3T and 4T (at a given V_{G2}) configurations. Similarly to the collected charge, the bipolar amplification is found to be higher in 3T-FinFET than in 3T-MC-NWFET for all LET values. These results are in good agreement with the results of the previous study (Munteanu et al. 2007) described in Section 12.3.2, which have shown that increasing the number of gates reduces the bipolar gain, because the floating-body effects are reduced by better control of the silicon film by the gate. In the same manner, at a given V_{G2}, the bipolar amplification is higher in 4T-FinFET than in 4T-MC-NWFET. This remark is valid for all V_{G2} values simulated here.

Particular attention was given in Munteanu and Autran (2012b) to the gain variation as a function of V_{G2} at low values of LET, shown in Figure 12.36a for 4T-FinFET and in Figure 12.36b for 4T-MC-NWFET. In these figures, the bipolar gain is minimum at $V_{G2} = 0$ V;

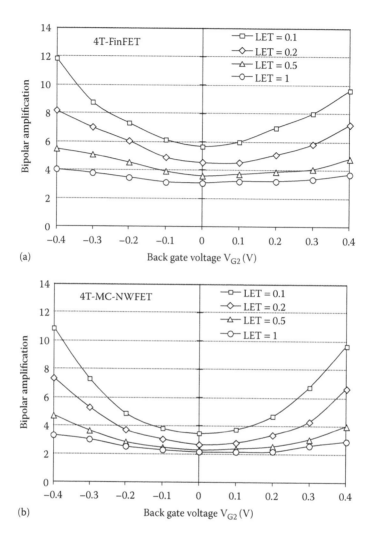

FIGURE 12.36
Bipolar amplification at low LET values as function of V_{G2} in 4T-FinFET (a) and 4T-MC-NWFET (b). LET values are expressed in MeV cm² mg⁻¹. (Reprinted from Munteanu, D., Autran, J.L., Simulation analysis of bipolar amplification in independent-gate FinFET and multi-channel NWFET submitted to heavy-ion irradiation. *IEEE Trans. Nucl. Sci.* 59, 3249–3257. © (2012) IEEE. With permission.)

it increases when the positive V_{G2} increases from 0 to 0.4 V and when the negative V_{G2} decreases from 0 to −0.4 V. The increase in bipolar gain is more important for negative V_{G2} than for positive V_{G2} (Munteanu and Autran 2012b). As shown in Figure 12.36, this behavior of bipolar gain versus V_{G2} is similar in FinFET and MC-NWFET. In order to better understand this point and gain insight into device operation, the SRH recombination rate was analyzed in Munteanu and Autran (2012b). The results have indicated that the SRH rate is higher for positive V_{G2} than for negative V_{G2}. Generally, the recombination rate increases when the body potential increases (Munteanu and Cristoloveanu 1999; Sze 1985) or when the carrier lifetime is reduced (Munteanu et al. 1998). In Munteanu and Autran (2012b), the same carrier lifetimes have been considered in simulation for all back-gate biases. But, in devices operated with independent gates, a positive V_{G2} induces a higher body potential than a negative V_{G2}. This leads to a higher recombination rate at the source–body junction for positive V_{G2}. The consequence is that the deposited charge is more quickly evacuated for positive V_{G2}. The collected charge and the bipolar amplification are, then, lower for positive V_{G2} than for negative V_{G2}. The same behavior of the recombination rate is found in MC-NWFET.

Finally, the 3T-FinFET simulated in Munteanu and Autran (2012b) exhibits a low bipolar amplification, lower than that found in Munteanu and Autran (2009a). This is mainly due to the better control of the film potential by the gate due to the thinner silicon film used in the present work. Similarly to the 3T-FinFET, the 4T-FinFET studied here has lower bipolar gain values than those presented in Munteanu and Autran (2009a) for both negative and positive V_{G2}. However, the bipolar gain in 3T-FinFET is found to be higher than that obtained in Munteanu et al. (2007). This was explained by the different geometric parameters of the structures simulated in Munteanu et al. (2007), where the 3T-FinFET was in fact a triple-gate MOSFET configuration, designed with a very narrow width and a fin height equal to the silicon film thickness ($h_{fin} = 10$ nm in Munteanu et al. [2007] compared with $h_{fin} = 62$ nm in Munteanu and Autran [2012b]). Then, the floating-body effects induced by the presence of the buried oxide were reduced in the devices simulated in Munteanu et al. (2007), which have decreased collected charge and bipolar amplification compared with those obtained in Munteanu and Autran (2012b). Also, the bipolar gain of the 3T-MC-NWFET simulated in Munteanu and Autran (2012b) is higher than the bipolar amplification of the 3T-NWFET (GAA device) simulated in Munteanu et al. (2007). This is probably due to the presence of a triple-gate device in the nanowire stack of the 3T-MC-NWFET; the triple-gate device has an intrinsically higher bipolar gain than the GAA device, which leads to an increase of the total bipolar amplification of the three-device stack contained in the 3T-MC-NWFET.

12.7 Junctionless Devices

In recent years, a new concept of field-effect MOS transistor without junctions (called junctionless MOSFET) has been proposed (Colinge et al. 2009, 2010; Lee et al. 2009, 2010; Chen et al. 2010; Kranti et al. 2010), experimentally validated, and analyzed in detail (Parihar et al. 2013; Barbut et al. 2013). A junctionless MOSFET (Figure 12.37a) is a transistor with the same type of semiconductor throughout the entire silicon film, including the source, channel, and drain regions. A double-gate junctionless MOSFET (JL-DGFET), then, contains a heavily doped silicon film sandwiched between two gate electrodes connected together. The two gates are used to deplete the silicon film and then turn off the device, or to accumulate majority carriers from the doped silicon layer and then turn on the device. This structure presents a real advantage, since its fabrication process is simplified compared with the conventional process (there are no doping gradients in the device [Kranti et al. 2010] and no semiconductor-type inversion). The off-state current of these devices is no longer degraded by the leakage current of the reversely biased source-channel and channel-drain diodes, but is uniquely controlled by the gate. This could be very attractive for ultrashort devices, typically for decananometer channel lengths, for which the off-state current could be reduced.

Although standard DGFETs with junctions (Figure 12.37b), also called inversion-mode (IM) DGFETs, and JL-DGFETs are very similar, the operating principle of junctionless

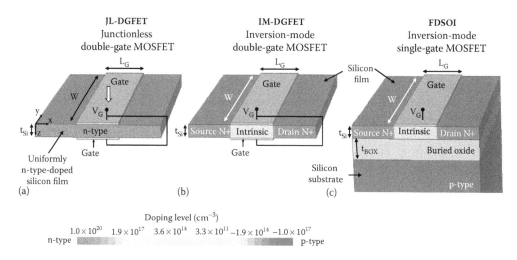

FIGURE 12.37
Schematic description of the simulated JL-DGFET (a), IM-DGFET (b), and FDSOI (c) structures considered in this work. The doping-level distribution in each device is shown and the main geometric parameters are defined. For a better view, the spacers and isolation oxide are not shown. The position of the ion strike is indicated by the *arrow*; the ion strikes vertically in the middle of the channel and in a direction parallel to the z axis.

devices is quite different from that of IM-DGFETs. The conventional IM-DGFET is normally in the off state under $V_G = 0$ V, and the source-channel and channel-drain junctions are reversely biased. A voltage must be applied on the gate to turn on the transistor. The vertical electric field created across the gate insulator attracts minority carriers at the silicon/insulator interface to create an inversion (conduction) channel, and then these carriers flow from source to drain regions through this channel. Thus, in the IM-DGFET transistor, the electric field is highest when the transistor is in the on state and the lowest in the off state. For JL-DGFET, the situation is reversed: the electric field is high in the off state (at $V_G = 0$ V) and very low in the on state (Kranti et al. 2010). In the junctionless transistor, the work-function difference between the gate and the doped silicon film leads to a positive flat-band voltage. Therefore, in the on state (high V_G), the junctionless transistor is under flat-band conditions and the transverse electric field is zero (Colinge et al. 2009). The current (majority carriers) flows into the channel, which extends throughout the entire silicon film (Colinge et al. 2009), unlike conventional devices, in which conduction takes place at the silicon/insulator interface. Thus, at high gate voltages, the conduction takes place in the film volume and not at the interface. This can be beneficial for carrier mobility because the impact of the surface roughness is reduced. In order to switch off the transistor, a low gate voltage has to be applied on the gates; the vertical electric field increases and depletes the film, which cuts off the transistor (in contrast to IM-DGFET, in which the high electric field is used to create an inversion layer at the interface).

The JL-DGFET technology is potentially interesting for future ultrascaled devices, due to the simplified technological process, reduced leakage current, and potential enhanced performance resulting from volume conduction and better carrier mobility. In the context of microelectronics, characterized by industrial needs for highly reliable circuits in a wide area of applications (medical, aerospace, automotive, networking, and nuclear engineering), it is important to investigate in detail the sensitivity to radiation of the JL-DGFET technology with respect to the radiation sensitivity of other more conventional technologies envisaged for high-reliability concerns, such as IM-DGFET and single-gate FDSOI (Figure 12.37c).

From a radiation-hardness point of view, the high doping level in the silicon film of a JL-DGFET could have a negative impact on its immunity to single events, because floating-body effects are expected to be strong. Then, in spite of its double-gate configuration, JL-DGFET should be more sensitive to radiation than IM-DGFET, for which the channel is intrinsic. The transient response of JL-DGFET has been analyzed in different studies using 3D numerical simulation at both device and circuit levels (Munteanu and Autran 2012a, 2014a,b). The behavior of JL-DGFET has been also systematically compared with that of more conventional IM devices: IM-DGFET and FDSOI. At device level, the bipolar amplification of JL-DGFET was studied in Munteanu and Autran (2012a) and compared with that of IM-DGFET and FDSOI with similar geometric parameters. However, the bipolar gain estimation on individual devices is not sufficient to evaluate the radiation hardness of a given technology. For this reason, additional simulations have been performed at circuit level in Munteanu and Autran (2014b) in order to carefully determine the SEU-threshold LET of a six-transistor (6T) SRAM cell based on JL-DGFET.

12.7.1 Simulation Details

Figure 12.37 shows a schematic 3D description of the devices simulated in Munteanu and Autran (2014a). Planar IM-DGFET structures were based on real devices reported in Vinet et al. (2005). These devices have been designed with 20 nm channel length, 100 nm gate width, 6 nm-thick silicon film, and 1 nm-thick gate oxide. The channel is intrinsic; source and drain regions are highly n-type doped, and the doping profile in these regions is uniform. Additionally, the transition between the channel and the highly doped source/drain regions is characterized by an abrupt doping profile. JL-DGFET have exactly the same geometric dimensions as IM-DGFET structures, with the notable exception that all the silicon film is uniformly n-type doped at 10^{19} cm^{-3}. In other words, there are no highly doped source/drain regions, as shown in Figure 12.37a. It should be noted that the channel thickness has to be sufficiently small in order to make possible the complete depletion of the silicon film and to be able to cut off the device (Lee et al. 2009). This condition is satisfied for the doping level and the film thickness considered in Munteanu and Autran (2014a). The two gates are connected together in both IM-DGFET and JL-DGFET architectures. Finally, for FDSOI devices, the silicon film has the same geometric parameters and doping profiles as the silicon film of IM-DGFET. However, only a single gate controls the electrostatic potential and the current flow in the film of FDSOI. A 10 nm-thick buried oxide and a thick silicon substrate lowly doped at 10^{16} cm^{-3} have also been considered. The very thin buried oxide is necessary to minimize the short-channel effects in this FDSOI device. All devices were calibrated to meet the requirements of the International Technology Roadmap for Semiconductors (ITRS) low-power technology node corresponding to the year 2015. To facilitate comparison, the gates' work function was refined to achieve the same off-state current (I_{OFF}) for all devices. The simulator and models used in this study are the same as in Section 12.2.2.2.

The steady-state drain-current characteristics of JL-DGFET, IM-DGFET, and FDSOI have been simulated in Munteanu and Autran (2014a). As shown in Table 12.4, the three devices have the same off-state current, but different SSs and on-state currents (Munteanu and Autran 2014a). While double-gate devices (both JL-DGFET and IM-DGFET) exhibit near-ideal SSs (65 mV dec^{-1}), FDSOI has a much higher SS (90 mV dec^{-1}) because of the single-gate configuration, which reduces the control by the gate of the channel potential and

TABLE 12.4

Off-State Current, Subthreshold Swing and On-State Current
for JL-DGFET, IM-DGFET, and FDSOI MOSFETs ($V_D = 0.75$ V)

Structure	JL-DGFET	IM-DGFET	FDSOI
Off-state current, I_{OFF} (nA μm^{-1})	2.2	2.2	2.2
Subthreshold swing, S (mV dec^{-1})	65	65	90
On-state current, I_{ON} (μA μm^{-1})	576	1641	657

increases the parasitic short-channel effects compared with a double-gate configuration. The highest on-state current is obtained in IM-DGFET, due to the combination of a double-gate structure and an intrinsic channel; this structure has the advantage of maximizing carrier mobility. In JL-DGFET, the highly doped silicon film degrades the mobility, and the on-state current is the lowest in spite of a double-gate configuration. The on-state current related to the FDSOI device is intermediate between those of IM-DGFET and JL-DGFET: it is lower than in IM-DGFET because only a single gate controls the channel, but it is higher than in JL-DGFET because the channel is intrinsic and the mobility is enhanced.

The devices and physical models described above have been further used for the simulation of drain-current transients induced by an energetic particle striking the sensitive area of the device (Munteanu and Autran 2014a). The transistors are simulated in the off state (most sensitive case) under $V_G = 0$ V and $V_D = 0.75$ V. Transient simulations are performed considering an ion track with a Gaussian shape with a characteristic radius of 20 nm. The ion track has a Gaussian time dependence centered on $t = 10$ ps, with a characteristic time of 2 ps. The LET value is kept constant along the ion track; this is justified by the very short ion-track length (equal to the silicon film thickness, 6 nm).

The ion strikes the device in a vertical incidence perpendicular to the gate, as shown in Figure 12.37. In a first step, it was considered that the ion strikes in the middle of the channel (i.e., at equal distance from the source and drain contacts). In a second step, several locations were considered for the ion impact between the source and the drain, in order to investigate the sensitivity of the device as a function of the ion-strike position. In most cases, the ion track (3D structure) is not entirely contained in the active area of the device, which requires the charge deposited by the ion in the device to be accurately calculated. The deposited charge is obtained in each case by taking into account the Gaussian distribution of the ion track, the 3D geometry of the silicon film, and the exact location of the ion strike. The collected charge is calculated using Equation 12.2, and, finally, the bipolar amplification of the charge deposited by the ion is obtained from Equation 12.1.

12.7.2 Radiation Sensitivity of Individual Devices

Simulated drain-current transients due to the ion strike are shown in Figure 12.38 for an LET value of 0.2 MeV cm^2 mg^{-1}. The prompt components of the current transients are almost identical for the three devices; however, the transient tails, representing the slow-discharge component (due to floating-body effects and carrier-recombination mechanisms), are significantly different. FDSOI shows the longest transient tail, indicating the presence of stronger floating-body effects than in double-gate devices. This is confirmed by the bipolar amplification, which is plotted as a function of the ion-strike LET in Figure 12.39. For an LET of 0.2 MeV cm^2 mg^{-1}, the bipolar gain is higher in FDSOI than in JL-DGFET and IM-DGFET.

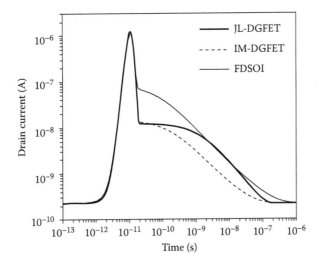

FIGURE 12.38
Drain-current transients in JL-DGFET, IM-DGFET, and FDSOI. All the transistors are biased in the off state. LET = 0.2 MeV cm² mg⁻¹.

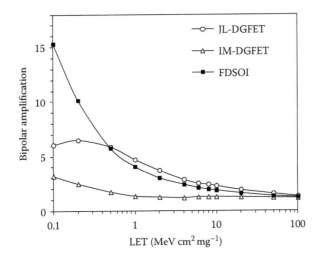

FIGURE 12.39
Bipolar gain versus ion LET in JL-DGFET, IM-DGFET, and FDSOI. The ion strikes vertically in the middle of the channel between the source and drain contacts.

IM-DGFET shows the lowest bipolar gain, which is the result of a double-gate configuration combined with an intrinsic channel. In spite of its double-gate structure, JL-DGFET is found to have a higher bipolar gain than IM-DGFET, essentially due to the high doping level of the silicon film, which enhances floating-body effects; however, at low LET values the bipolar amplification of JL-DGFET is lower than that of single-gate FDSOI.

12.7.2.1 *Impact of Channel Doping*

The impact of channel doping level on the drain-current transient and bipolar amplification was investigated in Munteanu and Autran (2012a). Three channel doping levels have been considered in simulation: 1×10^{19}, 2×10^{19}, and 3×10^{19} cm^{-3}. As explained before, in JL-DGFET the channel thickness has to be sufficiently small in order to be able to fully deplete the channel of carriers and to turn the device off. The higher the channel doping level, the smaller the film thickness needs to be. This condition is satisfied for all the doping levels and the film thickness considered here. The off-state current was considered to be the same for all doping levels (by changing the gate work function).

When the channel doping increases, the floating-body effects are enhanced and the drain-current transient is longer. Both collected charge and bipolar amplification (Figure 12.40) increase with channel doping. Impact ionization is also larger for higher doping levels, which additionally contribute to enhancing bipolar amplification. Very high values of bipolar gain are found for a channel doping of 3×10^{19} cm^{-3}, but these values are reduced when a larger ion-track radius is considered in simulation (as will be shown in Section 12.7.2.2). Finally, at very high LET the electric field is collapsed and the bipolar gain decreases below 2.5 for all devices.

12.7.2.2 Impact of Ion-Strike Location

Until now, it has been considered that the ion hits the device in the middle of the channel. Other ion-strike locations along the channel (x axis) have been explored in Munteanu and Autran (2014a) in order to study their impact on the radiation sensitivity of JL-DGFET compared with that of IM-DGFET and FDSOI. Several locations of ion strike have been considered between the source contact (x = 0) and the drain contact (x = 60 nm), as shown in Figure 12.41. The 3D profile of the heavy-ion charge density in the silicon film (the same for all devices) is also shown in Figure 12.41 for an ion striking the film at x = 45 nm. For this particular location, as shown in Figure 12.41, the ion track is not entirely contained on the silicon film. This is also the case for other locations, and this requires a specific calculation of the deposited charge. For each location, the current transient is simulated and the

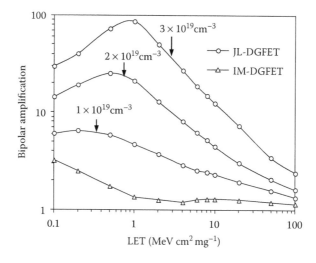

FIGURE 12.40
Bipolar amplification as a function of LET in IM-DGFET and JL-DGFET. $V_G = 0$ V and $V_D = 0.75$ V. (Reprinted from Munteanu, D., Autran, J.L. 3-D numerical simulation of bipolar amplification in junctionless double-gate MOSFETs under heavy-ion irradiation. *IEEE Trans. Nucl. Sci.* 59, 773–780. © (2012) IEEE. With permission.)

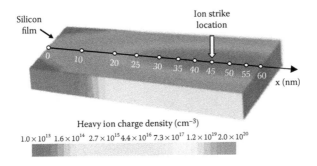

FIGURE 12.41
3D profiles of heavy-ion charge density in the silicon film of JL-DGFET for an ion strike at x=45 nm and LET=2 MeV cm^2 mg^{-1}. The values of the heavy-ion charge density are in cm^{-3}. For a better view of the film, gate material, spacers, and isolation oxide are not shown. The position of the ion strike is indicated by the *arrow*. Other positions for the ion strike considered in this work are also indicated.

collected charge is extracted from this transient. Finally, the bipolar gain is calculated at a given LET for each x value.

The collected charge as a function of the x location is shown in Figure 12.42 for very low and very high LET values. The deposited charge, calculated for each x location and each LET, is also shown in Figure 12.42. The deposited charge is highest in the middle of the channel and decreases toward the sides of the silicon film, because a reduced part of the ion track is contained in the active region when the ion-strike position moves toward the source and drain contacts (Munteanu and Autran 2014a). At very low LET=0.1 MeV cm^2 mg^{-1} and for all x locations, the lowest collected charge is obtained for IM-DGFET and the highest collected charge for FDSOI (Figure 12.43a). For all devices, the collected charge has a bell-shaped profile with a maximum around the middle of the channel (where the deposited charge is highest) and two minima at the source and drain contacts (where the deposited charge is lowest). The collected charge is always higher than the deposited charge for all x locations, which clearly indicates a strong bipolar amplification. The behavior is quite different for LET=100 MeV cm^2 mg^{-1}, as shown in Figure 12.42b. For ion strikes located between the source contact and the middle of the channel, the collected charge of JL-DGFET is higher than that of FDSOI and IM-DGFET. It is also interesting to note that for x<30 nm the collected charge is lower than the deposited charge. This indicates that the bipolar amplification is very low and that there is a strong recombination of the deposited charge in the device (Munteanu and Autran 2014a). Beyond x=30 nm, the collected charge for JL-DGFET decreases and becomes lower than that of IM-DGFET and FDSOI devices. For ion strikes beyond x=50 nm, the collected charge of JL-DGFET equals the deposited charge, which indicates that there is no longer any bipolar amplification of the deposited charge. This is not the case for IM-DGFET and FDSOI. These results show that, for a high LET, the trends obtained for ion strikes in the middle of the channel are no longer valid for ion strikes located in the vicinity of the drain region, beyond x=30 nm (Munteanu and Autran 2014a). As explained in Munteanu and Autran (2014a), these results show that JL-DGFET is able to collect a smaller amount of charge than IM-DGFET for these specific values of x location and LET.

To confirm these new findings and highlight the range of LET for which these observations are valid, the bipolar gain has been calculated in Munteanu and Autran (2014a) as a function of x for different LET values. Figure 12.43a shows, as expected, that at LET=0.1 MeV cm^2 mg^{-1} the bipolar gain of FDSOI is highest and the gain of JL-DGFET

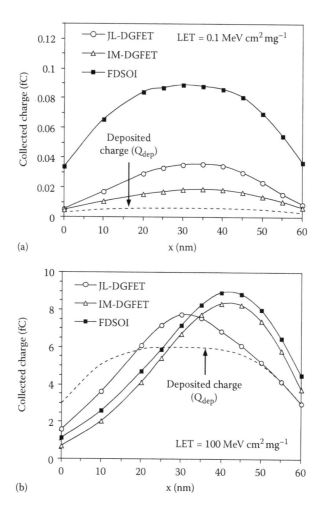

FIGURE 12.42
Collected charge as a function of x in IM-DGFET, JL-DGFET, and FDSOI. The deposited charge is also plotted for comparison. (a) LET = 0.1 MeV cm² mg⁻¹ and (b) LET = 100 MeV cm² mg⁻¹.

is intermediate between that of FDSOI and IM-DGFET. This trend continues when LET increases until ~0.5 MeV cm² mg⁻¹. For LET values approximately between 0.5 and 20 MeV cm² mg⁻¹, JL-DGFET shows the highest bipolar gain for all x locations, whereas IM-DGFET always exhibits the lowest gain. This can be shown in Figure 12.43b for the bipolar amplification at LET = 20 MeV cm² mg⁻¹. The bipolar gain of JL-DGFET is slightly higher than that of FDSOI for x locations beyond 40 nm. This trend changes for LET values above 20 MeV mg⁻¹ cm⁻². Although for x < 30 nm IM-DGFET has the lowest bipolar amplification, for x > 30 nm the bipolar gain of JL-DGFET falls below that of IM-DGFET and FDSOI. This is confirmed in Figure 12.44a and b, which show bipolar amplification as a function of x location for LET = 50 and LET = 100 MeV cm² mg⁻¹, respectively.

As indicated in Munteanu and Autran (2014a), these results show that the bipolar gains of JL-DGFET, IM-DGFET, and FDSOI have different dependence on strike location. In addition, results of Munteanu and Autran (2014a) have shown that there are LET ranges and specific ion-strike locations for which JL-DGFET has lower bipolar amplification than more conventional IM-DGFET and FDSOI devices. However, from a circuit point of view,

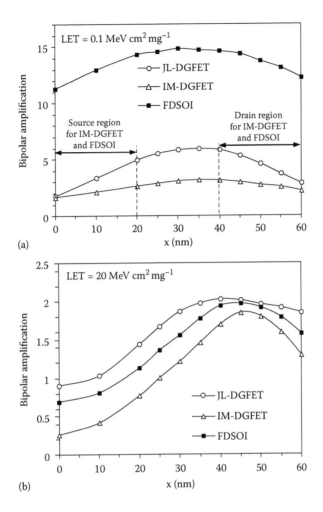

FIGURE 12.43
Bipolar amplification as a function of the x position of the ion strike in JL-DGFET, IM-DGFET, and FDSOI. (a) LET = 0.1 MeV cm² mg⁻¹ and (b) LET = 20 MeV cm² mg⁻¹.

these results are not sufficient to conclude on the radiation hardness of JL-DGFET technology. Additional simulations at circuit level are needed to understand the impact of different transistor responses on circuit radiation sensitivity, as will be shown in the following.

12.7.3 SEU Sensitivity of SRAM Cells

In Munteanu and Autran (2014b), the transient response and the sensitivity to heavy-ion irradiation of SRAM cells using JL-DGFETs have been investigated. JL-DGFETs with the same geometric parameters and doping levels as in Section 12.7.1 were considered for both cell inverters and cell-access control of the SRAM cell (for the schema of a SRAM cell, see Figure 5.18 in Chapter 5). All devices were built and simulated in the 3D domain with the Synopsys/DESSIS module and were connected via the Mixed-Mode module (Synopsys 2014). Only the OFF-state NMOS transistor was struck by an ionizing particle (see Figure 5.18), this particular case corresponding to the most effective scenario to disturb the cell and to flip its logical state. In order to compare the SEU sensitivity of JL-DGFET SRAM

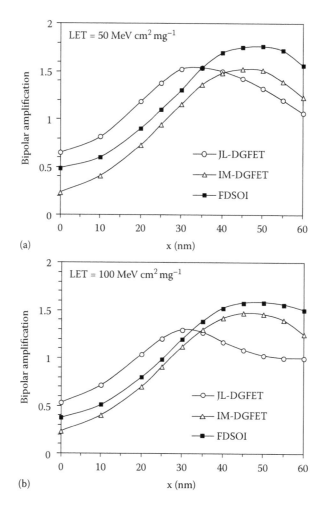

FIGURE 12.44
Bipolar amplification as a function of the x position of the ion strike in JL-DGFET, IM-DGFET, and FDSOI. (a) LET = 50 MeV cm² mg⁻¹ and (b) LET = 100 MeV cm² mg⁻¹.

cells with other technologies, we also constructed and simulated SRAM cells based on IM-DGFET and FDSOI devices. The six transistors of these cells were also entirely simulated in the 3D-device domain; the individual IM-DGFET and FDSOI transistors have the same geometric parameters and doping levels as in Section 12.7.1.

As a preliminary, the worst-case condition in terms of x location of the ion strike in the OFF-state NMOS transistor was determined in Munteanu and Autran (2014b) for each technology. The worst-case location is the x location along the channel where the bipolar amplification (and the collected charge) is highest. Then, the SEU LET threshold obtained in this point will be the smallest LET for which the SRAM cell flips. To find the worst case, we previously plotted bipolar gain as a function of x (according to a similar analysis to that presented in Section 12.7.2.2). The worst-case x locations were found to be x = 40 nm for JL-DGFET and FDSOI and x = 35 nm for IM-DGFET. In the following, we used these worst-case locations for all SRAM simulations.

Figure 12.45 shows the variation with time of the current measured at the drain (node "1," see Figure 5.18) of the impacted transistor for two values of the ion LET and for the

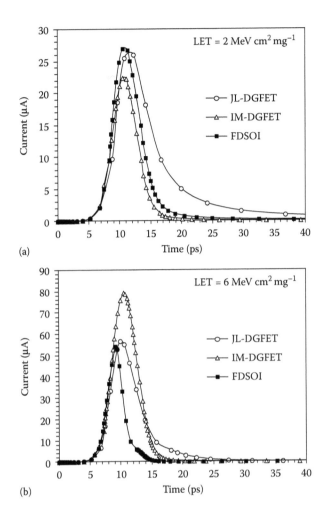

FIGURE 12.45
Current transient at node "1" (see Figure 5.18 in Chapter 5) of the JL-DGFET, IM-DGFET, and FDSOI SRAM cells after the ion strike for (a) LET = 2 MeV cm² mg⁻¹ and (b) LET = 6 MeV cm² mg⁻¹.

three SRAM cells made up of JL-DGFET, IM-DGFET, and FDSOI transistors. As explained in Munteanu and Autran (2014b), when the ion LET increases, the peak of the current transient increases, because the capacitive effect induced by the ion is stronger (Ferlet-Cavrois et al. 2003). The charge collected at node "1" also increases (in proportion to the ion LET). The current transient induced by the ionizing particle disturbs the voltages of nodes "1" and "2." The disturbance is stronger when the ion LET increases. In Figures 12.46 and 12.47, the time variations of voltages extracted at nodes "1" and "2" (V_1 and V_2) are plotted for the three SRAM cells; the responses to the ion strike of these different cells are compared in these figures. As shown in Figure 12.46a, for LET = 2 MeV cm² mg⁻¹, V_1 and V_2 are disturbed for all cells, but their values do not change at the end of the transient (i.e., the state of the SRAM cell is not modified; the cell did not flip). For LET = 3.5 MeV cm² mg⁻¹, the values of V_1 and V_2 for the FDSOI SRAM cell change with respect to their initial value; therefore the FDSOI SRAM cell flips (Figure 12.46b). In this case, the ion LET is sufficiently strong to induce a collected charge larger than the critical charge of the cell and to cause the SRAM cell upset. This is not the case for JL-DGFET and IM-DGFET SRAM cells, which

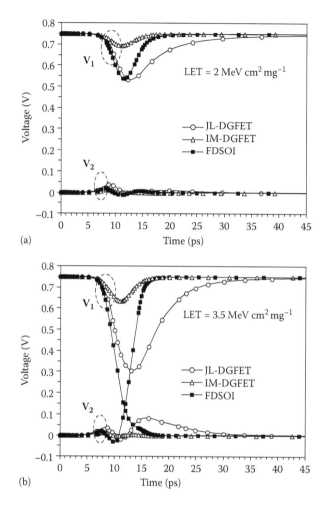

FIGURE 12.46

Time variation of voltages V_1 and V_2 in JL-DGFET, IM-DGFET, and FDSOI SRAM cells for (a) LET = 2 MeV cm^2 mg^{-1} and (b) LET = 3.5 MeV cm^2 mg^{-1}.

are disturbed but recover their initial state. For LET = 6 MeV cm^2 mg^{-1} (Figure 12.47a), the IM-DGFET SRAM cell does not flip while the two others flip, and for LET = 8 MeV cm^2 mg^{-1} (Figure 12.47b) all three cells flip. As indicated in Munteanu and Autran (2014b), this preliminary study gives a first indication of the radiation hardness of the considered technologies: the IM-DGFET SRAM cell is less sensitive to radiation than JL-DGFET and FDSOI cells, the FDSOI technology being the most sensitive to the ion strike.

In order to study in more depth the radiation hardness of these SRAM cells, the SEU-threshold LET (LET$_{th}$) of each cell was found in Munteanu and Autran (2014b), by varying the ion-strike LET until the SRAM was observed to upset. The SEU-threshold LET and the corresponding critical charge (Q_{crit}) are shown in Table 12.5. As expected, the highest critical charge and SEU-threshold LET are found for the IM-DGFET SRAM cell. The JL-DGFET SRAM cell exhibits lower critical charge and threshold LET than the IM-DGFET SRAM cell, but higher than the FDSOI SRAM cell. These results are explained in Munteanu and Autran (2014b) as follows. The critical charge depends on the equivalent capacitance of the struck node (C_N), on the supply voltage (V_{DD}), and on the maximum current of the

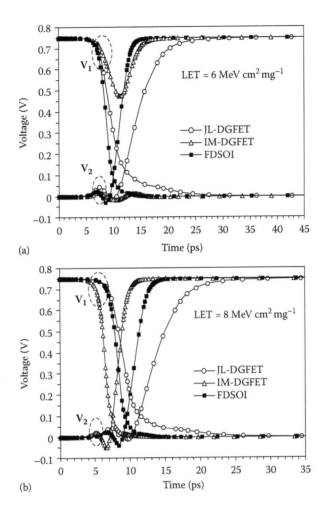

FIGURE 12.47
Time variation of voltages V_1 and V_2 in JL-DGFET, IM-DGFET, and FDSOI SRAM cells for (a) LET = 6 MeV cm^2 mg^{-1} and (b) LET = 8 MeV cm^2 mg^{-1}.

TABLE 12.5

SEU-Threshold LET (LET$_{th}$) and Critical Charge (Q$_{crit}$) for JL-DGFET, IM-DGFET, and FDSOI SRAM Cells

Structure	JL-DGFET	IM-DGFET	FDSOI
LET$_{th}$ (MeV cm^2 mg^{-1})	3.64	7.48	3.15
Q$_{crit}$ (fC)	0.309	0.51	0.205

on-state PMOS transistor (I$_{PMOS}$), as previously highlighted in literature (Roche et al. 1999; Jahinuzzaman et al. 2009). These previous works showed that the critical charge increases with C$_N$, V$_{DD}$, and I$_{PMOS}$. In simulations presented in Munteanu and Autran (2014b), all cells are operating at the same supply voltage. The equivalent capacitance of the struck node (node "1") is higher for double-gate devices (JL-DGFET and IM-DGFET) than for single-gate devices (FDSOI), due to the double-gate configuration (Colinge 2007). The maximum current of the PMOS transistor is highest for the IM-DGFET PMOS transistor, while it is almost the same for JL-DGFET and FDSOI. From these observations, the authors concluded in Munteanu and Autran (2014b) that the critical charge of IM-DGFET SRAM is higher than that of JL-DGFET SRAM because the current of the IM-DGFET PMOS is higher, while C_N is essentially the same for both cells. For JL-DGFET and FDSOI SRAM cells, the maximum current of the PMOS transistor is almost the same, but the node capacitance is higher for JL-DGFET (due to its double-gate configuration) than for FDSOI. Then, the critical charge of JL-DGFET SRAM cell is higher than that of FDSOI SRAM cell. As explained in Munteanu and Autran (2014b), this means that a higher charge is needed to flip the JL-DGFET SRAM cells than the FDSOI SRAM cell, indicating that the JL-DGFET SRAM cell is more immune to ion irradiation than its FDSOI counterpart. The results obtained here are very encouraging for junctionless double-gate technology.

Finally, it is important to note that, in the SRAM simulation presented in Munteanu and Autran (2014b), the ionizing particle strikes the OFF-state NMOS transistor. Generally, NMOS transistors are serially connected by sharing drain and source nodes, and the adjacent NMOS transistor may be affected by the ionizing particle induced to the adjacent transistors. Especially in the JL-DGFET, there are no junctions between channel and source or drain region, and the particle at the adjacent FET may have a strong effect. Additional extensive simulations are necessary to address this issue, which may be the object of future investigations.

12.8 III–V FinFET and Tunnel FET

In order to improve CMOS performance, especially on-state current, according to ITRS recommendations for future technological nodes, high-mobility materials have recently been investigated to potentially replace silicon in the MOSFET channel (Krishnamohan et al. 2006). Due to low conductivity, effective masses in both valence and conduction bands (which leads to higher hole and electron mobilities than in Si), Ge and III–V materials such as GaAs, InAs, InSb, or ternary compounds have been given detailed consideration in recent investigations to enhance the performance of p-MOSFETs and n-MOSFETs for high-performance CMOS logic applications. The SEE sensitivity of III–V materials technologies was first investigated by McMorrow et al. (1996, 2004) and Weaver et al. (2003). As explained in Dodd et al. (2010), III–V materials technologies are often even more sensitive to SEU than their silicon counterparts due to internal gain mechanisms, which often result in high charge collection efficiency and hence enhanced SEE sensitivity. In addition, similarly to advanced CMOS technologies, higher-speed operation results in increased sensitivity to SETs (Dodd et al. 2010).

Steep switching devices such as Tunnel FET (Choi et al. 2007; Mookerjea et al. 2009; Wang et al. 2004) have been proposed in recent years for low-voltage application. Tunneling FETs (TFETs) enable the fundamental SS value of 60 mV dec^{-1}, an inherent limitation of conventional MOSFETs, to be overcome. This limitation is due to the use of a particular mechanism for carrier injection based on the band-to-band tunneling (BTBT) that appears in gated p-i-n diodes operating under reverse bias as the building block of existing TFET designs (Padilla et al. 2012; Verhulst et al. 2010). This allows a reduction in the supply voltage (needed for low-voltage applications) and strongly reduces OFF-state currents in TFETs, which makes these devices among the most promising novel devices currently

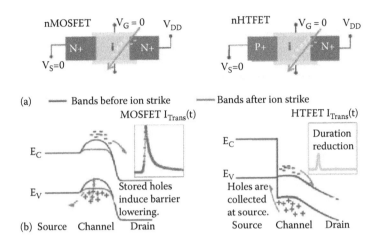

(a)

(b) Source   Channel   Drain   Source   Channel   Drain

FIGURE 12.48
(a) Schematic description of the simulated nMOSFET and nHTFET; (b) Band diagram of nMOSFET (*left*) and nHTFET (*right*) before and after ion strike. (Reprinted from Liu, H. et al., Technology assessment of Si and III–V FinFETs and III–V tunnel FETs from soft error rate perspective, In *Proceedings of the International Electron Devices Meeting Technical Digest*, pp. 25.5.1–25.5.4, 10–13 December, San Francisco, CA. © (2012) IEEE. With permission.)

under study (Khatami and Banerjee 2009; Jeon et al. 2010; Padilla et al. 2012). Combining high-mobility materials with TFETs could be seriously considered to reach predicted performances for low-power applications.

Transient error generation and transient drain current in III–V FinFET and III–V heterojunction tunnel FET (HTFET) have been investigated by Liu et al. (2012) using both device and circuit simulations. The generic structure of the HTFET is schematically represented in Figure 12.48a. As explained in Liu et al. (2012), in nMOSFET the radiation-induced holes are stored in the body (due to the source barrier), which induces barrier lowering (see Figure 12.48b) and increases the channel potential. Additional electrons flow into the channel due to the parasitic bipolar amplification mechanism, which further increases the drain node charge collection (Liu et al. 2012). On the contrary, in HTFET, due to the asymmetric source and drain doping, both electrons and holes can be collected through ambipolar transport (Liu et al. 2012); by this mechanism, the hole density in HTFET decreases rapidly due to ambipolar transport and the channel barrier is unchanged (Figure 12.48b). Holes and electrons can be collected at the source and drain, respectively, which greatly reduces body-charge-storage-induced bipolar gain and further reduces collected charge, transient-current magnitude, and transient duration (Liu et al. 2012).

Drain-current transients and collected charges have been compared in Liu et al. (2012); these results have shown that, at LET = 0.1 pC μm^{-1}, transient duration in HTFET is reduced by 80% and collected charge is reduced by 90% compared with Si FinFET. It is also indicated that a twofold charge-collection enhancement is obtained in III–V FinFET compared with Si FinFET due to high carrier mobility. The results obtained in Liu et al. (2012) also indicate significant reduction in bipolar gain in HTFET compared with MOSFET. Also, HTFET shows reduced current magnitude and a tenfold reduction in charge collection compared with Si FinFET (Liu et al. 2012).

SRAM SER and logic SER projections have also been investigated in Liu et al. (2012) using a simplified analytical model. As explained in Liu et al. (2012), III–V FinFET shows increased charge deposition due to low ionization energy, which increases the SER for

SRAM cells for all V_{DD} compared with Si FinFET (Liu et al. 2012). For logic circuits, III–V FinFET shows reduced SER compared with Si FinFET below 0.5 V due to improved latching-window masking (Liu et al. 2012). Concerning HTFET, these devices show superior radiation resilience compared with both Si and III–V FinFET over the voltage range of 0.3–0.6 V for both SRAM and logic (Liu et al. 2012). As concluded in Liu et al. (2012), this fundamental advantage stems from bipolar gain reduction, on-state enhanced Miller capacitance effect, and improved latching-window masking, which makes HTFET desirable for radiation-resilient ultralow-power application.

References

Autran, J.L. and Munteanu, D. 2008. Simulation of electron transport in nanoscale independent-gate DG devices using a full 2D Green's function approach. *Journal of Computational and Theoretical Nanoscience* 5:1120–1127.

Ball, D.R., Alles, M.L., Schrimpf, R.D., and Cristoloveanu, S. 2010. Comparing single event upset sensitivity of bulk vs. SOI based FinFET SRAM cells using TCAD simulations. In *Proceedings of the IEEE International SOI Conference*, pp. 1–2. 11–14 October, San Diego, CA, IEEE.

Barbut, L., Jazaeri, F., Bouvet, D., and Sallese, J.-M. 2013. Transient off-current in junctionless FETs. *IEEE Transactions on Electron Devices* 60:2080–2083.

Benedetto, J., Eaton, P., Avery, K., Mavis, D., Gadlage, M., and Turflinger, T. 2004. Heavy ion induced digital single-event transients in deep submicron processes. *IEEE Transactions on Nuclear Science* 51:3480–3485.

Bernard, E., Ernst, T., Guillaumot, B., et al. 2007. Impact of the gate stack on the electrical performances of 3D multi-channel MOSFET (MCFET) on SOI. In *Proceedings of the 37th European Solid-State Device Research Conference*, pp. 147–150. 11–13 September, Munich, IEEE.

Brisset, C., Dollfus, P., Musseau, O., Leray, J.L., and Hesto, P. 1994. Theoretical study of SEU's in 0.25-pm fully-depleted CMOS/SOI technology. *IEEE Transactions on Nuclear Science* 41:2297–2303.

Chen, C-Y., Lin, J-T., Chiang, M-H., and Kim, K. 2010. High-performance ultra-low power junctionless nanowire FET on SOI substrate in subthreshold logic application. In *Proceedings of the IEEE International SOI Conference*, pp. 1–2. 11–14 October, San Diego, CA, IEEE.

Choi, W., Park, B., Lee, J., and Liu, T. 2007. Tunneling field-effect transistors (TFETs) with subthreshold swing (SS) less than 60 mV/dec. *IEEE Electron Device Letters* 28:743–745.

Choi, Y., Lindert, N., Xuan, P., et al. 2001. Sub-20 nm CMOS FinFET technologies. In *Proceedings of the International Electron Devices Meeting Technical Digest*, pp. 421–424, 2–5 December, Washington, DC.

Colinge, J.P. 1997. *Silicon-On-Insulator Technology: Materials to VLSI*. Norwell, MA: Kluwer Academic.

Colinge, J.P. 2007. Multi-gate SOI MOSFETs. *Microelectronic Engineering* 84:2071–2076.

Colinge, J.P., Gao, M.H., Romano-Rodríguez, A., Maes, H., and Claeys, C. 1990. Silicon-on-insulator "Gate-all-around device". In *Proceedings of the International Electron Devices Meeting Technical Digest*, pp. 595–598. 9–12 December, San Francisco, CA, IEEE.

Colinge, J.P., Lee, C.-W., Afzalian, A., et al. 2010. Nanowire transistors without junctions. *Nature Nanotechnology* 5:225–229.

Colinge, J.P., Lee, C.-W., Ferain, I., et al. 2009. Reduced electric field in junctionless transistors. *Applied Physics Letters* 96:073510.

Cristoloveanu, S. and Li, S.S. 1995. *Electrical Characterization of Silicon-On-Insulator Materials and Devices*. Norwell, MA: Kluwer Academic.

Dodd, P.E., Shaneyfelt, M.R., Felix, J.A., and Schwank, J.R. 2004. Production and propagation of single-event transients in high-speed digital logic ICs. *IEEE Transactions on Nuclear Science* 51:3278–3284.

Dodd, P.E., Shaneyfelt, M.R., Schwank, J.R., and Felix, J.A. 2010. Transient response of III–V field-effect transistors to heavy-ion irradiation. *IEEE Transactions on Nuclear Science* 57:1747–1763.

Dupre, C., Ernst, T., Arvet, C., Aussenac, F., Deleonibus, S., and Ghibaudo, G. 2007. Stacked nanowires ΦFET with independent gates: A novel device for ultra-dense low-power applications. In *Proceedings of the IEEE International SOI Conference*, pp. 95–96. 1–4 October, Indian Wells, CA, IEEE.

El-Mamouni, F., Zhang, E.X., Ball, D.R., et al. 2012. Heavy-ion-induced current transients in bulk and SOI FinFETs. *IEEE Transactions on Nuclear Science* 59:2674–2681.

El-Mamouni, F., Zhang, E.X., Pate, N.D., et al. 2011a. Laser- and heavy ion-induced charge collection in bulk FinFETs. *IEEE Transactions on Nuclear Science* 58:2563–2569.

El-Mamouni, F., Zhang, E.X., Schrimpf, R.D., et al. 2011b. Pulsed laser-induced transient currents in bulk and silicon-on-insulator FinFETs. In *Proceedings of the International Reliability Physics Symposium*, pp. SE.4.1–SE.4.4. 10–14 April, Monterey, CA, IEEE.

Ernst, T., Dupré, C., Isheden, C., et al. 2006. Novel 3D integration process for highly scalable Nano-Beam stacked-channels GAA (NBG) FinFETs with HfO$_2$/TiN gate stack. In *Proceedings of the International Electron Devices Meeting Technical Digest*, pp. 1–4. 11–13 December, San Francisco, CA, IEEE.

Ferlet-Cavrois, V. 2004. Comportement des technologies SOI sous irradiations [Behavior of SOI technologies under irradiation]. HDR Thesis (in French).

Ferlet-Cavrois, V., Gasiot, G., Marcandella, C., et al. 2002a. Insights on the transient response of fully and partially depleted SOI technologies under heavy-ion and dose-rate irradiations. *IEEE Transactions on Nuclear Science* 49:2948–2956.

Ferlet-Cavrois, V., Marcandella, C., Giraud, G., et al. 2002b. Characterization of the parasitic bipolar amplification in SOI technologies submitted to transient irradiation. *IEEE Transactions on Nuclear Science* 49:1456–1461.

Ferlet-Cavrois, V., Paillet, P., McMorrow, D., et al. 2005. Direct measurement of transient pulses induced by laser irradiation in deca-nanometer SOI devices. *IEEE Transactions on Nuclear Science* 52:2104–2113.

Ferlet-Cavrois, V., Paillet, P., Schwank, J.R., et al. 2003. Charge collection by capacitive influence through isolation oxides. *IEEE Transactions on Nuclear Science* 50:2208–2218.

Ferlet-Cavrois, V., Vizkelethy, G., Paillet, P., et al. 2004. Charge enhancement effect in NMOS bulk transistors induced by heavy ion irradiation-comparison with SOI. *IEEE Transactions on Nuclear Science* 51:3255–3262.

Fischetti, M.V. and Laux, S.E. 2001. Long-range coulomb interactions in small Si devices. Part I: Performance and reliability. *Journal of Applied Physics* 89:1205–1231.

Frank, D.J., Laux, S.E., and Fischetti, M.V. 1992. Monte Carlo simulation of a 30 nm dual-gate MOSFET: How short can Si go? In *Proceedings of the International Electron Devices Meeting Technical Digest*, pp. 553–556. 13–16 December, San Francisco, CA, IEEE.

Gaillardin, M., Paillet, P., Ferlet-Cavrois, V., et al. 2007. Transient radiation response of single- and multiple-gate FD SOI transistors. *IEEE Transactions on Nuclear Science* 54:2355–2362.

Guarini, K.W., Solomon, P.M., Zhang, Y., et al. 2001. Triple-self-aligned, planar double-gate MOSFETs: Devices and circuits. In *Proceedings of the International Electron Devices Meeting Technical Digest*, pp. 19.2.1–19.2.4. 2–5 December, Washington, DC, IEEE.

Gusev, E.P., Narayanan, V., and Frank, M.M. 2006. Advanced high-κ dielectric stacks with polySi and metal gates: Recent progress and current challenges. *IBM Journal of Research and Development* 50:387–410.

Haensch, W., Nowak, E.J., Dennard, R.H., et al. 2006. Silicon CMOS devices beyond scaling. *IBM Journal of Research and Development* 50:339–361.

Harrison, S., Coronel, P., Leverd, F., et al. 2003. Highly performant double gate MOSFET realized with SON process. In *Proceedings of the International Electron Devices Meeting Technical Digest*, pp. 18.6.1–18.6.4. 8–10 December, Washington, DC, IEEE.

Hiramoto, T., Saitoh, M., and Tsutsui, G. 2006. Emerging nanoscale silicon devices taking advantage of nanostructure physics. *IBM Journal of Research and Development* 50:411–418.

Hisamoto, D., Kaga, T., Kawamoto, Y., and Takeda, E. 1989. A fully depleted lean-channel transistor (DELTA)—A novel vertical ultra thin SOI MOSFET. In *Proceedings of the International Electron Devices Meeting Technical Digest*, pp. 833–836. 3–6 December, Washington, DC, IEEE.

Hite, L.R., Lu, H., Houston, T.W., Hurta, D.S., and Bailey, W.E. 1992. An SEU resistant 256 K SOI SRAM. *IEEE Transactions on Nuclear Science* 39:2121–2125.

Houssa, M. 2004. *Fundamental and Technological Aspects of High-κ Gate Dielectrics*. London: Institute of Physics.

Huang, X., Lee, W.-C., Chang, D.H.L., et al. 2001. Sub 50 nm P channel FinFETs. *IEEE Transactions on Electron Devices* 48:880–886.

ITRS (International Technology Roadmap for Semiconductors). 2013. International Technology Roadmap for Semiconductors (ITRS). Available at: http://public.itrs.net/.

Jahinuzzaman, S.M., Sharifkhani, M., and Sachdev, M. 2009. An analytical model for soft error critical charge of nanometric SRAMs. *IEEE Transactions on Very Large Scale Integration (VLSI) Systems* 17:1187–1195.

Jeon, K., Loh, W.-Y., Patel, P., et al. 2010. Si tunnel transistors with a novel silicided source and 46 mV/dec swing. In *Proceedings of the VLSI Technology Symposium*, pp. 121–122. 15–17 June, Honolulu, IEEE.

Jiao, Z. and Salama, C.A.T. 2001. A fully depleted Δ-channel SOI nMOSFET. *Proceedings of the Electrochemical Society* 3:403–408.

Jiménez, D., Iniguez, B., Suné, J., et al. 2004. Continuous analytic I–V model for surrounding-gate MOSFETs. *IEEE Electron Device Letters* 25:571–573.

Kado, Y. 1997. The potential of ultra-thin SOI devices for low-power and high-speed applications. *IEICE Transactions on Electronics* E-80-C:443–454.

Kedzierski, J., Nowak, E., Kanarsky, T., et al. 2002. Metal-gate FinFET and fully-depleted SOI devices using total gate silicidation. In *Proceedings of the International Electron Devices Meeting Technical Digest*, pp. 247–250. 8–11 December, San Francisco, CA, IEEE.

Kerns, S.E., Massengill, L.W., Kerns Jr, D.V., et al. 1989. Model for CMOS/SOI single-event vulnerability. *IEEE Transactions on Nuclear Science* 36:2305–2310.

Khatami, Y. and Banerjee, K. 2009. Steep subthreshold slope n- and p-type tunnel-FET devices for low-power and energy-efficient digital circuits. *IEEE Transactions on Electron Devices* 56:2752–2761.

Kobayashi, D., Aimi, M., Saito, H., and Hirose, K. 2006. Time-domain component analysis of heavy-ion-induced transient currents in fully-depleted SOI MOSFETs. *IEEE Transactions on Nuclear Science* 53:3372–3378.

Kranti, A., Yan, R., Lee, C.-W., et al. 2010. Junctionless nanowire transistor (JNT): Properties and design guidelines. In *Proceedings of the European Solid-State Circuits Conference*, pp. 357–360. 14–16 September, Sevilla, IEEE.

Krishnamohan, T., Jungemann, C., Kim, D., et al. 2006. Theoretical investigation of performance in uniaxially- and biaxially-strained Si, SiGe and Ge double-gate p-MOSFETs. In *Proceedings of the International Electron Devices Meeting Technical Digest*, pp. 1–4. 11–13 December, San Francisco, CA, IEEE.

Lee, C.-W., Afzalian, A., Akhavan, N.D., Yan, R., Ferain, I., and Colinge, J.P. 2009. Junctionless multi-gate field-effect transistor. *Applied Physics Letters* 94:053511.

Lee, C.-W., Borne, A., Ferain, I., et al. 2010. High temperature performance of silicon junctionless MOSFETs. *IEEE Transactions on Electron Devices* 53:620–625.

Liu, H., Cotter, M., Datta, S., and Narayanan, V. 2012. Technology assessment of Si and III–V FinFETs and III–V tunnel FETs from soft error rate perspective. In *Proceedings of the International Electron Devices Meeting Technical Digest*, pp. 25.5.1–25.5.4. 10–13 December, San Francisco, CA, IEEE.

Majkusiak, B., Janik, T., and Walczak, J. 2002. Semiconductor thickness effects in the double-gate SOI MOSFET. *IEEE Transactions on Electron Devices* 45:1127–1134.

Manoj, C.R., Nagpal, M., Varghese, D., and Rao, V.R. 2008. Device design and optimization considerations for bulk FinFETs. *IEEE Transactions on Electron Devices* 55:609–615.

Masahara, M., Liu, Y., Sakamoto, K., et al. 2005. Demonstration, analysis, and device design considerations for independent DG MOSFETs. *IEEE Transactions on Electron Devices* 52:2046–2051.

Massengill, L.W., Kerns, D.V., Kerns, S.E., and Alles, M.L. 1990. Single-event charge enhancement in SOI devices. *IEEE Electron Device Letters* EDL-11:98–99.

Mathew, L., Du, Y., Thean, A.V., et al. 2004. CMOS vertical multiple independent gate field effect transistor (MIGFET). In *Proceedings of the IEEE International SOI Conference*, pp. 187–189. 4–7 October, Charleston, SC, IEEE.

Mavis, D.G. and Eaton, P.H. 2002. Soft error rate mitigation techniques for modern microcircuits. In *Proceedings of the International Reliability Physics Symposium*, pp. 216–225, 7–11 April, Dallas, TX, IEEE.

McMorrow, D., Boos, J.B., Knudson, A.R., et al. 2004. Transient response of III–V field-effect transistors to heavy-ion irradiation. *IEEE Transactions on Nuclear Science* 51:3324–3331.

McMorrow, D., Weatherford, T.R., Buchner, S., et al. 1996. Single-event phenomena in GaAs devices and circuits. *IEEE Transactions on Nuclear Science* 43:628–644.

Mookerjea, S., Mohata, D., Krishnan, R., et al. 2009. Experimental demonstration of 100 nm channel length In0.53Ga0.47As-based vertical inter-band tunnel field effect transistors (TFETs) for ultra low-power logic and SRAM applications. In *Proceedings of the International Electron Devices Meeting Technical Digest*, pp. 1–3. 7–9 December, Baltimore, MD, IEEE.

Moore, G.E. 1965. Cramming more components onto integrated circuits. *Electronics* 38:19. See also: http://www.intel.com/research/silicon/mooreslaw.htm.

Moreau, M., Munteanu, D., and Autran, J.L. 2008. Simulation analysis of quantum confinement and short-channel effects in independent double-gate metal-oxide-semiconductor field-effect transistors. *Japanese Journal of Applied Physics* 47:7013–7018.

Morishita, F., Yamaguchi, Y., Eimori, T., et al. 1999. Analysis and optimization of floating body cell operation for high-speed SOI-DRAM. *IEICE Transactions on Electronics* E-82-C:544–550.

Munteanu, D. and Autran, J.L. 2003. Two-dimensional modeling of quantum ballistic transport in ultimate double-gate SOI devices. *Solid State Electronics* 47:1219–1225.

Munteanu, D. and Autran, J.L. 2008. Modeling and simulation of single-event effects in digital devices and ICs. *IEEE Transactions on Nuclear Science* 55:1854–1878.

Munteanu, D. and Autran, J.L. 2009a. 3-D simulation analysis of bipolar amplification in planar double-gate and FinFET with independent gates. *IEEE Transactions on Nuclear Science* 56:2083–2090.

Munteanu, D. and Autran, J.L. 2009b. Transient response of 3D multi-channel nanowire MOSFETs submitted to heavy ion irradiation: A 3D simulation study. *IEEE Transactions on Nuclear Science* 56:2042–2049.

Munteanu, D. and Autran, J.L. 2012a. 3-D numerical simulation of bipolar amplification in junctionless double-gate MOSFETs under heavy-ion irradiation. *IEEE Transactions on Nuclear Science* 59:773–780.

Munteanu, D. and Autran, J.L. 2012b. Simulation analysis of bipolar amplification in independent-gate FinFET and multi-channel NWFET submitted to heavy-ion irradiation. *IEEE Transactions on Nuclear Science* 59:3249–3257.

Munteanu, D. and Autran, J.L. 2014a. Investigation of sensitivity to heavy-ion irradiation of junctionless double-gate MOSFETs by 3-D numerical simulation. In J. Awrejcewicz (ed.), *Computational and Numerical Simulations*, pp. 227–249. Vienna: InTech.

Munteanu, D. and Autran, J.L. 2014b. Radiation sensitivity of junctionless double-gate 6T SRAM cells investigated by 3-D numerical simulation. *Microelectronics Reliability* 54:2284–2288.

Munteanu, D., Autran, J.L., Ferlet-Cavrois, V., Paillet, P., Baggio, J., and Castellani, K. 2007. 3-D quantum numerical simulation of single-event transients in multiple-gate nanowire MOSFETs. *IEEE Transactions on Nuclear Science* 54:994–1001.

Munteanu, D., Autran, J.L., Harrison, S., Nehari, K., Tintori, O., and Skotnicki, T. 2005. Compact model of the quantum short-channel threshold voltage in symmetric double-gate MOSFET. *Molecular Simulation* 31:831–837.

Munteanu, D. and Cristoloveanu, S. 1999. Model for the extraction of the recombination lifetime in partially depleted SOI MOSFET. *Microelectronic Engineering* 48:355–358.

Munteanu, D., Ferlet-Cavrois, V., Autran, J.L., et al. 2006. Investigation of quantum effects in ultra-thin body single- and double-gate devices submitted to heavy ion irradiation. *IEEE Transactions on Nuclear Science* 53:3363–3371.

Munteanu, D., Weiser, D., Cristoloveanu, S., Faynot, O., Pelloie, J.L., and Fossum, J.G. 1998. Generation-recombination transient effects in partially depleted SOI transistors: Systematic experiments and simulations. *IEEE Transactions on Electron Devices* 45:1678–1683.

Musseau, O., Ferlet-Cavrois, V., Pelloie, J.L., Buchner, S., McMorrow, D., and Campbell, A.B. 2000. Laser probing of bipolar amplification in 0.25-μm MOS/SOI transistors. *IEEE Transactions on Nuclear Science* 47:2196–2203.

Musseau, O., Leray, J.L., Ferlet-Cavrois, V., Coic, Y.M., and Giffard, B. 1994. SEU in SOI SRAMS—A static model. *IEEE Transactions on Nuclear Science* NS-41:607–612.

Nowak, E.J., Aller, I., Ludwig, T., et al. 2004. Turning silicon on its edge. *IEEE Circuits and Devices Magazine* 20:20–31.

Padilla, J.L., Gámiz, F., and Godoy, A. 2012. Impact of quantum confinement on gate threshold voltage and subthreshold swings in double-gate tunnel FETs. *IEEE Transactions on Electron Devices* 59:3205–3211.

Paillet, P., Gaillardin, M., Ferlet-Cavrois, V., et al. 2005. Total ionizing dose effects on deca-nanometer FD SOI devices. *IEEE Transactions on Nuclear Science* 52:2345–2352.

Parihar, M.S., Ghosh, D., and Kranti, A. 2013. Single transistor latch phenomenon in junctionless transistors. *Journal of Applied Physics* 113:184503.

Park, J.T. and Colinge, J.P. 2002. Multiple-gate SOI MOSFETs: Device design guidelines. *IEEE Transactions on Electron Devices* 49:2222–2229.

Park, J.T., Colinge, J.P., and Diaz, C.H. 2001. Pi-gate SOI MOSFET. *IEEE Electron Device Letters* 22:405–406.

Pei, G. and Kan, E.C. 2004. Independently driven DG MOSFETs for mixed-signal circuits: Part I—quasi-static and nonquasi-static channel coupling. *IEEE Transactions on Electron Devices* 51:2086–2093.

Rim, K., Hoyt, J.L., and Gibbons, J.F. 1998. Transconductance enhancement in deep submicron strained Si nMOSFETs. In *Proceedings of the International Electron Devices Meeting Technical Digest*, pp. 707–710. 6–9 December, San Francisco, CA, IEEE.

Roche, P., Autran, J.L., Gasiot, G., and Munteanu, D. 2013. Technology downscaling worsening radiation effects in bulk: SOI to the rescue. In *Proceedings of the International Electron Devices Meeting Technical Digest*, pp. 31.1.1–31.1.4, 9–11 December, Washington, DC.

Roche, P., Palau, J.M., Bruguier, G., Tavernier, C., Ecoffet, R., and Gasiot, J. 1999. Determination of key parameters for SEU occurrence using 3-D full cell SRAM simulations. *IEEE Transactions on Nuclear Science* 46:1354–1362.

Schwank, J.R., Ferlet-Cavrois, V., Dodd, P.E., et al. 2004. Analysis of heavy-ion induced charge collection mechanisms in SOI circuits. *Solid-State Electronics* 48:1027–1044.

Schwank, J.R., Ferlet-Cavrois, V., Shaneyfelt, M.R., Paillet, P., and Dodd, P.E. 2003. Radiation effects in SOI technologies. *IEEE Transactions on Nuclear Science* 50:522–538.

Seifert, N., Gill, B., Jahinuzzaman, S., et al. 2012. Heavy-ion-induced current transients in bulk and SOI FinFETs. *IEEE Transactions on Nuclear Science* 59:2666–2673.

Synopsys. 2014. Synopsys Sentaurus TCAD tools, manual. Available at: http://www.synopsys.com/tools/tcad/Pages/default.aspx.

Sze, S.M. 1985. *Semiconductor Devices. Physics and Technology*. New York: Wiley.

Taur, Y., Buchanan, D., Chen, W., et al. 1997. CMOS scaling into the nanometer regime. *Proceedings of the IEEE* 85:486–504.

Taur, Y. and Ning, T.H. 1998. *Fundamentals of Modern VLSI Devices*. Cambridge, UK: Cambridge University.

Verhulst, A., Soree, B., Leonelli, D., Vandenberghe, W., and Groeseneken, G. 2010. Modeling the single-gate, double-gate and gate-all-around tunnel field effect transistor. *Journal of Applied Physics* 107:024518-1–024518-8.

Vinet, M., Poiroux, T., Widiez, J., et al. 2005. Bonded planar double-metal-gate NMOS transistors down to 10 nm. *IEEE Electron Device Letters* 26:317–319.

Wang, P., Hilsenbeck, K., Nirschl, T., et al. 2004. Complementary tunneling transistor for low-power applications. *Solid State Electronics* 48:2281–2286.

Weaver, B.D., McMorrow, D., and Cohn, L.M. 2003. Radiation effects in III–V semiconductor electronics. *International Journal of High Speed Electronics and Systems* 13:293–326.

Wettstein, A., Schenk, A., and Fichtner, W. 2002. Quantum device-simulation with density-gradient model. *IEEE Transactions on Electron Devices* 48:279–284.

Yang, F., Chen, H., Chen, F., et al. 2002. 25 nm CMOS Omega FETs. In *Proceedings of the International Electron Devices Meeting Technical Digest*, pp. 255–258. 8–11 December, San Francisco, CA, IEEE.

Yang, J.-W. and Fossum, J.G. 2005. On the feasibility of nanoscale triple-gate CMOS transistors. *IEEE Transactions on Electron Devices* 52:1159–1164.

Young, K.K. and Burns, J.A. 1988. Avalanche-induced drain-source breakdown in silicon-on-insulator n-MOSFETs. *IEEE Transactions on Electron Devices* 35:426–431.

Zhang, W., Fossum, J.G., Mathew, L., and Du, Y. 2005. Physical insights regarding design and performance of independent-gate FinFETs. *IEEE Transactions on Electron Devices* 52:2198–2206.

Index

Milton Keynes UK
Ingram Content Group UK Ltd.
UKHW051943071024
449327UK00026B/2139